Bornologies
and Lipschitz Analysis

Gerald Beer
President's Distinguished Professor of Mathematics Emeritus
California State University, Los Angeles, California, USA

CRC Press
Taylor & Francis Group
Boca Raton London New York

CRC Press is an imprint of the
Taylor & Francis Group, an **informa** business

A SCIENCE PUBLISHERS BOOK

First edition published 2023
by CRC Press
6000 Broken Sound Parkway NW, Suite 300, Boca Raton, FL 33487-2742

and by CRC Press
4 Park Square, Milton Park, Abingdon, Oxon, OX14 4RN

Library of Congress Cataloging-in-Publication Data (applied for)

ISBN: 978-0-367-49787-3 (hbk)
ISBN: 978-0-367-49821-4 (pbk)
ISBN: 978-1-003-04737-7 (ebk)

DOI: 10.1201/9781003047377

Typeset in Times New Roman
by Radiant Productions

Preface

As a 21-year-old student at U.C.L.A. I had my first exposure to general topology, taking an introductory course from Robert Sorgenfrey who adopted for the class a sparse text written by his colleague Sze-Tsen Hu. But Hu chose to do something novel in his text. He included a section on topological spaces equipped with an auxiliary family of subsets $\mathscr{B}$ which he called a *boundedness*. While this family was only required to be hereditary and stable under finite unions, for almost all applications, $\mathscr{B}$ also formed a cover of the underlying set, and would be called a *bornology* by subsequent authors. This is the case for the family of subsets of finite diameter in a metric space, i.e., the metrically bounded sets. Hu in fact characterized those bornologies in a metrizable spaces that could be so realized with respect to some metric compatible with the topology. While he made this notable discovery 20 years before he wrote his textbook and wanted to expose it to a wider audience, Hu clearly felt that the perspective of a space being a set equipped with a topology and such an auxiliary family of subsets deserved further study. The focus of general topology is myopic, and the large structure of the space and its interaction with the small structure deserved more systematic attention. What such a study might produce was unforseen to him. For example, how was Hu to know that in a metric space, important bornologies $\mathscr{B}$ arise as families of subsets such that for some geometric functional ϕ, a subset A belongs to $\mathscr{B}$ if and only if each sequence in A along which ϕ goes to zero must cluster?

While I was consumed by convex analysis and hyperspace topology during the first part of my career, I turned to the study of topological spaces equipped with a bornology - especially metric spaces - as the turn of the century approached. My first partner in crime in this endeavor was Sandro Levi, and we initially worked on bornological convergence of nets of closed sets in a metric space, which provided an overall framework in which to consider convergence in Hausdorff distance, Attouch-Wets convergence, and convergence with respect to the Fell topology. We studied approximation of sets by members of a bornology in two different senses, and characterized when bornologies on a metrizable space could be realized as certain classical bornologies under an appropriate choice of metric. We introduced variational notions of uniform continuity and uniform convergence for functions relative to a bornology. Important colleagues in these efforts were Camillo Costantini, Giuseppe Di Maio, Alois Lechicki, Som Naimpally and Jesus Rodríguez-López.

Working separately with Cristina Vipera, we saw how all regular T_1-extensions of a topological space (as a dense subset) arise from a family of bornologies - one for each point in the remainder - on the underlying space. While our approach builds on the simple Alexandroff one-point compactification, our point of view was heretical and has not attracted so much attention, much in the way that Hu's initial discoveries did not. Bornologies and extensions is a promising area of research for young mathematicians.

The latest phase involves collaboration with Maribel Garrido, who encouraged me to think about various aspects of Lipschitz analysis and their relation to large structure. Indeed, our work now concerns more generally the interplay between bornologies and various classes of continuous functions. The most notable result in this area arguably belongs to Javi Cabello-Sánchez: the uniformly continuous real-valued functions on a metric space form a ring if and only if each subset of the metric space either belongs to the bornology of Bourbaki-bounded sets or contains an infinite uniformly isolated subset. This interplay is a now a hot topic internationally, attracting particular attention from researchers in India.

This monograph has as its genesis notes for a short course that I gave at Universidad Complutense de Madrid 8 years ago. I would like to thank Maribel Garrido and Jesus Jaramillo for receiving me so frequently at Complutense, and before that, Sandro Levi for his parallel hospitality at Milano Bicocca. It has taken so long to complete the monograph because of my technological incompetence. I need to acknowledge the staff at CRC Press - particularly Vijay Primlani - for their help in this regard. Vijay, I so appreciate your kindness and patience.

The book is divided into 39 sections and is admittedly discursively written, in part because the subject as a whole is still in its infancy. Some sections treat bornologies alone, some treat Lipschitzian behavior alone, while others are a curious melange of things. Of course, there is an attempt made to study in some detail specific bornologies and specific classes of Lipschitz-type functions. Certain themes are revisited throughout the manuscript, e.g., (1) vector lattices of real-valued functions that are either stable under pointwise product or reciprocation of its non-vanishing members, and (2) the density of Lipschitz-type functions in larger classes of continuous functions. I do not argue that the study of bornologies should be part of general topology; rather, the subject seems to fit better within the framework of set-valued analysis. In any case, there is no shortage of things to be done here going forward.

October, 2022

Gerald Beer
Burbank, California

Contents

List of Symbols

A^-	for $A \subseteq X$, those subsets of X that hit A
A^+	for $A \subseteq X$, those subsets of X that are contained in A
$\mathrm{aco}(A)$	the absolutely convex hull of a subset A of a real linear space $= \mathrm{co}(A \cup -A)$
$A \in (\mathscr{B}, \mathrm{d}) - \lim A_\lambda$	the net of closed sets $\langle A_\lambda \rangle$ is bornologically convergent to A
$A = AW_d - \lim A_\lambda$	the net of closed sets $\langle A_\lambda \rangle$ is Attouch-Wets convergent to A
$A = W_d - \lim A_\lambda$	the net of closed sets $\langle A_\lambda \rangle$ is Wijsman convergent to A
$\mathscr{A}^*$	the weakly totally bounded subsets determined by a family of subsets of $\langle X, d \rangle$
$\mathscr{A}_*$	the totally bounded subsets determined by a family of subsets of $\langle X, d \rangle$
$\alpha(\cdot)$	the Kuratowski measure of noncompactness functional for a metric space
$\mathrm{born}(\mathscr{A})$	the smallest bornology containing a family of subsets $\mathscr{A}$ of a set X
$\mathscr{B}(A)$	the principal bornology determined by a subset A, that is, $\mathrm{born}(\{A\})$
$\mathscr{B}_d(X)$	the bornology of metrically bounded subsets on a metric space $\langle X, d \rangle$
$\mathscr{B}\mathscr{B}_d(X)$	the bornology of Bourbaki bounded subsets on a metric space $\langle X, d \rangle$
$\mathscr{B}(\phi)$	the bornology of subsets B such that each sequence in B along which $\phi \to 0$ clusters
$\mathscr{B}_{\mathrm{box}}$	the box bornology on a product of sets each equipped with a bornology
$\mathscr{B}_{\mathrm{prod}}$	the product bornology on a product of sets each equipped with a bornology
$\mathscr{B}_d^{uc}(X)$	the bornology of UC-subsets on a metric space $\langle X, d \rangle$
$\mathscr{B}_d^{up}(X)$	the bornology of uniformly paracompact subsets on a metric space $\langle X, d \rangle$
$\overline{\mathscr{B}}$	for a bornology $\mathscr{B}$ on a topological space, the bornology with base $\{\mathrm{cl}(B) : B \in \mathscr{B}\}$
$\mathscr{B}^f$	the family of subsets of the domain of f to which the restriction of f is uniformly continuous
$\mathscr{B}_f$	the family of subsets of the domain of f on which f is strongly uniformly continuous
$\mathscr{B}_d^{ni}(X)$	the bornology of infinitely nonuniformly isolated subsets on $\langle X, d \rangle$
$\mathscr{B}_i(\sigma)$	for an extension $\langle X \cup I, \sigma \rangle$ of $\langle X, \tau \rangle$ in which X is dense, the ideal $\{E \subseteq X : i \notin \mathrm{cl}_\sigma (E)\}$
$\mathscr{B}^{\mathrm{small}}$	for a bornology $\mathscr{B}$, the smallest bornology whose totally bounded sets agree with $\mathscr{B}_*$
$[B, d, \varepsilon]$	for a metric space $\langle X, d \rangle$, $\{(A, C) \in \mathscr{C}(X)^2 : A \cap B \subseteq S_d(C, \varepsilon)$ and $C \cap B \subseteq S_d(A, \varepsilon)\}$
$[B, d, \varepsilon][A]$	for a closed subset A of $\langle X, d \rangle$, $\{C \in \mathscr{C}(X) : (A, C) \in [B, d, \varepsilon]\}$
$\mathbf{B}(X, Y)$	the continuous linear transformations between normed linear spaces X and Y
$\mathrm{cl}(A)$	the closure of a subset A in a topological space
$\mathrm{co}(A)$	the convex hull of a subset A of a real linear space
$C(X, Y)$	the continuous functions between topological spaces X and Y

$C_b(X, Y)$	the bounded continuous functions from $\langle X, d \rangle$ to $\langle Y, \rho \rangle$
$C_{\mathscr{B}}(X, Y)$	the family of functions uniformly continuous when restricted to members of $\mathscr{B}$
$C^s_{\mathscr{B}}(X, Y)$	the family of strongly uniformly continuous functions on a bornology $\mathscr{B}$
$CC(X, Y)$	the Cauchy continuous functions from $\langle X, d \rangle$ to $\langle Y, \rho \rangle$
$\mathscr{C}(X)$	the family of closed subsets of a topological space
$\mathscr{C}_0(X)$	the family of nonempty closed subsets of a topological space
$\mathrm{diam}_d(A)$	the diameter of a set A with respect to a metric d
$\mathrm{dom}(f)$	for $f : X \to (-\infty, \infty]$, those $x \in X$ with $f(x) < \infty$
$d(x, A)$	the distance between a point x and a subset A of a metric space $\langle X, d \rangle$
d_{box}	the box metric on a finite product of metric spaces
$D_d(A, B)$	the gap between subsets A and B of a metric space $\langle X, d \rangle$
$\mathbf{D}(X)$	the family of compatible metrics for a metrizable space $\langle X, \tau \rangle$
$\Delta_{\mathscr{B}}$	the standard uniformity for uniform convergence on a bornology $\mathscr{B}$ for Y^X
$\Delta^s_{\mathscr{B}}$	the standard uniformity for strong uniform convergence on a bornology $\mathscr{B}$ for Y^X
$\mathrm{epi}(f)$	the epigraph of an extended real-valued function f, i.e., $\{(x, \alpha) \in X \times \mathbb{R} : \alpha \geq f(x)\}$
$e_d(A, B)$	the excess of a set A over a set B in a metric space $\langle X, d \rangle$
$f\vert_A$	the restriction of f to a subset A of its domain
f^{conj}	the conjugate of a convex function with values in $(-\infty, \infty]$ with $\mathrm{epi}(f)$ closed, nonempty.
$\mathscr{F}(X)$	the bornology of finite subsets on a set X
$\mathrm{Gr}(f)$	the graph of a function f
$\mathrm{hypo}(f)$	the hypograph of an extended real-valued function f, i.e., $\{(x, \alpha) \in X \times \mathbb{R} : \alpha \leq f(x)\}$
$H_d(A, B)$	the Hausdorff distance between subsets A and B of $\langle X, d \rangle$
$\mathrm{hc}(\mathscr{A})$	the hereditary core of a family of subsets $\mathscr{A}$, i.e., $\{A \in \mathscr{A} : \mathscr{P}(A) \subseteq \mathscr{A}\}$
$\mathrm{int}(A)$	the interior of a subset A in a topological space
$I(\cdot)$	the isolation function for a metric space
I_X	the identity function on a set X
$\mathrm{Ker}(\phi)$	for $\phi \in C(X, [0, \infty))$, the set $\{x : \phi(x) = 0\}$
$\mathscr{K}(X)$	the bornology of relatively compact subsets
ℓ_1	the normed linear space of real absolutely summable sequences equipped with its usual norm
ℓ_2	the normed linear space of real square summable sequences equipped with its usual norm
$L(f)$	the smallest Lipschitz constant for a Lipschitz function f
$\mathrm{Lip}(X, Y)$	the Lipschitz functions from $\langle X, d \rangle$ to $\langle Y, \rho \rangle$
$\mathrm{Lip}_b(X, Y)$	the bounded Lipschitz functions from $\langle X, d \rangle$ to $\langle Y, \rho \rangle$
$LLD(X, Y)$	the functions that are Lipschitz for large distances from $\langle X, d \rangle$ to $\langle Y, \rho \rangle$
A'	the set of limit points of a subset A of a topological space
$m_f(\cdot)$	the modulus of continuity for a function f between metric spaces
$\mathrm{nlc}(X)$	for a topological space $\langle X, \tau \rangle$ those points with no compact neighborhood

$\|\cdot\|_\infty$	the supremum norm $\sup\{|f(x)| : x \in X\}$ where $f \in C_b(X, \mathbb{R})$
$\|\cdot\|_{\mathrm{Lip}}$	the Lipschitz norm $\max\{|f(x_0)|, L(f)\}$ where $f \in \mathrm{Lip}(X, \mathbb{R})$
$\|\cdot\|_{\mathrm{op}}$	the operator norm for continuous linear transformations or functionals
$\|\cdot\|_S$	the Sherbert norm $\|f\|_\infty + L(f)$ where $f \in \mathrm{Lip}_b(X, \mathbb{R})$
$\|\cdot\|_W$	the Weaver norm $\max\{\|f\|_\infty, L(f)\}$ where $f \in \mathrm{Lip}_b(X, \mathbb{R})$
$v(\cdot)$	the measure of local compactness functional on a metric space
0_X	the origin of a real linear space X
$o(\mathscr{B}, p)$	for a nontrivial bornology with closed base on $\langle X, \tau \rangle$, the one-point extension with ideal point p and topology $\tau \cup \{(X \cup \{p\}) \setminus B : B \in \mathscr{B} \text{ and } B = \mathrm{cl}(B)\}$
$\omega(f, x)$	the oscillation of a function f between metric spaces at a point x in the domain
$\Omega(f, A)$	the oscillation of a function f between metric spaces at a subset A of the domain
$\bar{\phi}(A)$	for $\phi \in C(X, [0, \infty))$ and $A \subseteq X$; the value $\sup_{a \in A} \phi(a)$
$\underline{\phi}(A)$	for $\phi \in C(X, [0, \infty))$ and $A \subseteq X$; the value $\inf_{a \in A} \phi(a)$
$\mathscr{P}(X)$	the power set of a set X
$\mathscr{P}_0(X)$	the nonempty members of the power set of a set X
$\mathrm{span}(A)$	in a vector space, the smallest linear subspace containing A
$\mathrm{st}(A, \mathscr{U})$	the union of those $U \in \mathscr{U}$ that hit A
$\mathrm{star}(A; 0_X)$	the union of all line segments from points of A to the origin of the linear space X
$S_d(p, \alpha)$	the open ball with center p and radius α with respect to a metric d
$S_d(A, \alpha)$	the open enlargement of radius α of a subset A with respect to a metric d
$S_d^n(A, \alpha)$	the nth-iterated open enlargement of radius α of A with respect to d
τ_d	the topology determined by a metric d on a set
τ_{H_d}	the topology determined on $\mathscr{P}(X)$ by Hausdorff distance
τ_V	the Vietoris topology on $\mathscr{P}(X)$
$\tau_{\{f_i : i \in I\}}$	the weak topology on the common domain of a family of functions $\{f_i : i \in I\}$
$\tau_0(\{\mathscr{B}_i : i \in I\})$	the topology on $X \cup I$ having as base $\{i \in I : E \in \mathscr{B}_i\} \cup (X \setminus E)$ $(E \in \mathscr{C}(X))$
$\tau_0(\sigma)$	for an extension $\langle X \cup I, \sigma \rangle$ of $\langle X, \tau \rangle$, this is short for $\tau_0(\{\mathscr{B}_i(\sigma) : ii \in I\})$
$\tau_{\mathscr{B}, d}$	the topology of bornological convergence on $\mathscr{C}(X)$ when $\mathscr{B}$ is shielded from closed sets
$\mathscr{T}\mathscr{B}_d(X)$	the bornology of totally bounded subsets on a metric space $\langle X, d \rangle$
$\mathscr{T}_{\mathscr{B}}$	the topology of uniform convergence on a bornology $\mathscr{B} \subseteq \mathscr{P}(X)$
$\mathscr{T}_{\mathscr{B}}^s$	the topology of strong uniform convergence on a bornology $\mathscr{B} \subseteq \mathscr{P}(X)$
U_X	the closed unit ball of a real normed linear space X
$UC(X, Y)$	the uniformly continuous functions from $\langle X, d \rangle$ to $\langle Y, \rho \rangle$
$UC_b(X, Y)$	the bounded uniformly continuous functions from $\langle X, d \rangle$ to $\langle Y, \rho \rangle$
$\mathscr{V}^-$	for $\mathscr{V} \subseteq \mathscr{P}(X)$, the subsets of X that hit each member of $\mathscr{V}$
X^*	the dual space of the normed linear space X equipped with the usual operator norm
$\langle X, d \rangle$	a set X equipped with a metric d, i.e., a metric space
$\langle \hat{X}, \hat{d} \rangle$	the completion of the metric space $\langle X, d \rangle$
$\langle X, \tau \rangle$	a set X equipped with a topology τ, i.e., a topological space

$\langle X, \tau, \mathscr{B} \rangle$	a topological space $\langle X, \tau \rangle$ equipped with a bornology $\mathscr{B}$, i.e., a bornological universe
$x \simeq_{\varepsilon} w$	in a metric space, there is an ε-chain of finite length from x to w
Y^X	the set of functions from a set X to a set Y
$\vee$	the least upper bound binary operation in a lattice
$\wedge$	the greatest lower bound binary operation in a lattice
$\downarrow \mathscr{A}$	the family of subsets of members of a family of sets $\mathscr{A}$
$\uparrow \mathscr{A}$	the family of supersets of members of a family of sets $\mathscr{A}$
$\Sigma(\mathscr{A})$	the family of finite unions of members of a family of sets $\mathscr{A}$

Introduction

Let X be a set with at least two points. Like a topology or a sigma algebra on X, a bornology on X is a family of subsets $\mathscr{B}$ satisfying certain properties. For a family of subsets $\mathscr{B}$, put $\downarrow \mathscr{B} := \{A : A \subseteq B \text{ for some } B \in \mathscr{B}\}$ and $\sum(\mathscr{B}) := \{A : A \text{ is a finite union of elements of } \mathscr{B}\}$. The family $\mathscr{B}$ is called an ideal of subsets provided both $\mathscr{B} = \downarrow \mathscr{B}$ and $\mathscr{B} = \sum(\mathscr{B})$. If in addition, $\mathscr{B}$ contains the singletons, then $\mathscr{B}$ is called a bornology. Thus a bornology is at once an ideal on X and a cover of X. The largest bornology is the power set of X (in which case the bornology is called trivial) and the smallest is the family of finite subsets $\mathscr{F}(X)$. We call $\mathscr{B}_0$ a base for the bornology $\mathscr{B}$ provided $\mathscr{B} = \downarrow \mathscr{B}_0$. Given a family $\mathscr{A}$ of subsets of X, the smallest bornology containing it is clearly $\text{born}(\mathscr{A}) := \downarrow \sum(\mathscr{A} \cup \mathscr{F}(X))$.

Hu [102, 103] conceived of a "space" as a triple: a nonempty set X equipped with a topology τ and a bornology $\mathscr{B}$, where there is to be some interplay between τ and $\mathscr{B}$. For example, each member B of the bornology of relatively compact subsets on a Hausdorff space has compact closure. We call a triple $\langle X, \tau, \mathscr{B} \rangle$ a bornological universe. Hu showed that for a metrizable space, a bornology $\mathscr{B}$ on X is the bornology of metrically bounded subsets with respect to some compatible metric d if and only if $\mathscr{B}$ has an open base, a closed base, and a countable base. For a noncompact metrizable space, there are uncountably many distinct bornologies of this kind [18]. Hogbe-Nlend [97] considered bornologies in the context of locally convex spaces, where he insisted on some interplay between the bornology and the vector operations.

Sections 11 and 12 introduce basic terminology and constructions relative to sets equipped with bornologies and more specifically bornological universes. For example, a net $\langle x_\lambda \rangle$ in a set X equipped with a bornology $\mathscr{B}$ is declared convergent to ∞ provided it is eventually outside each member of $\mathscr{B}$. Given a family $\{\langle X_i, \mathscr{B}_i \rangle : i \in I\}$ of sets equipped with bornologies, we describe two ways to equip $\prod_{i \in I} X_i$ with a bornology that respects the factor bornologies. Given a bornology $\mathscr{B}_1$ on X and a bornology $\mathscr{B}_2$ on Y we call $f : X \to Y$ bornological provided $f(B) \in \mathscr{B}_2$ whenever $B \in \mathscr{B}_1$. Dually, we call f coercive if $f^{-1}(B) \in \mathscr{B}_1$ whenever $B \in \mathscr{B}_2$. Our usage of coercive is motivated by its usage in optimization theory, as a coercive function maps nets convergent to ∞ with respect to $\mathscr{B}_1$ to nets convergent to ∞ with respect to $\mathscr{B}_2$. It is easy to see that these two properties are independent of one another in the context of continuous functions between metric spaces where the bornologies are metric bornologies.

A bornology $\mathscr{B}$ on $\langle X, \tau \rangle$ is called local if it contains a neighborhood of each point of X. On a metric space, it is called stable under small enlargments if for each $B \in \mathscr{B}$ there exists $\varepsilon > 0$ such that the ε-enlargement $S_d(B, \varepsilon)$ belongs to $\mathscr{B}$. The relatively compact subsets in a locally compact metric space evidently have this property. A weaker but very important property is that $\mathscr{B}$ be shielded from closed sets: each $B \in \mathscr{B}$ has a superset $B_1 \in \mathscr{B}$ such that each neighborhood of B_1 contains an enlargement of B. The relatively compact subsets in an arbitrary metric space are shielded from closed sets, as for each relatively compact set, its closure serves as a shield. A bornology that is shielded from closed sets must have a closed base. Bornologies describe the large structure of a space.

In a metric space $\langle X, d \rangle$ there are other important bornologies besides the metrically bounded subsets and the relatively compact subsets: the totally bounded subsets, the Bourbaki bounded subsets, the UC-subsets, the uniformly paracompact subsets, etc. All of these can be characterized sequentially and in terms of the behavior of particular families of continuous functions on them. For example, (1) a subset B is metrically bounded if and only if each Lipschitz function on X is bounded restricted to B; (2) a subset B is totally bounded if and only if each sequence in it has a Cauchy subsequence; (3) a subset B is Bourbaki bounded if and only if each uniformly continuous function on X restricted to B is bounded [119]; (4) a subset B is a UC-subset if and only if whenever $\langle b_n \rangle$ is a sequence in B with $\lim_{n \to \infty} d(b_n, X \backslash \{b_n\}) = 0$, then the sequence must cluster; (5) a subset B is uniformly paracompact if and only if for each continuous function f on X, there exists $\delta > 0$ such that f restricted to the ball of radius δ about each point of B is bounded. Following Hu, we try to identify conditions for a bornology on a metrizable space so that it is of a particular type with respect to an appropriately chosen compatible metric.

As with the bornology of UC-subsets, membership to the bornology of uniformly paracompact subsets is characterized by the clustering of sequences in it along which a certain geometric functional ϕ goes to zero - in this second case, it is the measure of local compactness functional [32]. In Section 28, we consider bornologies induced by such a signature functional collectively, obtaining as particular cases known results for UC-subsets and uniformly paracompact subsets. We explain how such functionals ϕ often arise from hereditary properties of open sets.

We are interested in when particular bornologies coincide. A classical result in this direction: a metrizable space has a compatible metric for which the relatively compact subsets coincide with the metrically bounded subsets if and only if X is locally compact and separable [149]. The relatively compact subsets coincide with the Bourbaki bounded subsets in $\langle X, d \rangle$ if and only if each so-called Bourbaki-Cauchy sequence in X clusters, as discovered by Garrido and Meroño [86].

As expected, attention is given to spaces where X itself belongs to a given bornology, in which case the bornology is trivial. Variational aspects of membership to the bornology become hidden and statements become misleadingly simple. For example, $\langle X, d \rangle$ is a UC-space if and only if each open cover of the space has a Lebesgue number. As another example, a metrizable space $\langle X, \tau \rangle$ has a compatible metric so that the space becomes uniformly paracompact if and only if the set of points in X with no compact neighborhood is compact, as first observed by Romaguera [137].

If $\langle X, d \rangle$ and $\langle Y, \rho \rangle$ are metric spaces and $\mathscr{A}$ is a cover of X, then the topology of uniform convergence for Y^X on each member of $\mathscr{A}$ is made no stronger by replacing $\mathscr{A}$ by $\mathrm{born}(\mathscr{A})$. Thus we may assume that all classical topologies of uniform convergence on X containing the topology of pointwise convergence are determined by a bornology $\mathscr{B}$ on X. There is a useful variational strengthening of uniform convergence on $\mathscr{B}$ called strong uniform convergence on $\mathscr{B}$: if $\langle f_\lambda \rangle$ is a net in Y^X, then it is strongly uniformly convergent to $f \in Y^X$ relative to $\mathscr{B}$ if $\forall B \in \mathscr{B}\ \forall \varepsilon > 0$, there exists λ_0 such for all $\lambda \succeq \lambda_0$ there exists $\delta > 0$ (dependent on λ as well as ε) such that $x \in S_d(B, \delta) \Rightarrow \rho(f_\lambda(x), f(x)) < \varepsilon$ [45]. Convergence so described is uniformizable in a way that parallels the classical case. Strong uniform convergence comports well with functions g that are strongly uniformly continuous on members of $\mathscr{B}$: $\forall B \in \mathscr{B}\ \forall \varepsilon > 0$, there exists $\delta > 0$ such that $\max \{d(x, B), d(w, B), d(x, w)\} < \delta \Rightarrow \rho(g(x), g(w)) < \varepsilon$. It can be shown B is a UC-subset if and only if each continuous function on X is strongly uniformly continuous on B, and thus the interest in this particular bornology. Strong uniform convergence on $\mathscr{B}$ agrees with the classical notion for Y^X (resp. $C(X, Y)$) if and only if $\mathscr{B}$ is stable under small enlargements (resp. $\downarrow \{\mathrm{cl}(B) : B \in \mathscr{B}\}$ is shielded from closed sets) [46]. Falling out of our analysis is the fact that a pointwise convergent net of continuous functions has a continuous limit if and only if the net is strongly uniformly convergent relative to $\mathscr{F}(X)$ [45, 55].

We are interested in vector lattices of real-valued functions - especially Lipschitz and locally Lipschitz functions - and their interplay with bornologies. For a vector lattice Ω that contains the constant functions, it was shown by Beer, García-Lirola and Garrido that if the reciprocal of each nonvanishing member of Ω is back in Ω, then Ω is stable under pointwise product [33]. The most notable pair of results here involve $UC(X, \mathbb{R})$, the uniformly continuous real-valued functions on $\langle X, d \rangle$: $UC(X, \mathbb{R})$ is stable under reciprocation if and only if X is a UC-space [38] and stable under pointwise product if and only if each subset of X is either Bourbaki bounded or contains an infinite uniformly isolated subset, as shown by Cabello-Sánchez [61].

The locally Lipschitz functions on $\langle X, d \rangle$ with values in a second metric space $\langle Y, \rho \rangle$ are exactly those functions that are Lipschitz when restricted to each relatively compact subset. Within the locally Lipschitz functions, we pay special attention to the functions that are Lipschitz when restricted to the range of each Cauchy sequence and to functions f that are Lipschitz in the small: $\exists \delta > 0 \ \exists \lambda > 0$ such that $d(x, w) < \delta \Rightarrow \rho(f(x), f(w)) \leq \lambda d(x, w)$. For real-valued functions, the locally Lipschitz functions are uniformly dense in the continuous functions $C(X, \mathbb{R})$; the Cauchy-Lipschitz functions are uniformly dense in the functions that map Cauchy sequences in X to Cauchy sequences in $\mathbb{R}$; the Lipschitz in the small functions are uniformly dense in $UC(X, \mathbb{R})$ [35, 36, 65, 84, 123]. The family of subsets on which the Cauchy-Lipschitz functions are all Lipschitz is exactly the bornology of totally bounded subsets. The family of subsets on which each Lipschitz in the small function is Lipschitz - which need not be a bornology - has been characterized by by Leung and Tan [114].

An intriguing result of Garrido and Jaramillo that has a functional analytic proof says that if $h : \langle X, d \rangle \to \langle Y, \rho \rangle$ has the property that if whenever $f : \mathbb{R} \to \mathbb{R}$ is Lipschitz, $f \circ h$ remains Lipschitz, then h itself must be Lipschitz [83]. We show through a series of results that this is no isolated phenomenon. For example, a function between metric space is uniformly continuous provided whenever it is followed by a real-valued Lipschitz function in a composition, the result remains uniformly continuous. We do other functional analysis here and there. Following Michael [122], we show that each metric space can be isometrically embedded in the continuous dual of the space $\mathrm{Lip}(X, \mathbb{R})$ of real-valued Lipschitz functions on X equipped with a natural norm, and related to that, give his short proof of the Arens-Eells theorem. We introduce the Weaver norm and Sherbert norm on the bounded real-valued Lipschitz functions, but do not deal with them significantly.

Section 31 through Section 36 are devoted to the notion of bornological convergence of nets of closed subsets in a metric space $\langle X, d \rangle$ as introduced by Lechicki, Levi and Spakowski [113], but which has as its point of departure Attouch-Wets convergence, of fundamental importance in convex and variational analysis. Let $\mathscr{B}$ be a bornology on a metric space $\langle X, d \rangle$ and let $\mathscr{C}(X)$ be the family of all closed subsets, writing $\mathscr{C}_0(X)$ for the nonempty closed subsets. A net $\langle A_\lambda \rangle$ in $\mathscr{C}(X)$ is declared $(\mathscr{B}, d)$-convergent to $A \in \mathscr{C}(X)$ provided for each $B \in \mathscr{B}$ and $\varepsilon > 0$, eventually we have both

$$A_\lambda \cap B \subseteq S_d(A, \varepsilon) \quad \text{and} \quad A \cap B \subseteq S_d(A_\lambda, \varepsilon).$$

When $\mathscr{B}$ is the bornology of metrically bounded subsets, this is Attouch-Wets convergence. For the trivial bornology, and for nets in $\mathscr{C}_0(X)$, this is nothing but convergence in Hausdorff distance. For the bornology of relatively compact subsets, this coincides with convergence in the Fell topology on $\mathscr{C}(X)$, which is curiously always compact.

As shown by Beer, Costantini and Levi [29], the convergence is topological if and only if $\mathscr{B}$ is shielded from closed sets. The topology is uniformizable if and only if $\mathscr{B}$ is stable under small enlargements, and it is metrizable if and only if in addition $\mathscr{B}$ has a countable base. In this case, there is an equivalent metric ρ so that $(\mathscr{B}, d)$-convergence becomes Attouch-Wets convergence as determined by ρ [42].

For nets of nonempty closed sets, necessary and sufficient conditions are given due to Beer, Naimpally, and Rodríguez-López for $(\mathscr{B}, d)$-convergence of $\langle A_\lambda \rangle$ to A to coincide with the uniform convergence of $\langle d(\cdot, A_\lambda) \rangle$ to $d(\cdot, A)$ on members of $\mathscr{B}$ [48]. This of course is the case for Attouch-Wets convergence and convergence in Hausdorff distance.

As an application, we reconcile various modes of convergence for continuous linear transformations between normed linear spaces with the bornological convergence of their graphs. In our analysis, completeness is irrelevant while starshapedness plays a basic role. As an example, convergence in the operator norm corresponds to Attouch-Wets convergence of graphs, which in this setting are closed linear subspaces of the product of the domain and the codomain [22, 131].

The last two sections detail the intimate connection between families of bornologies on a topological space and extensions of the space in which the base space is dense. Suppose $\langle X, \tau \rangle$ is a topological space and $\{\mathscr{B}_i : i \in I\}$ is a family of nontrivial bornologies on X each with closed base and where $X \cap I = \emptyset$. Actually a family of ideals will do. We describe an associated extension topology on $X \cup I$ such that X is dense in $X \cup I$: take as a base all sets of the form $\{i \in I : E \in \mathscr{B}_i\} \cup (X \backslash E)$ where E runs over $\mathscr{C}(X)$. Remarkably, all regular extensions $\langle X \cup I, \sigma \rangle$ in which X is dense arise in this way. More precisely, if X is dense in the extension $\langle X \cup I, \sigma \rangle$, then a family of ideals on X indexed by I exists so determining the extension topology if and only if $\{(X \cup I) \backslash \mathrm{cl}_\sigma (E) : E \in \mathscr{C}(X)\}$ is a base for σ.

Following Beer and Vipera [54], we call such an extension bornological whether or not each ideal is actually a bornology. For a bornological extension $\langle X \cup I, \sigma \rangle$, there is a unique family of ideals on X indexed by I that determine the extension in the way described: each is of the form $\{E \subseteq X : i \notin \mathrm{cl}_\sigma(E)\}$. From this description, each of the ideals must be a bornology provided the bornological extension is T_1, better justifying our terminology.

1. Background Material

Let X be a set with at least two points and without assumed auxiliary structure. If A and B are subsets of X, we write $A \subseteq B$ if A is a subset of B and $A \subset B$ if the inclusion is proper. If $A \subseteq X$, we denote its complement by A^c or by $X \backslash A$. The power set of X is denoted by $\mathscr{P}(X)$ while the family of nonempty subsets will be denoted by $\mathscr{P}_0(X)$. The family of finite subsets of X will be denoted by $\mathscr{F}(X)$. If $\mathscr{A}$ is a family of subsets of X, by $\downarrow \mathscr{A}$ we mean the *hereditary hull* of $\mathscr{A}$, i.e., $\{E \subseteq X : \exists A \in \mathscr{A} \text{ with } E \subseteq A\}$ [44]. Of course, if $\mathscr{A} = \downarrow \mathscr{A}$, we call $\mathscr{A}$ *hereditary*. The set of all finite unions of members of $\mathscr{A}$ will be denoted by $\sum(\mathscr{A})$.

Our first result collects some basic facts about these operators.

Proposition 1.1. *Let $\mathscr{A}, \mathscr{B}$ be families of subsets of a set X. Then*

(1) $\mathscr{A} \subseteq \mathscr{B} \Rightarrow \downarrow \mathscr{A} \subseteq \downarrow \mathscr{B}$ *and* $\sum(\mathscr{A}) \subseteq \sum(\mathscr{B})$.

(2) $\mathscr{A} \subseteq \downarrow \mathscr{A}$ *and* $\mathscr{A} \subseteq \sum(\mathscr{A})$.

(3) $\downarrow \sum(\mathscr{A}) = \sum(\downarrow \mathscr{A})$.

(4) $\downarrow\downarrow (\mathscr{A}) = \downarrow (\mathscr{A})$ *and* $\sum(\sum(\mathscr{A})) = \sum(\mathscr{A})$.

By an *ideal* $\mathscr{A}$ of subsets of X, we mean a family satisfying $\mathscr{A} = \downarrow \sum(\mathscr{A})$. An ideal $\mathscr{A}$ is called *nontrivial* if it does not contain X; equivalently, the ideal is a proper subfamily of the power set of X. Dually, $\uparrow \mathscr{A}$ will denote $\{E \subseteq X : \exists A \in \mathscr{A} \text{ with } A \subseteq E\}$. The downarrow operator will be much more important to us here than the uparrow operator.

By a *neighborhood* of subset A of a topological space $\langle X, \tau \rangle$, we mean some $V \in \tau$ with $A \subseteq V$. Topological spaces in this text will be assumed to contain at least two points and to satisfy the *Hausdorff separation axiom*: distinct points have disjoint neighborhoods. A Hausdorff space is called *regular* if whenever $A \subseteq X$ is closed and $p \in A^c$, then p and A have disjoint neighborhoods. It is called *normal* if disjoint closed subsets have disjoint neighborhoods. Thus, in Hausdorff spaces, each normal space is regular.

If A is a subset of a Hausdorff space $\langle X, \tau \rangle$ we write $\text{cl}(A)$ for its closure, $\text{int}\,(A)$ for its interior, and A' for its set of limit points. If $A \subseteq B \subseteq X$ we say that A is *dense* in B provided $B \subseteq \text{cl}(A)$. We will denote the closed subsets of $\langle X, \tau \rangle$ by $\mathscr{C}(X)$ and the nonempty closed subsets by $\mathscr{C}_0(X)$. If A is a nonempty subset of X, the relative topology on A will be denoted by τ_A. The product topology on the product of an indexed family of Hausdorff spaces will be denoted by τ_{prod}.

Let $\mathscr{V}$ be a family of subsets of $\langle X, \tau \rangle$. The family is called *locally finite* if each point $x \in X$ has a neighborhood that intersects at most finitely many members of $\mathscr{V}$. The family is called *discrete* if each point has a neighborhood that intersects at most one member of the family. A second family $\mathscr{U}$ is said to *refine* $\mathscr{V}$ if for each $U \in \mathscr{U}$ there exists $V \in \mathscr{V}$ with $U \subseteq V$. We say that $\mathscr{V}$ is a *cover* of $A \subseteq X$ if $A \subseteq \cup \mathscr{V}$. We call A *compact* if each open cover of A has a finite subcover. A Hausdorff space is called *paracompact* if for each open cover $\mathscr{V}$ of X, there is a locally finite open cover of X that refines $\mathcal{V}$. Each compact Hausdorff space is paracompact and each paracompact space is normal [155, p. 147]. Normality of course is equivalent to the existence of a Urysohn function for each pair A and B of disjoint nonempty closed subsets of X: a continuous function on X with values in $[0, 1]$ mapping one set to zero and other to one.

If $\mathcal{U}$ is a cover of X and $A \subseteq X$, we put $\mathrm{St}(A, \mathcal{U}) := \cup\{U \in \mathcal{U} : U \cap A \neq \emptyset\}$, writing $\mathrm{St}(x, \mathcal{U})$ for $\mathrm{St}(\{x\}, \mathcal{U})$. A cover $\mathcal{V}$ is said to *star-refine* $\mathcal{U}$ provided the associated cover $\{\mathrm{St}(V, \mathcal{V}) : V \in \mathcal{V}\}$ refines $\mathcal{U}$. As shown by A. Stone [141], the existence of an open star-refinement for each open cover of X is equivalent to paracompactness of $\langle X, \tau \rangle$.

We write Y^X for the set of all functions with domain X and codomain Y. If $f \in Y^X$, we write $\mathrm{Gr}(f)$ for its graph. If $A \in \mathscr{P}_0(X)$ and $f \in Y^X$, we write $f|_A$ for the restriction of f to A. The function space $\mathbb{R}^X$ is a vector space under ordinary scalar multiplication and addition of functions. Given $f, g \in \mathbb{R}^X$ we put $f \vee g := \max\{f, g\}$ and $f \wedge g := \min\{f, g\}$. In this way, the ordinary pointwise partial order on real-valued functions makes $\mathbb{R}^X$ a *lattice*, i.e., each pair of elements $\mathbb{R}^X$ have both a least upper bound and a greatest lower bound. Consistent with the language of lattice theory, we call these the *join* and *meet* of f and g, respectively.

We call a subfamily Ω of $\mathbb{R}^X$ a *vector lattice* if it is both a vector subspace of $\mathbb{R}^X$ and a sublattice of $\mathbb{R}^X$ under $\vee$ and $\wedge$ as we have defined them. For a vector subspace Ω of $\mathbb{R}^X$, being a vector lattice is equivalent to the requirement $f \in \Omega \Rightarrow |f| \in \Omega$. This follows from the following two observations.

(a) $f \vee g = \frac{f+g}{2} + \frac{|f-g|}{2}$ and $f \wedge g = \frac{f+g}{2} - \frac{|f-g|}{2}$;

(b) $|f| = f \vee -f$.

Two other properties that we might look for in a vector lattice Ω of real-valued functions is that it be stable under pointwise product or that whenever $f \in \Omega$ is never zero, then its reciprocal is back in Ω. We call the second property *stability under reciprocation*. Provided Ω contains the constant functions, the second property implies the first, as observed by Beer, García-Lirola, and Garrido [33, Theorem 1.1]. Of course, if Ω contains the constant functions, then stability under pointwise product means that Ω is a ring with multiplicative identity.

Notice that the condition that a vector space Ω be stable under pointwise product is equivalent to the formally weaker condition $f \in \Omega \Rightarrow f^2 \in \Omega$, because

$$fg = \frac{1}{4}[(f+g)^2 - (f-g)^2].$$

Theorem 1.2. *Let Ω be a vector lattice of real-valued functions on a nonempty set X containing the constant functions. If Ω is stable under reciprocation, then Ω is stable under pointwise product.*

Proof. It suffices to show that Ω is stable under squaring. We first claim that if $g \in \Omega$ and there exists $\alpha > 0$ such that for all $x \in X$, $|g(x)| \neq \alpha$, then $g^2 \in \Omega$. Indeed, note that

$$\frac{1}{g^2 - \alpha^2} = \frac{1}{2\alpha(g - \alpha)} - \frac{1}{2\alpha(g + \alpha)}$$

from which $g^2 - \alpha^2$ and then g^2 belong to Ω.

Now let $f \in \Omega$ be arbitrary and put $g := |f| + 2$ which fulfills the above condition with $\alpha = 1$. Then $g^2 \in \Omega$ and $f^2 = |f|^2 = g^2 - 4|f| - 4$. This implies that $f^2 \in \Omega$ as well. $\square$

One focus of this monograph is to identify necessary and sufficient conditions for particular vector lattices of real-valued functions to be stable under pointwise product and to be stable under reciprocation.

Corollary 1.3. *Let Ω be a vector lattice of real-valued functions on a nonempty set X containing the constant functions. If Ω is stable under reciprocation, then whenever $f \in \Omega$ is nonvanishing and $g \in \Omega$, we have $\frac{g}{f} \in \Omega$.*

Recall that a function f between Hausdorff spaces $\langle X, \tau \rangle$ and $\langle Y, \sigma \rangle$ is called *continuous* at $p \in X$ if whenever $f(p) \in V \in \sigma$ there exists a neighborhood U of p with $f(U) \subseteq V$; f is globally continuous if and only if the inverse image of each open subset of Y is open in X. Each globally continuous function has closed graph. If $f : X \to Y$ and $g : X \to Y$ are continuous and A is a dense subset such that $f|_A = g|_A$, then $f = g$. This means that the values of a globally defined continuous function between two Hausdorff spaces are determined by its restriction to any dense subset.

We write $C(X, Y)$ for the space of (globally) continuous functions from X to Y. To say that f is a *homeomorphism* or *bicontinuous* means that f is a continuous bijection and $f^{-1} : Y \to X$ is continuous as well. If $f \in C(X, Y)$ is merely one-to-one, we say that f is an *embedding* provided that f determines a homeomorphism between X and $f(X)$, equipped with its relative topology.

If $\{f_i : i \in I\}$ is a family of functions on a nontrivial set X where $f_i : X \to \langle Y_i, \tau_i \rangle$, the topology generated by $\{f_i^{-1}(V) : i \in I, V \in \tau_i\}$ is the weakest topology on X with respect to which each f_i is continuous. We denote this topology by $\tau_{\{f_i : i \in I\}}$ and call it the *weak topology* determined by the family. A function h from another topological space into X equipped with $\tau_{\{f_i : i \in I\}}$ is continuous if and only if $f_i \circ h$ is continuous for each index i [155, p. 56]. The weak topology is Hausdorff provided the family *separates points*, i.e., for $x \neq w$ in X, there exists f_i for which $f_i(x) \neq f_i(w)$.

For example, the product topology is the weak topology determined by the projection maps. A topological space $\langle X, \tau \rangle$ is called *completely regular* if for each nonempty closed subset A of $\langle X, d \rangle$ and each $p \notin A$ there exists $f \in C(X, [0, 1])$ with $f(p) = 0$ and $f(A) = \{1\}$. In the context of Hausdorff spaces, such spaces are also called *Tychonoff spaces*. A Hausdorff space $\langle X, \tau \rangle$ is completely regular if and only if the weak topology on X determined by $C(X, \mathbb{R})$ agrees with τ [155, p. 96]. Thus, a function h into a completely regular space is continuous if and only if $f \circ h$ is continuous for each $f \in C(X, \mathbb{R})$.

Each open cover $\mathscr{V}$ of a paracompact space $\langle X, \tau \rangle$ has a *continuous partition of unity* $\{p_i : i \in I\}$ *subordinated to it*. This means that (i) each p_i belongs to $C(X, [0, 1])$; (ii) $\{\{x \in X : p_i(x) > 0\} : i \in I\}$ is a locally finite (open) refinement of $\mathscr{V}$; and (iii) $\forall x \in X$, $\sum_{i \in I} p_i(x) = 1$ [74, 155]. These are useful in constructing globally defined functions meeting certain requirements.

Occasionally, we will work with a function f defined on a Hausdorff space with values in $(-\infty, \infty]$; in this case, its *effective domain* is the set $\mathrm{dom}(f) := \{x \in X : f(x) < \infty\}$ [17]. By the *epigraph* of such an extended real-valued function, we mean

$$\mathrm{epi}(f) := \{(x, \alpha) : x \in X, \alpha \in \mathbb{R}, \text{ and } \alpha \geq f(x)\}.$$

Dually, the *hypograph* of f is given by

$$\mathrm{hypo}(f) := \{(x, \alpha) : x \in X, \alpha \in \mathbb{R}, \text{ and } \alpha \leq f(x)\}.$$

Such a function is called *lower* (resp. *upper*) *semicontinuous* if $\{x : f(x) > \alpha\}$ (resp. $\{x : f(x) < \alpha\}$) is open for each $\alpha \in \mathbb{R}$.

Proposition 1.4. *Let $\langle X, \tau \rangle$ be a Hausdorff space and let $f : X \to (-\infty, \infty]$. Then f is lower semicontinuous if and only if its epigraph is a closed subset of $X \times \mathbb{R}$.*

Proof. For necessity, suppose f is lower semicontinuous and $(p, \alpha) \notin \mathrm{epi}(f)$. Choose β with $f(p) > \beta > \alpha$. Then $\{x : f(x) > \beta\} \times (-\infty, \beta)$ is a neighborhood of (p, α) disjoint from $\mathrm{epi}(f)$. Conversely, suppose $\alpha \in \mathbb{R}$ and $f(p) > \alpha$. Since we are assuming $\mathrm{epi}(f)$ is closed, there exists $V \in \tau$ and $\varepsilon > 0$ such that $V \times (\alpha - \varepsilon, \alpha + \varepsilon) \cap \mathrm{epi}(f) = \emptyset$. Since epigraphs recede upwards, we have $\forall x \in V$, $f(x) > \alpha$, and so $\{x : f(x) > \alpha\}$ contains a neighborhood of p. $\quad\square$

Corollary 1.5. *Let $\{f_i : i \in I\}$ be a family of lower semicontinuous functions on a Hausdorff space $\langle X, \tau \rangle$. Then $\sup_{i \in I} f_i$ is lower semicontinuous.*

Proof. The epigraph of a supremum of a family of extended real-valued functions is the intersection of their epigraphs. $\quad\square$

Dually f is upper semicontinuous if its hypograph is a closed subset of $X \times \mathbb{R}$. Of course, a real-valued function with closed graph need not be continuous; instead, continuity for a real-valued function means that both its epigraph and hypograph are closed subsets.

By a *metric* d on X we mean a function from $X \times X$ to $[0, \infty)$ satisfying the following conditions for all x, w, p in X: (1) $d(x, w) = 0$ if and only if $x = w$; (2) $d(x, w) = d(w, x)$; (3) $d(x, w) \leq d(x, p) + d(p, w)$. We write $\langle X, d \rangle$ for X equipped with the metric d. All metric spaces will be assumed to contain at least two points.

Example 1.6. The *Euclidean metric* d on $\mathbb{R}^n$ is defined as follows: if $x = (\alpha_1, \alpha_2, \ldots, \alpha_n)$ and $w = (\beta_1, \beta_2, \ldots, \beta_n)$, then $d(x, w) := \sqrt{\sum_{i=1}^{n} (\alpha_i - \beta_i)^2}$.

Example 1.7. The *discrete metric* or *zero-one metric* on X is defined by $d(x, w) = 0$ if $x = w$ and $d(x, w) = 1$ if $x \neq w$.

We call a metric d on a set X *convex* provided whenever $x \neq w$ in X and $\alpha \in (0, d(x, w))$, then there exists $z \in X$ with $d(x, z) = \alpha$ and $d(z, w) = d(x, w) - \alpha$ [5, 53]. The Euclidean metric is convex while the discrete metric is not. We call the metric space $\langle X, d \rangle$ *metrically convex* if the metric d is convex.

On occasion, we will allow our metrics to assume the value ∞; such distances are called *extended metrics*. If condition (1) in the definition is weakened to allow distinct points to have distance zero between them, the distance we get is called a *pseudometric*. For example, if X is a set with at least two points and $f : X \to \mathbb{R}$ is a function that fails to be one-to-one, then $d(x_1, x_2) := |f(x_1) - f(x_2)|$ is a pseudometric but not a metric. Adding a pseudometric to a metric produces a metric.

If X is a real linear space, we denote its origin by 0_X, and if A is a nonempty subset of X, we write $\mathrm{span}(A)$ for the smallest vector subspace of X containing A. This consists of all linear combinations of elements of A. We call a subset A of a real linear space *convex* if whenever $a_1 \in A$ and $a_2 \in A$ and $t \in [0, 1]$, then $ta_1 + (1 - t)a_2 \in A$. We call A *absolutely convex* if in addition $A = -A$. By a *convex combination* of a finite set of elements in X, we mean a linear combination using nonnegative scalars that sum to one. The set of all convex combinations of elements of a subset A, called the *convex hull* of A, is the smallest convex subset of X containing A. We denote the convex hull of A by $\mathrm{co}(A)$. By a *polytope*, we mean the convex hull of a finite set.

On the other hand, $\mathrm{aco}(A) := \mathrm{co}(A \cup -A)$ is evidently the smallest absolutely convex set containing A. This consists of all linear combinations of the form

$$\alpha_1 a_1 + \alpha_2 a_2 + \cdots + \alpha_n a_n \quad (\{a_1, a_2, \ldots, a_n\} \subseteq A \text{ and } \sum_{j=1}^{n} |\alpha_j| = 1)$$

Naturally, $\mathrm{aco}(A)$ is called the *absolutely convex hull* of A; clearly, $\mathrm{co}(A) \subseteq \mathrm{aco}(A)$.

A function T between real linear spaces X and Y is called a *linear transformation* if it preserves linear combinations of elements of X. If $y_0 \in Y$ is fixed and $T : X \to Y$ is a linear transformation, then $x \mapsto T(x) + y_0$ $(x \in X)$ is called an *affine function*. Such functions are the ones that preserve linear combinations whose coefficients sum to one.

By a *norm* $|| \cdot ||$ on a real linear space X, we mean a function from X to $[0, \infty)$ such that for all $x, w \in X$ and $\alpha \in \mathbb{R}$, (1) $||x|| = 0$ if and only if $x = 0_X$; (2) $||\alpha x|| = |\alpha| \cdot ||x||$; (3) $||x + w|| \leq ||x|| + ||w||$. If $|| \cdot ||$ is a norm on X, then $d(x, w) := ||x - w||$ is called *the metric induced by the norm*. The Euclidean metric is determined by the *Euclidean norm* on $\mathbb{R}^n$: $||x|| := \sqrt{\sum_{i=1}^{n} \alpha_i^2}$ where $x = (\alpha_1, \alpha_2, \ldots, \alpha_n)$. Another important normed linear space for our purposes is the vector space $C_b(X, \mathbb{R})$ of bounded real-valued continuous functions on a Hausdorff space $\langle X, \tau \rangle$ equipped with the *supremum norm* $|| \cdot ||_\infty$ defined by

$$||f||_\infty := \sup_{x \in X} |f(x)|.$$

The distance between two such functions in this normed linear space is called *uniform distance*, and convergence of a sequence of functions with respect to $|| \cdot ||_\infty$ means uniform convergence. More generally, if $\langle Y, \rho \rangle$ is a metric space and $f, g : X \to Y$, we call $\sup_{x \in X} \rho(f(x), g(x))$ the uniform distance between f and g, whether or not the supremum is finite. This extended distance is induced by an extended norm on Y^X. A *seminorm* p on a real linear space X replaces condition (1) by the weaker condition (1') $p(0_X) = 0$, retaining (2) $p(\alpha x) = |\alpha| p(x)$ and (3) $p(x + w) \leq p(x) + p(w)$. So for example, the rule of assignment $f \mapsto \int_a^b |f(x)| dx$ is a seminorm on the Riemann integrable real-valued functions on $[a, b]$.

We denote the *dual space* of continuous linear functionals on a normed linear space X by X^*. By linearity, f belongs to X^* provided it is just continuous at the origin, in which case $f(\{x : ||x|| \leq 1\})$ is a bounded subset of $\mathbb{R}$ and conversely. We equip X^* with the *operator norm* defined by

$$||f||_{\mathrm{op}} := \sup\{|f(x)| : ||x|| \leq 1\}.$$

Of course, this is nothing but $||\tilde{f}||_\infty$ where $\tilde{f}$ is the restriction of f to $\{x : ||x|| \leq 1\}$, the closed unit ball U_X of the normed linear space. The distance between two linear functionals f and g determined by $|| \cdot ||_{\mathrm{op}}$ is given by $\sup_{||x|| \leq 1} |f(x) - g(x)|$. More generally, if $\langle Y, || \cdot ||_0 \rangle$ is a second real normed linear space, then a linear transformation $T : X \to Y$ is globally continuous if and only if it is continuous at O_X, in which case we put $||T||_{\mathrm{op}} := \sup\{||T(x)||_0 : x \in U_X\}$. We denote the continuous linear transformations from X to Y by $\mathbf{B}(X, Y)$; note that $X^* = \mathbf{B}(X, \mathbb{R})$.

If $B \subseteq \langle X, d \rangle$ and $\varepsilon > 0$, we call B ε-*discrete* if whenever $b_1 \neq b_2$ in B, we have $d(b_1, b_2) \geq \varepsilon$. More generally, we call B *uniformly discrete* if it is ε-discrete for some positive ε. Notice that any finite subset is uniformly discrete. If $a \in X$, we define *the distance from a to $B \subseteq X$* by

$$d(a, B) := \inf\{d(a, b) : b \in B\}.$$

The definition is consistent with our definition of distance between points and gives $d(a, \emptyset) = \infty$. We denote $d(p, \{p\}^c)$ by $I(p)$ and call this the *isolation* of p in X. Note that $I(p) = 0$ means that p is a limit point of X. We call a nonempty subset A of X *uniformly isolated* if $\inf_{a \in A} I(a) > 0$.

If A and B are subsets of X, at least one of which is nonempty, we define the *gap* between them by
$$D_d(A, B) := \inf\{d(a, b) : a \in A, b \in B\}.$$

As a special case, when $A = \{a\}$, we get the distance from a to B, and if $A \neq \emptyset$, then $D_d(A, \emptyset) = D_d(\emptyset, A) = \infty$. Two nonempty subsets are called *near* if the gap between them is zero; intuitively, this means they either intersect or are asymptotic.

If $A \neq \emptyset$ and B are subsets of X, we define the *excess of A over B* by
$$e_d(A, B) := \sup\{d(a, B) : a \in A\}.$$

In particular, when $B = \emptyset$, we get $e(A, \emptyset) = \infty$. We adopt the convention $e_d(\emptyset, B) = 0$ whatever $B \subseteq X$ may be.

Given $p \in X$ and $\alpha > 0$, the *open ball* with center p and radius α is defined by
$$S_d(p, \alpha) := \{x \in X : d(x, p) < \alpha\}.$$

A subset A of X is called *d-bounded* or *metrically bounded* if it is contained in an open ball. For a nonempty subset, A is d-bounded if and only if its *diameter* $\operatorname{diam}_d(A) := \sup\{d(a_1, a_2) : a_1 \in A, a_2 \in A\}$ is finite. A subset A of a metric space $\langle X, d \rangle$ is called *open* if it contains an open ball about each of its points. A subset is called *closed* if its complement is open in this sense. A compact subset of a metric space is both closed and bounded; if each closed and bounded subset of a metric space is compact, then the metric space is called *boundedly compact*.

- The open sets as determined by a metric d form a Hausdorff, first countable, paracompact topology that we will denote by τ_d.

- Each open ball is an open set.

- Metrics d and ρ on a set X are called *equivalent* if $\tau_d = \tau_\rho$.

We say a sequence $\langle x_n \rangle$ in $\langle X, d \rangle$ is *convergent* to $p \in X$ if $\forall \varepsilon > 0, \exists k \in \mathbb{N}$ such that $n \geq k \Rightarrow d(x_n, p) < \varepsilon$. There is at most one point p in X to which a given sequence can converge which is called the *limit* of the sequence, and we write any of the following to express this convergence: $\langle x_n \rangle \to p$ or $\lim_{n \to \infty} x_n = p$ or $\lim x_n = p$. We say that p is a *cluster point* of $\langle x_n \rangle$ if $\forall \varepsilon > 0, \forall k \in \mathbb{N}, \exists n \geq k$ with $d(x_n, p) < \varepsilon$. To say that p is a cluster point is the same as saying that $\langle x_n \rangle$ has a subsequence convergent to p.

- A point p of X is a cluster point of $\langle x_n \rangle$ if and only if $p \in \cap_{n=1}^{\infty} \operatorname{cl}(\{x_k : k \geq n\})$; as such, the set of cluster points of a sequence forms a closed set.

- The set of terms of a sequence without a cluster point forms a closed set.

- A subset A of $\langle X, d \rangle$ is closed if and only if whenever $\langle a_n \rangle$ in A is convergent, the limit stays in A.

- Two metrics d and ρ for X are equivalent if and only if they determine the same convergent sequences with the same limits.

- A subset A of $\langle X, d \rangle$ is compact if and only if each sequence in A has a cluster point in A.

- Each convergent sequence $\langle x_n \rangle$ in X has the *Cauchy property*: $\forall \varepsilon > 0, \exists k \in \mathbb{N}$ such that $n > j \geq k \Rightarrow d(x_n, x_j) < \varepsilon$.

- The set of terms of a Cauchy sequence forms a d-bounded set.

- If each Cauchy sequence in $\langle X, d \rangle$ is convergent, then d is called a *complete metric*. Completeness of a metric is characterized by Cantor's theorem: d is complete if and only if each decreasing sequence of nonempty closed sets whose diameters tend to zero has nonempty intersection.

- If the metric induced by a norm is complete, the normed linear space is called a *Banach space*. The spaces X^* for a normed linear space X and $C_b(X, \mathbb{R})$ for a Hausdorff space $\langle X, \tau \rangle$ are both Banach spaces.

Convergence and clustering of a sequence can be formulated in a topological space in the obvious ways, but as these notions cannot in general be used to characterize topological properties, we are reluctant to do so formally, apart from a more inclusive discussion of nets that we undertake later.

We call a Hausdorff space $\langle X, \tau \rangle$ *metrizable* if there is a metric d on X such that $\tau = \tau_d$. For a metrizable space, we write $\mathbf{D}(X)$ for the family of its compatible metrics. A major accomplishment of general topology was to identify necessary and sufficient conditions for metrizability of a Hausdorff topological space. Early on came the Alexandroff-Urysohn metrization theorem [9]: a Hausdorff space $\langle X, \tau \rangle$ is metrizable if and only if there exists a sequence of open covers $\langle \mathscr{U}_n \rangle$ of X with the following two properties:

(1) $\forall n \in \mathbb{N}$, $\mathscr{U}_{n+1}$ star-refines $\mathscr{U}_n$;

(2) $\forall x \in X, \{ \mathrm{St}(x, \mathscr{U}_n) : n \in \mathbb{N} \}$ is a neighborhood base for τ at x.

A sequence of open covers satisfying both (1) and (2) is called a *star-development* for X. More precisely, the existence of a star-development $\langle \mathscr{U}_n \rangle$ allows for the construction of a compatible metric d such that $\forall n \geq 2$, both of the following conditions hold [155, p. 167]:

(i) $\mathscr{U}_n$ refines $\{ S_d(x, \frac{1}{2^{n-1}}) : x \in X \}$;

(ii) $\{ S_d(x, \frac{1}{2^n}) : x \in X \}$ refines $\mathscr{U}_{n-1}$.

We will call on these particular properties of the constructed metric at one point in the monograph.

If $\langle X, d \rangle$ and $\langle Y, \rho \rangle$ are metric spaces, then $X \times Y$ equipped with the product topology is metrizable; a compatible metric for the product topology is the *box metric*:

$$d_{\mathrm{box}}((x_1, y_1), (x_2, y_2)) := \max\{ d(x_1, x_2), \rho(y_1, y_2) \}.$$

If d and ρ are complete metrics, so is the box metric. This construction has an obvious generalization to any finite product of metric spaces, and the associated metric on the product is still called the box metric.

If $\{\langle X_n, \tau_n \rangle : n \in \mathbb{N}\}$ is a denumerable family of metrizable spaces, then $\prod_{n \in \mathbb{N}} X_n$ equipped with product topology is metrizable. Here is a standard way to define the metric on the product. If d_n is a compatible metric for $\langle X_n, \tau_n \rangle$, and $x = \langle x_n \rangle$ and $w = \langle w_n \rangle$ are two points of the product, then

$$\rho(x, w) := \sum_{n=1}^{\infty} 2^{-n} \min\{1, d_n(x_n, w_n)\}$$

is compatible with the product topology. It is routine to verify that if each d_n is a complete metric, then so is the standard product metric. In summary: a countable product of (completely) metrizable spaces is (completely) metrizable.

If $A \subseteq \langle X, d \rangle$ and $\varepsilon > 0$, we define the ε-*enlargement* of A by

$$S_d(A, \varepsilon) := \cup_{a \in A} S_d(a, \varepsilon) = \{x \in X : d(x, A) < \varepsilon\}.$$

This notation is consistent with our notation for open balls.

If A and B are subsets of $\langle X, d \rangle$, we define the *Hausdorff distance* between them by

$$H_d(A, B) := \inf\{\varepsilon > 0 : B \subseteq S_d(A, \varepsilon) \text{ and } A \subseteq S_d(B, \varepsilon)\} = \max\{e_d(A, B), e_d(B, A)\}.$$

If either no enlargement of A contains B or no enlargement of B contains A, then $H_d(A, B) = \infty$. This occurs, for example, in the plane equipped with the Euclidean metric if A and B are two lines that intersect in exactly one point. It also occurs if exactly one of the two sets is empty.

We list some well-known facts relative to Hausdorff distance without proof, all of which are established in [17].

- Whenever $\{a, c\} \subseteq X$, we have $d(a, c) = H_d(\{a\}, \{c\})$.

- If X is not d-bounded, then H_d must assume infinite values between some pair of nonempty subsets.

- Whenever A and B are subsets of X, $H_d(A, B) = H_d(\mathrm{cl}(A), \mathrm{cl}(B))$.

- Hausdorff distance as we have defined it when restricted to $\mathscr{C}_0(X)$ is an extended metric.

- Hausdorff distance restricted to the nonempty closed and bounded subsets of X is a bona fide metric; compactness and completeness of $\langle X, d \rangle$ are separately inherited by the closed and bounded subsets equipped with the Hausdorff metric.

- Hausdorff distance defines a pseudo-metrizable topology τ_{H_d} on $\mathscr{P}(X)$ where $\mathscr{A} \subseteq \mathscr{P}_0(X)$ is declared open if for each $A \in \mathscr{A}$ $\exists \alpha \in (0, \infty)$ such that $\{E \in \mathscr{P}_0(X) : H_d(A, E) < \alpha\} \subseteq \mathscr{A}$. Further, the empty set $\emptyset$ is an isolated point in this topology.

- Hausdorff distance between nonempty subsets is the uniform distance between their associated distance functionals: $H_d(A, B) = \sup_{x \in X} |d(x, A) - d(x, B)|$.

There is a second topology on $\mathscr{P}(X)$ that is almost as well known as the Hausdorff metric topology called the Vietoris or finite topology [17, 109, 122, 150]. If $E \subseteq X$ put

$$E^- := \{A \in \mathscr{P}(X) : A \cap E \neq \emptyset\} \text{ and } E^+ := \{A \in \mathscr{P}(X) : A \subseteq E\}.$$

Note that $X^+ = \mathscr{P}(X)$, $X^- = \mathscr{P}_0(X)$, $\emptyset^+ = \{\emptyset\}$, and $\emptyset^- = \emptyset$. Whenever E_1 and E_2 are subsets of X, we have $(E_1 \cap E_2)^+ = E_1^+ \cap E_2^+$. The *Vietoris topology* τ_V on $\mathscr{P}(X)$ is generated by all sets of the form V^- and W^+ where V and W run over the open subsets of X. As with the Hausdorff metric topology, $\emptyset$ is an isolated point of the hyperspace. It is easy to check that the relative

Vietoris topology on $\mathscr{P}_0(X)$ has as a base all strings of the form $(\cap_{i=1}^n V_i^-) \cap W^+ \cap \mathscr{P}_0(X)$ where $V_1, V_2, \ldots, V_n, W$ are nonempty open subsets of X.

Of course, this topology makes sense in any Hausdorff space. It is left to the reader to show that $x \mapsto \{x\}$ is an embedding of the underlying Hausdorff space into $\mathscr{P}(X)$ equipped with the Vietoris topology. Topologies on spaces of subsets of a Hausdorff space are often called *hyperspace topologies*. Hyperspace topologies can be defined on the compact subsets of a Hausdorff space, the nonempty closed convex subsets of a normed linear space, etc. A hyperspace topology defined on a family of subsets containing the singletons is called *admissible* provided $x \mapsto \{x\}$ is an embedding [17].

2. Continuous Functions on Metric Spaces

If $\langle X, d \rangle$ and $\langle Y, \rho \rangle$ are metric spaces, continuity of $f : X \to Y$ is achieved at $p \in X$ if either of the following conditions is satisfied:

(1) whenever $\langle x_n \rangle$ converges to p, then $\langle f(x_n) \rangle$ converges to $f(p)$, or

(2) $\forall \varepsilon > 0, \exists \delta > 0$ such that $d(p, x) < \delta \Rightarrow \rho(f(p), f(x)) < \varepsilon$.

The topology of a metric space $\langle X, d \rangle$ is of course normal, and in this setting, a concrete Urysohn function for a pair of disjoint nonempty closed subsets A and B is

$$f(x) = \frac{d(x, A)}{d(x, A) + d(x, B)}.$$

Let $\langle X, d \rangle$ and $\langle Y, \rho \rangle$ be metric spaces. We distinguish the following subclasses of $C(X, Y)$ in this setting:

- f is called *uniformly continuous* if $\forall \varepsilon > 0, \exists \delta > 0$ such that $\forall x, w \in X, d(x, w) < \delta \Rightarrow \rho(f(x), f(w)) < \varepsilon$;

- f is called λ-*Lipschitz* (for $\lambda \geq 0$) if $\forall x \in X, \forall w \in X,\ \rho(f(x), f(w)) \leq \lambda d(x, w)$;

- f is called an *isometry* if $\forall x \in X, \forall w \in X,\ \rho(f(x), f(w)) = d(x, w)$;

- f is called *Cauchy continuous* [34, 45, 52, 115, 140] if it maps Cauchy sequences in X to Cauchy sequences in Y.

A function between metric spaces is called *Lipschitz* if it is λ-Lipschitz for some $\lambda \geq 0$. We write $UC(X, Y), \text{Lip}(X, Y)$ and $CC(X, Y)$ for the classes of functions from $\langle X, d \rangle$ to $\langle Y, \rho \rangle$ that are uniformly continuous, Lipschitz and Cauchy continuous, respectively.

A 0-Lipschitz function is a constant function. We call a 1-Lipschitz function *nonexpansive*. Evidently, a real-valued function f is λ-Lipschitz if and only if for each x_1, x_2 in X, we have $f(x_1) \leq f(x_2) + \lambda d(x_1, x_2)$, a formulation that is often more convenient to use.

Our approach to the next few results is adapted from [92].

Lemma 2.1. *Let $\{f_j : j \in J\}$ be a family of λ-Lipschitz real-valued functions on a metric space $\langle X, d \rangle$ such that for some $p \in X$, $\inf_{j \in J} f_j(p) > -\infty$. Then $x \mapsto \inf_{j \in J} f_j(x)$ is a real-valued λ-Lipschitz function on X.*

Proof. This is clear if $\lambda = 0$. Otherwise, for each $x \in X$, put $\alpha_x = \inf_{j \in J} f_j(x)$. We claim that for each $x \in X$, $\alpha_x \geq \alpha_p - \lambda d(p, x)$. Suppose this fails for some point $w \in X$, that is, $\alpha_w < \alpha_p - \lambda d(p, w)$. This means that for some $j \in J$, we have

$$f_j(w) < \alpha_p - \lambda d(p, w) \leq f_j(p) - \lambda d(p, w),$$

and thus $f_j(p) > f_j(w) + \lambda d(p,w)$, contradicting f_j being λ-Lipschitz. This establishes the claim so that $\forall x \in X$, α_x is a real number. Exactly the same reasoning shows that whenever x_1, x_2 in X, we have $\alpha_{x_1} \leq \alpha_{x_2} + \lambda d(x_1, x_2)$ and this shows that the rule of assignment $x \mapsto \alpha_x$ is λ-Lipschitz as required. □

Replacing each function f_j by $-f_j$ in Lemma 2.1 allows us to replace infimum by supremum throughout in the statement, provided the supremum is finite at some point.

We immediately apply Lemma 2.1 to show that each lower bounded real-valued lower semicontinuous function g is a pointwise limit of an increasing sequence of Lipschitz functions. As is easy to check, lower semicontinuity of such a limit is necessary, for more generally, a real-valued function that is the supremum of a family of lower semicontinuous functions must itself be lower semicontinuous. However, lower boundedness is not necessary, rather what is only needed is that the initial function g majorize some Lipschitz function.

Proposition 2.2. *Let $\langle X, d \rangle$ be a metric space and let $g : X \to \mathbb{R}$ be a lower bounded and lower semicontinuous function. Then there exists a sequence $\langle f_n \rangle$ in $Lip(X, \mathbb{R})$ such that $\forall n \in \mathbb{N}$, $f_n \leq f_{n+1} \leq g$ and $\langle f_n \rangle$ converges pointwise to g. Furthermore, if f is in addition upper bounded and uniformly continuous then we can obtain uniform convergence.*

Proof. Put $\alpha = \inf \{g(x) : x \in X\}$. For each $n \in \mathbb{N}$ and each $w \in X$, let $h_{w,n} : X \to \mathbb{R}$ be the n-Lipschitz function defined by

$$h_{w,n}(x) := g(w) + nd(x,w) \quad (x \in X),$$

and then put $f_n := \inf \{h_{w,n} : w \in X\}$ for $n \in \mathbb{N}$. Clearly, $\alpha \leq f_n$ and so by Lemma 2.1, f_n is n-Lipschitz. By construction $f_n \leq f_{n+1}$ for each positive integer n and

$$\forall x \in X, \; f_n(x) \leq h_{x,n}(x) = g(x).$$

To show pointwise convergence, we show $\forall \varepsilon > 0$, $\forall x \in X$ $\exists n \in \mathbb{N}$ with $f_n(x) \geq g(x) - \varepsilon$. By lower semicontinuity of g, there exists $\delta > 0$ such that $d(x,w) < \delta \Rightarrow g(w) > g(x) - \varepsilon$. Choose $n \in \mathbb{N}$ so large that $\alpha + n\delta \geq g(x)$, and let $w \in X$ be arbitrary. If $d(x,w) \geq \delta$, then

$$h_{w,n}(x) = g(w) + nd(x,w) \geq \alpha + n\delta \geq g(x).$$

On the other hand, if $d(x,w) < \delta$, then $h_{w,n}(x) \geq g(w) > g(x) - \varepsilon$. From these two estimates, we get $f_n(x) = \inf \{h_{w,n}(x) : w \in X\} \geq g(x) - \varepsilon$.

For the last statement, let $\beta = \sup \{g(x) : x \in X\}$, and given $\varepsilon > 0$ choose $\delta > 0$ such that $|f(x_1) - f(x_2)| < \varepsilon$ whenever $d(x_1, x_2) < \delta$. Now simply repeat the last argument where $n \in \mathbb{N}$ is chosen so that $\alpha + n\delta \geq \beta$. □

Corollary 2.3. *Let $\langle X, d \rangle$ be a metric space and let $g : X \to \mathbb{R}$ be bounded and uniformly continuous. Then for each $\varepsilon > 0$ there exists a bounded member f of $Lip(X, \mathbb{R})$ such that $\sup_{x \in X} |f(x) - g(x)| < \varepsilon$.*

Proof. A function that has finite uniform distance from a bounded function is itself bounded. □

Given a lower bounded function $g : X \to \mathbb{R}$ on a metric space $\langle X, d \rangle$, the function $f_n : X \to \mathbb{R}$ defined by

$$f_n(x) := \inf_{w \in X} g(w) + nd(w,x)$$

is called the *n-Lipschitz regularization* of g. This is the largest n-Lipschitz function that g majorizes. For this reason, f_n is sometimes called the lower n-Lipschitz envelope of g. We also note that it is not possible in general to uniformly approximate a lower bounded uniformly continuous function by Lipschitz functions, e.g., consider $f : \mathbb{N} \to \mathbb{N}$ defined by $g(n) = n^2$ where $\mathbb{N}$ is equipped with the usual metric of the line. Later in this monograph we will characterize those metric spaces such that the each member of $UC(X, \mathbb{R})$ can be uniformly approximated by members of $\mathrm{Lip}(X, \mathbb{R})$.

In the sequel, Lipschitz functions and Lipschitz-type functions will be particularly important. If $f : X \to Y$ is λ-Lipschitz for some $\lambda \geq 0$, we put

$$L(f) := \min\{\lambda \geq 0 : f \text{ is } \lambda - \text{Lipschitz}\} = \sup_{x \neq w} \frac{\rho(f(x), f(w))}{d(x, w)}.$$

Clearly $L(f + g) \leq L(f) + L(g)$ and $L(\alpha f) = |\alpha| L(f)$ for f, g in $\mathrm{Lip}(X, \mathbb{R})$. However, $f \mapsto L(f)$ fails to be a norm on $\mathrm{Lip}(X, \mathbb{R})$ because $L(f) = 0$ if and only if f is a constant function.

Example 2.4. If A is an interval, a ray, or all of $\mathbb{R}$ and $f : A \to \mathbb{R}$ is differentiable and $\forall a \in A, |f'(a)| \leq \lambda$, then by the mean-value theorem, f is λ-Lipschitz. Moreover, $L(f) = \sup\{|f'(a)| : a \in A\}$.

Example 2.5. If $T : X \to Y$ is a continuous linear transformation between normed linear spaces $\langle X, || \cdot ||_1 \rangle$ and $\langle Y, || \cdot ||_2 \rangle$, then T is Lipschitz with $L(T) = \sup\{||T(x)||_2 : ||x||_1 \leq 1\}$;

Example 2.6. Given any nonempty subset A of a metric space $\langle X, d \rangle$, $x \mapsto d(x, A)$ is nonexpansive.

Example 2.7. For each metric space $\langle X, d \rangle$, its isolation functional $x \mapsto I(x)$ is nonexpansive. For the proof we need to consider two separate cases for $x \neq w$:

(a) either $d(x, w) = I(x)$ or $d(x, w) = I(w)$;

(b) $\max\{I(x), I(w)\} < d(x, w)$.

In case (a) we can assume without loss of generality that $I(x) = d(x, w)$. Since $I(w) \leq d(x, w)$ we have $|I(x) - I(w)| = I(x) - I(w) \leq I(x) = d(x, w)$. In case (b), choose $\varepsilon > 0$ with $\varepsilon < d(x, w) - \max\{I(x), I(w)\}$ and then $p \neq x$ with $d(x, p) < I(x) + \varepsilon$. By the choice of ε, $p \neq w$ and we get

$$I(w) \leq d(w, p) \leq d(w, x) + d(x, p) < d(w, x) + I(x) + \varepsilon,$$

and letting ε go to zero, this yields $I(w) - I(x) \leq d(w, x)$. In the same way, $I(x) - I(w) \leq d(w, x)$ and we have shown that the isolation functional is nonexpansive.

Example 2.8. Given any metric space $\langle X, d \rangle$, $x \mapsto \{x\}$ is an isometry of the space into its nonempty closed and bounded subsets equipped with Hausdorff distance.

Example 2.9. The vector space ℓ_2 of square summable real sequences equipped with the norm $||\langle \alpha_n \rangle||_2 := (\sum_{n=1}^{\infty} \alpha_n^2)^{1/2}$ is a Banach space which contains a copy of $\mathbb{R}^n$ with the Euclidean metric isometrically.

Example 2.10. Given any metric space $\langle X, d \rangle$ and a fixed $p \in X$, $x \mapsto \phi_x : X \to \mathbb{R}$ where

$$\phi_x(w) = d(x, w) - d(p, w)$$

is an isometry of X into the Banach space $C_b(X, \mathbb{R})$ equipped with the supremum norm:

$$\|\phi_{x_1} - \phi_{x_2}\|_\infty = d(x_1, x_2) \quad \text{for } x_1, x_2 \in X.$$

More specifically, the image of X in the function space consists of 2-Lipschitz functions and $\forall x \in X$, $\|\phi_x\|_\infty = d(x, p)$. Thus, the metric space can be isometrically embedded as a dense subset of a complete metric space, namely the closure of $\{\phi_x : x \in X\}$ in $C_b(X, \mathbb{R})$.

With respect to the last example, the closure of $\phi(X)$ equipped the metric given by $\|\cdot\|_\infty$ is called *the completion* of $\langle X, d \rangle$, for if $\langle X, d \rangle$ sits as a dense subset of two complete metric spaces, then there is isometry between them whose restriction to the initial space is the identity [155, p. 176]. Another familiar way to get the completion of $\langle X, d \rangle$ is to identify Cauchy sequences $\langle x_n \rangle$ and $\langle y_n \rangle$ in X where $\lim_{n \to \infty} d(x_n, y_n) = 0$ and then equipping the equivalence classes with a natural complete metric, namely the limiting distance between corresponding terms of representatives from each equivalence class. Of course, each $x \in X$ is identified with the equivalence class of sequences convergent to x. As a third possibility, we can isometrically embed the metric space in the Banach space of continuous linear functionals on $\mathrm{Lip}(X, \mathbb{R})$ equipped with an appropriate norm as employed in the proof of Theorem 4.2 below.

Proposition 2.11. *For functions between metric spaces $\langle X, d \rangle$ and $\langle Y, \rho \rangle$, Lipschitz continuity $\Rightarrow$ uniform continuity $\Rightarrow$ Cauchy continuity $\Rightarrow$ global continuity.*

Proof. We only prove that if $f : X \to Y$ is Cauchy continuous, then it is continuous at each point $p \in X$. Let $\langle x_n \rangle$ be a sequence in X convergent to p; then the auxiliary sequence $p, x_1, p, x_2, p, x_3, p, \ldots$ is Cauchy, whence $f(p), f(x_1), f(p), f(x_2), f(p), f(x_3), f(p), \ldots$ is Cauchy as well, which gives $\lim_{n \to \infty} f(x_n) = f(p)$. $\qquad \square$

Example 2.12. $x \mapsto \sqrt{x}$ as a function from $[0, \infty)$ to $\mathbb{R}$ is uniformly continuous but not Lipschitz.

Example 2.13. Consider $X = \mathbb{N} \cup \{n + \frac{1}{n+1} : n \in \mathbb{N}\}$ equipped with the usual metric of the real line. Then the characteristic function of $\mathbb{N}$ is Cauchy continuous but not uniformly continuous.

Example 2.14. The function $x \mapsto \frac{1}{x}$ on $(0, \infty)$ is continuous but not Cauchy continuous.

Continuity, Cauchy continuity, uniform continuity, and Lipschitz continuity are all preserved under composition. For real-valued functions defined on a metric space, each class of functions forms a vector-lattice of functions containing the constant functions.

It is a standard exercise in a first course in analysis to construct a sequence of Lipschitz functions on $[0, 1]$ convergent pointwise to the characteristic function of the origin, so that there is no hope that any of our four function classes is in general stable under pointwise convergence. While continuity, Cauchy continuity and uniform continuity are all preserved under uniform convergence, this need not be so for Lipschitz continuity, even for real-valued functions. For example, $f(x) = \sqrt{x}$ on $[0, \infty)$ is the uniform limit of the sequence $\langle f_n \rangle$ where for each $n \in \mathbb{N}$, $f_n(x) = \sqrt{x}$ if $x \geq 1/n$ and $f_n(x) = \sqrt{n}x$ otherwise.

Theorem 2.15. *Let A be a nonempty subset of a metric space $\langle X, d \rangle$ and let $\langle Y, \rho \rangle$ be a complete metric space. Suppose $f : A \to Y$ is Cauchy (resp. uniformly) continuous. Then f has a Cauchy (resp. uniformly) continuous extension to $cl(A)$.*

Proof. To define our extension g, let $x \in \mathrm{cl}(A)$ be arbitrary. Let $\langle a_n \rangle$ be a sequence in A convergent to x; as $\langle a_n \rangle$ is Cauchy, the image sequence is Cauchy in Y, and by completeness of the target space, it is thus convergent to some $y_x \in Y$. We put $g(x) = y_x$. To verify that g is well-defined, let $\langle b_n \rangle$ be a second sequence in A convergent to x. Then the spliced sequence $a_1, b_1, a_2, b_2, \ldots$ is Cauchy and so the image sequence $f(a_1), f(b_1), f(a_2), f(b_2), \ldots$ is Cauchy as well. As this image sequence has a convergent subsequence to y_x, the entire sequence converges to y_x and so $\langle f(b_n) \rangle$ converges to y_x.

Clearly, g extends f, as given any $a \in A$, the sequence $a, a, a, \ldots$ converges to a. It remains to show that g is Cauchy continuous or uniformly continuous as the case may be. We only deal with the first scenario, as the second is covered in standard texts. Let $\langle x_n \rangle$ be a Cauchy sequence in $\mathrm{cl}(A)$. By our construction, for each $n \in \mathbb{N}$ we can find $a_n \in A$ with both $d(a_n, x_n) < \frac{1}{n}$ and ($\Diamond$) $d(f(a_n), g(x_n)) < \frac{1}{n}$. As $\langle a_n \rangle$ is Cauchy, so is $\langle f(a_n) \rangle$. It follows from ($\Diamond$) that $\langle g(x_n) \rangle$ is Cauchy. $\qquad\square$

The next result shows that $CC(X, \mathbb{R})$, like $C(X, \mathbb{R})$, is always well-behaved with respect to pointwise products. In the next section, we will show $CC(X, \mathbb{R})$ is stable under reciprocation if and only if the domain space is complete.

Theorem 2.16. *Let f and g be real-valued Cauchy continuous functions on a metric space $\langle X, d \rangle$. Then their pointwise product fg is Cauchy continuous.*

Proof. Suppose $\langle x_n \rangle$ is a Cauchy sequence in X and let $\varepsilon > 0$. Since Cauchy sequences are bounded, there exists $\alpha > 0$ such that $\forall n \in \mathbb{N}$, $|f(x_n)| \leq \alpha$ and $|g(x_n)| \leq \alpha$. Choose $k \in \mathbb{N}$ such that whenever $n \geq k$ and $j \geq k$ we have both $|f(x_n) - f(x_j)| < \frac{\varepsilon}{2\alpha}$ and $|g(x_n) - g(x_j)| < \frac{\varepsilon}{2\alpha}$. It follows that

$$|f(x_n)g(x_n) - f(x_j)g(x_j)| \leq |f(x_n)||g(x_n) - g(x_j)| + |g(x_j)||f(x_n) - f(x_j)| < 2\alpha \cdot \frac{\varepsilon}{2\alpha} = \varepsilon$$

whenever $n \geq k$ and $j \geq k$. $\qquad\square$

Clearly, the proof strategy we employed is a modification of the standard proof used to show that the product of two bounded real-valued uniformly continuous (resp. Lipschitz) functions is again uniformly continuous (resp. Lipschitz), using the inequality

$$|f(x_1)g(x_1) - f(x_2)g(x_2)| \leq |f(x_1)||g(x_1) - g(x_2)| + |g(x_2)|f(x_1) - f(x_2)| \quad (x_1, x_2 \in X).$$

However, neither $UC(X, \mathbb{R})$ nor $\mathrm{Lip}(X, \mathbb{R})$ is in general stable under pointwise product: consider $f(x) = x$ and $g(x) = \sin x$ as functions on $\mathbb{R}$ (note that one of these is bounded).

Theorem 2.17. *Let $\langle X, d \rangle$ be a metric space. Then $\mathrm{Lip}(X, \mathbb{R})$ is stable under pointwise product if and only if X is bounded.*

Proof. If X is bounded, then each Lipschitz function on X is bounded and so the product of any two members of $\mathrm{Lip}(X, \mathbb{R})$ remains there. If X is not bounded fix $x_0 \in X$; while $x \mapsto d(x, x_0)$ is nonexpansive, $x \mapsto [d(x, x_0)]^2$ is not Lipschitz. $\qquad\square$

In the sequel, we will give various necessary and sufficient conditions on the structure of a metric space so that $UC(X, \mathbb{R})$ is stable under pointwise product. These conditions are likely to be unfamiliar to the reader.

Equivalence of two metrics d and ρ on a set X means that the identity map $I_X : X \to X$ is bicontinuous. If the map is uniformly continuous in both directions, then d and ρ are called *uniformly equivalent*. Similarly, if the map is Cauchy continuous in both directions, then d and ρ are called *Cauchy equivalent*.

- Given a metric d on X, $\rho(x, w) := \min\{1, d(x, w)\}$ gives a metric uniformly equivalent to d for which X is ρ-bounded.

- Two norms on a common real vector space X are equivalent if and only if they are uniformly equivalent if and only if they determine the same metrically bounded sets.

- If $\langle X, \tau \rangle$ is compact and metrizable, then any two compatible metrics are uniformly equivalent.

- If d and ρ are metrics on a set X, then uniform equivalence of d and ρ on X is necesseary and sufficient for the agreement of the topologies determined by H_d and H_ρ on $\mathscr{C}_0(X)$ [17, pp. 92–93].

Example 2.18. On $\mathbb{N}$, the usual metric $d(n, j) = |n - j|$ and the metric $\rho(n, j) = |\frac{1}{n} - \frac{1}{j}|$ are equivalent but not Cauchy equivalent.

Example 2.19. Let $X = \{a_n : n \in \mathbb{N}\}$ where $a_{2n-1} = n$ and $a_{2n} = n + \frac{1}{n+1}$. Consider the metrics d and ρ on X defined by $d(a_n, a_k) = |a_n - a_k|$ and $\rho(a_n, a_k) = |n - k|$. Then d and ρ are Cauchy equivalent, as the only Cauchy sequences with respect to either metric are those that are eventually constant. However, $I_X : \langle X, d \rangle \to \langle X, \rho \rangle$ is not uniformly continuous because $d(a_{2n-1}, a_{2n}) = \frac{1}{n+1}$ while $\rho(a_{2n-1}, a_{2n}) = 1$ for $n = 1, 2, 3, \ldots$.

Example 2.20. Uniformly equivalent metrics can determine different classes of metrically bounded sets: on $\mathbb{N}$ the usual metric of the line and the zero-one metric are uniformly equivalent.

The following proposition spells out a useful way to produce a metric equivalent or uniformly equivalent or Cauchy equivalent to a given metric with desired properties. The simple proof is left to the reader.

Proposition 2.21. *Let $\langle X, d \rangle$ be a metric space and let $f \in C(X, \mathbb{R})$ (resp. $UC(X, \mathbb{R})$; resp. $CC(X, \mathbb{R})$). Then $(x, w) \mapsto d(x, w) + |f(x) - f(w)|$ is a metric on X that is equivalent (resp. uniformly equivalent; resp. Cauchy equivalent) to d.*

As metric spaces are paracompact, given an open cover of the space, there is a continuous partition of unity subordinated to it. But one can do better in the metric setting: one can choose the functions in the partition of unity to be Lipschitz, a result established by Frolik [81, Theorem 1] using some complicated set-theoretic topology (see in the same vein [65, pp. 120–126]). We omit the proof.

Theorem 2.22. *Let $\{U_i : i \in I\}$ be an open cover of a metric space $\langle X, d \rangle$. Then we can find a family of nonnegative Lipschitz real-valued functions $\mathscr{P}$ with these properties:*

(1) *the family of cozero sets of $\mathscr{P}$ form a locally finite open refinement of $\{U_i : i \in I\}$;*

(2) *for each $x \in X$, $\sum_{p \in \mathscr{P}} p(x) = 1$.*

We will use the Frolik result in a subsequent section to prove the fundamental uniform density theorem for $C(X, \mathbb{R})$ for a general metric space $\langle X, d \rangle$. Concurrently with Frolik, Fried [80] proved a slightly weaker version of Theorem 2.22 where the cozero sets of $\mathscr{P}$ are only shown to be a *point finite* family, that is, each point of X belongs to at most finitely many cozero sets. While his proof involves a particularly clever application of the well-ordering principle and is simpler than the known proofs of the last result, Fried's weaker result is not adequate to prove our density theorem.

3. Extension of Real-Valued Continuous Functions on Subsets of a Metric Space

The classical Tietze extension theorem is a key result in a standard course in general topology that we shall not reprove here. Recall that Tietze's theorem says that if A is a closed nonempty subset of a normal space, then each continuous real-valued function f on A can be extended continuously to the entire space [155, p. 103]. Further, any bounds that apply to f can be preserved by the extension. For example, if $f(A) \subseteq [\alpha, \beta]$, say, and f_0 were such an extension, then a continuous extension that preserves those bounds is given by

$$
\tilde{f}_0(x) = \begin{cases} f_0(x) & \text{if } \alpha \leq f_0(x) \leq \beta \\ \alpha & \text{if } f_0(x) < \alpha \\ \beta & \text{if } f_0(x) > \beta. \end{cases}
$$

In particular, the Tietze theorem is valid in metrizable spaces. Not only is the classical Tietze extension theorem available to us here, but we can also extend continuous functions defined on closed subsets with values in a normed linear space, as established by Dugundji [74, p. 188].

Theorem 3.1. *Let A be a nonempty closed subset of a metric space $\langle X, d \rangle$ and let Y be a normed linear space. Suppose $f \in C(A, Y)$. Then f has a continuous extension to X whose function values lie in the convex hull of $f(A)$.*

The proof of the Dugundji extension theorem not unexpectedly uses partitions of unity which produce convex combinations of elements of the range. We do not include it here.

In this section, we give two applications of Tietze's theorem with respect to Cauchy continuity. Our first result characterizes those metric spaces on which continuity forces Cauchy continuity [140]. We will describe when Cauchy continuity forces uniform continuity in a subsequent section.

Proposition 3.2. *Let $\langle X, d \rangle$ be a metric space. Then the metric d is complete if and only if each continuous function from X to an arbitrary metric space is Cauchy continuous.*

Proof. Suppose the metric is complete. As each Cauchy sequence in X is convergent, a continuous function on X must map each Cauchy sequence to a convergent one, which of course must be Cauchy. Conversely, if $\langle X, d \rangle$ fails to be a complete metric space, let $\langle x_n \rangle$ be a Cauchy sequence in X with distinct terms that is nonconvergent. As $\langle x_n \rangle$ fails to cluster as well, $\{x_n : n \in \mathbb{N}\}$ is an infinite closed set without limit points. Thus, any real-valued function on the set is continuous with respect to the relative topology. As a result, by the Tietze extension theorem, there is $f \in C(X, \mathbb{R})$ mapping each x_n to n, so that f cannot be Cauchy continuous. $\square$

Corollary 3.3. *Let $\langle X, d \rangle$ be a metric space. Then $CC(X, \mathbb{R})$ is stable under reciprocation if and only if d is complete.*

Proof. Sufficiency is immediate from the last result. For necessity, if the domain space is not complete, let $\langle \hat{X}, \hat{d} \rangle$ be its completion. For $\hat{p} \in \hat{X} \backslash X$, consider $x \mapsto \hat{d}(x, \hat{p})$ as a Cauchy continuous function on X.   □

Extensions of real-valued Cauchy continuous functions and of real-valued Lipschitz functions to the entire space do not require that the initial domain be a closed subset of $\langle X, d \rangle$. For the first fact, we again use the Tietze extension theorem, and for the second fact, we use Lemma 2.1.

Theorem 3.4. *Let $\langle X, d \rangle$ be a metric space and let A be a nonempty subset. If $f : A \to \mathbb{R}$ is Cauchy continuous, then f has a Cauchy continuous extension to X.*

Proof. Let $\langle \hat{X}, \hat{d} \rangle$ be the completion of $\langle X, d \rangle$; by Theorem 2.15, we can extend f to a Cauchy continuous function on the closure of A relative to $\langle \hat{X}, \hat{d} \rangle$ and then by the Tietze extension theorem, we can find $\hat{g} \in C(\hat{X}, \mathbb{R})$ that extends f. By Proposition 3.2, $\hat{g}$ is Cauchy continuous and so is $\hat{g}|_X$.   □

Theorem 3.5. *Let A be a nonempty subset of a metric space $\langle X, d \rangle$. If $f : \langle A, d \rangle \to \mathbb{R}$ is λ-Lipschitz, then f has a λ-Lipschitz extension to X.*

Proof. For each $a \in A$, define $f_a : X \to \mathbb{R}$ by $f_a(x) = f(a) + \lambda d(x, a)$. Evidently for each $a \in A, f_a$ is λ-Lipschitz. For each $p \in A$, we have $f(p) = \inf_{a \in A} f_a(p)$ because $f(p) = f_p(p)$ and since f is λ-Lipschitz on $A, \forall a \in A$,

$$f(p) \leq f(a) + \lambda d(p, a) = f_a(p).$$

By Lemma 2.1, $\inf_{a \in A} f_a$ is the desired extension.   □

It is of interest to note that if $\lambda > 0$ and A is a nonempty closed subset of X and f is a strictly positive function on A, then the extension above is strictly positive as well because whenever $x \notin A$, we have for each $a \in A$,

$$0 < \lambda d(x, A) \leq f(a) + \lambda d(x, a).$$

It is easy to verify that $x \mapsto \inf_{a \in A} f(a) + \lambda d(x, a)$ is the largest global λ-Lipschitz extension h of f as asserted by Theorem 3.5. Furthermore, if A is a nonempty convex subset of a normed linear space $\langle X, || \cdot || \rangle$ and f is a λ-Lipschitz *convex function* on A, that is, for each $a_1, a_2 \in A$ and each $\alpha \in [0, 1]$ we have

$$f(\alpha a_1 + (1 - \alpha)a_2) \leq \alpha f(a_1) + (1 - \alpha)f(a_2),$$

then that extension will be convex as well [66].

The smallest λ-Lipschitz extension of f is given by $g(x) := \sup_{a \in A} f(a) - \lambda d(x, a)$. To appreciate the difference, consider $f(x) = |x|$ defined on $[-1, 1]$. The largest 1-Lipschitz extension to $\mathbb{R}$ is given by $h(x) = |x|$ while the smallest 1-Lipschitz extension is given by $g(x) = \min\{|x|, 2 - |x|\}$.

Finally, we remark that we can find also find Lipschitz constant preserving extensions provided the domain and target space are both Hilbert spaces [147].

It is not in general possible to find a uniformly continuous extension of a real-valued uniformly continuous function defined on a subset A of $\langle X, d \rangle$, even if A is closed. For example, $f(x) = x^2$ as a function on $\mathbb{N}$ has no uniformly continuous extension defined on $\mathbb{R}$. However, each bounded uniformly continuous real-valued function f defined on an arbitrary nonempty subset A has a uniformly continuous extension to X, as first proved by McShane [120]. As with Tietze's theorem, any bounds applicable to the values of f can be preserved in the extension.

We obtain McShane's extension theorem in a somewhat different way than he: we create a new metric ρ on X that is uniformly equivalent to d with respect to which f is nonexpansive and then use our Lipschitz extension result [25]. This takes some development, but it is all quite transparent.

Let C be a nonempty convex subset of a normed linear space $\langle X, ||\cdot||\rangle$. We say that $g : C \to \mathbb{R}$ is *concave* if whenever x_1 and x_2 lie in C and $\alpha \in [0,1]$, we have

$$\alpha g(x_1) + (1 - \alpha)g(x_2) \leq g(\alpha x_1 + (1 - \alpha)x_2).$$

Obviously, g is concave if and only if $-g$ is convex. There is a much broader literature on convex functions. Facts about convex functions can be dualized to obtain corresponding facts about concave functions. For example, h is convex if and only if its epigraph $\{(x, \alpha) \in C \times \mathbb{R} : \alpha \geq h(x)\}$ is a convex set [134, p. 80], while g is concave if and only if its hypograph $\{(x, \alpha) \in C \times \mathbb{R} : \alpha \leq g(x)\}$ is a convex set.

Actually we will only be looking at nonnegative nonconstant concave functions defined on $[0, \infty)$. We collect some facts about these in the following result. We call $g : [0, \infty) \to [0, \infty)$ *nondecreasing* if whenever $t_2 > t_1 \geq 0$, we have $g(t_2) \geq g(t_1)$.

Proposition 3.6. *Let* $g : [0, \infty) \to [0, \infty)$ *be concave and nonconstant. Then the following statements are valid.*

(1) *g is a subadditive function;*

(2) *g is a nondecreasing function;*

(3) *whenever $t > 0$, we have $g(t) > 0$.*

Proof. For the first statement, we are asked to show that whenever $t_1 \geq 0$ and $t_2 \geq 0$ then $g(t_1 + t_2) \leq g(t_1) + g(t_2)$. By nonnegativity of values, this is clearly true if $t_1 = 0 = t_2$. Otherwise, $t_1 + t_2 > 0$ and we compute

$$g(t_1) = g\left(\frac{t_1}{t_1 + t_2}(t_1 + t_2) + \frac{t_2}{t_1 + t_2}(0)\right) \geq \frac{t_1}{t_1 + t_2}g(t_1 + t_2) + \frac{t_2}{t_1 + t_2}g(0)$$

$$\geq \frac{t_1}{t_1 + t_2}g(t_1 + t_2).$$

In the same way, $g(t_2) \geq \frac{t_2}{t_1 + t_2}g(t_1 + t_2)$. Adding these inequalities yields our subadditivity.

For the second statement, suppose $g(t_2) < g(t_1)$ where $t_2 > t_1$. Put $\gamma = g(t_1) - g(t_2)$. By concavity it is routine to show using mathematical induction that for each $n \in \mathbb{N}$,

$$g(t_1 + n(t_2 - t_1)) \leq g(t_1) - n\gamma.$$

As a result, $g(t_1 + n(t_2 - t_1))$ will be negative for large n which is impossible.

For the third statement, suppose $t_1 > 0$ and $g(t_1) = 0$. Since g is nondecreasing and nonnegative, g is zero throughout $[0, t_1]$. If $g(t_2) > 0$ for some $t_2 > t_1$, then concavity fails, as

$$g(t_1) = g\left(\frac{t_2 - t_1}{t_2}(0) + \frac{t_1}{t_2}(t_2)\right) < \frac{t_2 - t_1}{t_2}g(0) + \frac{t_1}{t_2}g(t_2) = \frac{t_1}{t_2}g(t_2).$$

We conclude that g is the zero function on $[0, \infty)$, a contradiction. $\qquad\square$

We now let Δ denote the family of nonnegative concave functions g on $[0, \infty)$ that are continuous at the origin, and for which $g^{-1}(\{0\}) = \{0\}$. Members of Δ actually belong to $C([0, \infty), \mathbb{R})$ as a concave function is automatically continuous at each point of the interior of its domain [134, p. 4]. Of course, $g(t) = \arctan t$ belongs to Δ as does $g(t) = \sqrt{t}$. As an example of a member of Δ that is eventually constant, consider g defined by $g(t) = \frac{1}{2}t$ for $t \in [0, 2]$ and $g(t) = 1$ otherwise.

Let A be a nonempty subset of $\langle X, d \rangle$ and let $f : A \to \mathbb{R}$. For the purposes of the next two results, we introduce the auxiliary function $m_f : [0, \infty) \to [0, \infty]$ defined by

$$m_f(t) := \sup \{|f(a_1) - f(a_2)| : \{a_1, a_2\} \subseteq A \text{ and } d(a_1, a_2) \leq t\}.$$

Uniform continuity of f is equivalent to the continuity of m_f at the origin.

Proposition 3.7. *Let A be a nonempty subset of $\langle X, d \rangle$ and let $f : A \to \mathbb{R}$ be nonconstant and uniformly continuous and such that m_f has an affine majorant $t \mapsto \alpha t + \beta$. Then there exists $g_f \in \Delta$ majorizing m_f.*

Proof. Clearly $m_f(0) = 0$ and m_f is nondecreasing. From these points, $\alpha \geq 0$ and $\beta \geq 0$. Let g_f be the infimum of the family of all nondecreasing affine functions that majorize m_f on $[0, \infty)$. Evidently g_f is concave [134, p. 80], its values are nonnegative, and g_f majorizes m_f. Since f is nonconstant, there exists $t_1 > 0$ such that $m_f(t_1) > 0$ so that $g_f(t_1) > 0$ as well. We will show that g_f is continuous at $t = 0$ and that $g_f(0) = 0$ at the same time.

If $\beta = 0$, it is clear that $g_f(0) = 0$ and g_f is continuous at the origin. Otherwise, let $\varepsilon \in (0, \beta)$ be arbitrary. By uniform continuity of f on A, the function m_f is continuous at the origin. This means that for some $t_2 > 0$ and all $t \in [0, t_2]$ we have $m_f(t) < \varepsilon$. Let v be the affine function whose graph contains $(0, \varepsilon)$ and $(t_2, \alpha t_2 + \beta)$. Clearly, v is nondecreasing and majorizes m_f and thus majorizes g_f. From these two facts, we have both $g_f(0) \leq \varepsilon$ and $\limsup_{t \to 0} g_f(t) \in [0, \varepsilon]$. Letting ε tend to zero gives $g_f(0) = 0 = \lim_{t \to 0} g_f(t)$. Finally, as $g_f(t_1) \geq m_f(t_1) > 0$ we see that g_f is nonconstant and so by Proposition 3.6, 0 is the only t-value for which $g_f(t) = 0$. We have now shown that g_f belongs to Δ. $\qquad\square$

Theorem 3.8. *Let A be a nonempty subset of $\langle X, d \rangle$ and let $f : A \to \mathbb{R}$ be nonconstant and uniformly continuous and where m_f has an affine majorant. Let g_f be the function of the last proposition. Then $g_f \circ d : X \times X \to [0, \infty)$ is a metric uniformly equivalent to d with respect to which f is a nonexpansive function on A.*

Proof. Put $\rho = g_f \circ d$. Since $g_f^{-1}(\{0\}) = \{0\}$ and d is a metric, $\rho(x_1, x_2) = 0$ if and only if $x_1 = x_2$. Clearly, $\rho(x_1, x_2) = \rho(x_2, x_1)$ for all x_1, x_2 in X. The triangle inequality follows from the subadditivity of g_f and its nondecreasing nature as established in Proposition 3.6: given x_1, x_2 and x_3 in X

$$\rho(x_1, x_3) = g_f(d(x_1, x_3)) \leq g_f(d(x_1, x_2) + d(x_2, x_3)) \leq \rho(x_1, x_2) + \rho(x_2, x_3).$$

To show uniform equivalence of d and ρ, we show that the identity function on X is uniformly continuous in both directions. Let $\varepsilon > 0$; by the continuity of g_f at $t = 0$ there exists $\delta > 0$ such that if $t = |t - 0| < \delta$ then $g_f(t) < \varepsilon$. As a result, if $d(x_1, x_2) < \delta$, then $\rho(x_1, x_2) < \varepsilon$. For uniform continuity in the other direction, again let $\varepsilon > 0$, and put $\delta = g_f(\varepsilon) > 0$. If $\rho(x_1, x_2) < \delta$ then since g_f is nondecreasing, we have $d(x_1, x_2) < \varepsilon$.

It remains to show that $f : A \to \mathbb{R}$ is 1-Lipschitz with respect to ρ. Suppose a_1 and a_2 are arbitrary points of A. We have

$$|f(a_1) - f(a_2)| \leq m_f(d(a_1, a_2)) \leq g_f(d(a_1, a_2)) = \rho(a_1, a_2)$$

as required. $\square$

Examining our argument that $g_f \circ d$ is a metric uniformly equivalent to d, we only used these properties of g_f: (i) g_f is nondecreasing; (ii) g_f is subadditive; (iii) $g_f^{-1}(\{0\}) = \{0\}$; (iv) g_f is continuous at $t = 0$. In the end, concavity was not called on at all. We refer the interested reader to the expository article of Corazza [67] that considers general functions $g : [0, \infty) \to [0, \infty)$ such that for any metric d, $g \circ d$ is a metric or better yet, $g \circ d$ is a metric equivalent to the initial metric.

Theorem 3.9. *Let A be a nonempty subset of a metric space $\langle X, d \rangle$ and suppose $f : A \to \mathbb{R}$ is a uniformly continuous function on A such that m_f has an affine majorant. Then f can be extended to a uniformly continuous function on $\langle X, d \rangle$.*

Proof. This is trivial if f is a constant function. Otherwise, by Theorem 3.8, there is a metric ρ on X that is uniformly equivalent to d such that with respect to ρ, $f : A \to \mathbb{R}$ is 1-Lipschitz. By Proposition 3.5, f has a 1-Lipschitz extension $\hat{f}$ to the entire space. Since $\hat{f}$ is uniformly continuous with respect to ρ and $I_X : \langle X, d \rangle \to \langle X, \rho \rangle$ is uniformly continuous, the extension $\hat{f}$ is uniformly continuous on X with respect to d. $\square$

McShane's extension theorem as we have stated it falls out of the last more general result.

Corollary 3.10. *Let A be a nonempty subset of a metric space $\langle X, d \rangle$ and suppose $f : A \to \mathbb{R}$ is a bounded uniformly continuous function on A. Then f can be extended to a bounded uniformly continuous function on $\langle X, d \rangle$.*

Proof. Since f is bounded, there exists $\mu > 0$ such that for all $a \in A$, $|f(a)| < \mu$. As a result, m_f has $t \mapsto 2\mu$ as an affine majorant. Taking a uniformly continuous extension $\hat{f}$ for f to X, $\min\{\max\{\hat{f}, -\mu\}, \mu\}$ is a bounded uniformly continuous extension. $\square$

Proceeding more generally, we can define the *modulus of continuity function* $m_f : [0, \infty) \to [0, \infty]$ for a function between metric spaces $f : \langle X, d \rangle \to \langle Y, \rho \rangle$ by the formula $m_f(t) := \sup\{\rho(f(x_1), f(x_2)) : d(x_1, x_2) \leq t\}$ [53]. Some simple examples:

- if $f : \mathbb{R} \to \mathbb{R}$ is defined by $f(x) = x^2$ then $m_f(0) = 0$ and $m_f(t) = \infty$ otherwise;

- if $f : \mathbb{R}^2 \to \mathbb{R}$ is defined by $f(x, y) = 3x + 4y - 2$ then $m_f(t) = 5t$ for all $t \geq 0$;

- if we equip the integers $\mathbb{Z}$ with the Euclidean metric, and $f : \mathbb{Z} \to \mathbb{R}$ is defined by $f(k) = |k|$, then $m_f(t) = n - 1$ when $n - 1 \leq t < n$ for some $n \in \mathbb{N}$.

We immediately list some basic properties of m_f:

(a) $m_f(0) = 0$ and m_f is a nondecreasing function on $[0, \infty)$;

(b) m_f is continuous at $t = 0$ if and only if f is uniformly continuous;

(c) m_f is finite valued if and only if for each $\alpha_1 \in (0, \infty)$ there exists $\alpha_2 \in (0, \infty)$ such that $\text{diam}_d(A) < \alpha_1 \Rightarrow \text{diam}_\rho f(A) < \alpha_2$;

(d) m_f is bounded if and only if f is bounded;

(e) f is λ-Lipschitz if and only if $t \mapsto \lambda t$ majorizes m_f on $[0, \infty)$.

We call a function between metric spaces a *coarse map* if its modulus of continuity is finite-valued [5, 58, 106].

We close this section with a result that gives alternate expressions for the modulus of continuity of a given function to have an affine majorant. Of course, such a function must be a coarse map. This property will arise again in this monograph.

Proposition 3.11. *Let* $\langle X, d \rangle$ *and* $\langle Y, \rho \rangle$ *be metric spaces and let* $f : X \to Y$. *The following statements are equivalent for its modulus of continuity function* m_f.

(1) m_f *has an affine majorant;*

(2) m_f *has a concave majorant;*

(3) m_f *has a nondecreasing subadditive majorant.*

Proof. (1) $\Rightarrow$ (2) is trivial while (2) $\Rightarrow$ (3) is a consequence of Proposition 3.6. We turn to (3) $\Rightarrow$ (1). Let g be a nondecreasing subadditive majorant for m_f. We intend to show that for all $t \geq 0$, we have $g(t) \leq (1+t)g(1)$. For $t \in [0, 1]$, this is true because g is nondecreasing and so $g(t) \leq g(1)$. Now suppose $t > 1$. Choose $n \in \mathbb{N}$ with $n \leq t < n+1$. By subadditivity, $g(n) \leq ng(1)$, and so

$$g(t) \leq g(n+1) \leq g(n) + g(1) \leq ng(1) + g(1) \leq (t+1)g(1).$$

We have shown that $t \mapsto (1+t)g(1)$ majorizes g and thus it majorizes m_f. $\square$

4. The Lipschitz Norm for the Vector Space of Lipschitz Real-Valued Functions

The approach in this section is more functional analytic than will be found in the rest of this monograph, except for Section 33. The material is presented deliberately, in part because it is difficult to find it presented as a coherent whole in any monograph. It also gives us the opportunity to employ the Dugundji extension theorem.

As we have already noted, we can isometrically embed a metric space $\langle X, d \rangle$ into $C_b(X, \mathbb{R})$. The image of X will be closed if and only if $\langle X, d \rangle$ is complete because $C_b(X, \mathbb{R})$ is a Banach space. Remarkably, we can still isometrically embed any noncomplete metric space as a closed subset of a normed linear space, as discovered by Arens and Eells [6]. Modifying a construction of Michael [122], we show that an appropriate choice for the target space is a linear subspace of the dual space of $\mathrm{Lip}(Y, \mathbb{R})$ equipped with a so-called Lipschitz norm where Y is a complete metric space containing X isometrically.

As we mentioned earlier the assignment $f \mapsto L(f)$ on $\mathrm{Lip}(X, \mathbb{R})$ fails to be a norm as $L(f) = 0$ provided f is any constant function. To fix this problem, let $x_0 \in X$ be arbitrary. The *Lipschitz norm* on $\mathrm{Lip}(X, \mathbb{R})$ with respect to the base point x_0 is defined by

$$\|f\|_{\mathrm{Lip}} := \max\{|f(x_0)|, L(f)\} \quad (f \in \mathrm{Lip}(X, \mathbb{R})).$$

Replacing x_0 by a different point yields an equivalent norm as does replacing the maximum by a sum. It is easy to see that convergence with respect to the Lipschitz norm implies uniform convergence on bounded subsets of X; the converse holds if and only if X is bounded and uniformly isolated [40, Theorem 3.7].

Proposition 4.1. *Let $\langle X, d \rangle$ be a metric space and let $x_0 \in X$. Then $\mathrm{Lip}(X, \mathbb{R})$ equipped with the Lipschitz norm with base point x_0 is a Banach space.*

Proof. Let $\langle f_n \rangle$ be a Cauchy sequence with respect to $\|\cdot\|_{\mathrm{Lip}}$. Then both $\langle f_n(x_0) \rangle$ and $\langle L(f_n) \rangle$ are Cauchy sequences of real numbers; put $L_0 = \lim_{n \to \infty} L(f_n)$. We claim that $\langle f_n \rangle$ is a uniformly Cauchy sequence of functions on each ball with with center x_0, and hence uniformly convergent on bounded subsets of X. Let $\varepsilon > 0$ and $\delta > 1$ be arbitrary. Choose $k \in \mathbb{N}$ so large that whenever $n > j \geq k$ we have both

$$L(f_j - f_n) < \frac{\varepsilon}{2\delta} \quad \text{and} \quad |f_j(x_0) - f_n(x_0)| < \frac{\varepsilon}{2}.$$

For $x \in S_d(x_0, \delta)$ we compute

$$|(f_j - f_n)(x)| \leq |(f_j - f_n)(x) - (f_j - f_n)(x_0)| + |(f_j - f_n)(x_0)| < L(f_j - f_n)d(x, x_0) + \frac{\varepsilon}{2}$$

$$< \tfrac{\varepsilon}{2\delta} \cdot d(x, x_0) + \tfrac{\varepsilon}{2} < \tfrac{\varepsilon}{2} + \tfrac{\varepsilon}{2} = \varepsilon.$$

Denote the pointwise limit of $\langle f_n \rangle$ by f_0. It remains to show that f_0 is Lipschitz and $||f_n - f_0||_{\text{Lip}} \to 0$. For the first assertion, let x_1 and x_2 be arbitrary points of X. We compute

$$|f_0(x_1) - f_0(x_2)| = \lim_{n \to \infty} |f_n(x_1) - f_n(x_2)| \leq \lim_{n \to \infty} L(f_n)d(x_1, x_2) = L_0 \cdot d(x_1, x_2).$$

This shows that f_0 is L_0-Lipschitz. For the second assertion, we already know that $|f_j(x_0) - f_0(x_0)| \to 0$. We will show that for all j sufficiently large, $L(f_j - f_0) < \varepsilon$. Let $x_1 \neq x_2$. Let k be as specified above, and fix $j \geq k$. Since $\delta > 1$, we get

$$\frac{|(f_j - f_0)(x_1) - (f_j - f_0)(x_2)|}{d(x_1, x_2)} = \lim_{n \to \infty} \frac{|(f_j - f_n)(x_1) - (f_j - f_n)(x_2)|}{d(x_1, x_2)}$$

$$\leq \limsup_{n \to \infty} L(f_j - f_n) \leq \frac{\varepsilon}{2\delta} < \frac{\varepsilon}{2}.$$

This shows that $L(f_j - f_0) < \varepsilon$ for all $j \geq k$ as required. While we did not need to show that $L(f_0) = L_0$, this follows from the fact that $L(\cdot)$ is a continuous seminorm with respect to $||\cdot||_{\text{Lip}}$.   □

Let us write $\text{Lip}_b(X, \mathbb{R})$ for the bounded real-valued Lipschitz functions on a metric space $\langle X, d \rangle$. Our norm $||\cdot||_{\text{Lip}}$ need not produce a complete metric on $\text{Lip}_b(X, \mathbb{R})$ because $\text{Lip}_b(X, \mathbb{R})$ is not in general a closed subspace of $\text{Lip}(X, \mathbb{R})$. To see this, let $X = [1, \infty)$ and $\forall n \in \mathbb{N}$, let $f_n(x) = \ln(x)$ if $1 \leq x \leq n$ and $f(x) = \ln(n)$ otherwise. Define $f : [1, \infty) \to \mathbb{R}$ by $f(x) = \ln(x)$, an unbounded 1-Lipschitz function. Whatever the base point for the Lipschitz norm may be, we have eventually

$$||f_n - f||_{\text{Lip}} = L(f_n - f) = \frac{1}{n}.$$

When working with bounded real-valued Lipschitz functions, one can replace our Lipschitz norm by a stronger norm $||\cdot||_W$ on $\text{Lip}_b(X, \mathbb{R})$ emphasized by Weaver in his monograph on Lipschitz algebras [153] which makes $\text{Lip}_b(X, \mathbb{R})$ a Banach space:

$$||f||_W := \max\{||f||_\infty, L(f)\} \qquad (f \in \text{Lip}_b(X, \mathbb{R})).$$

Alternatively, we can replace the maximum by a sum in the definition, a path followed by Sherbert [138, 139], and that makes $\text{Lip}_b(Y, \mathbb{R})$ a commutative Banach algebra with identity. The Weaver norm fails to be submultiplicative, requiring a factor of 2 in front of the product of the norms. We will not systematically study spaces of bounded real-valued Lipschitz functions in this monograph, referring the reader to [65, 153].

The next result is essentially due to Michael [122].

Theorem 4.2. *Let $\langle X, d \rangle$ be a metric space and equip $Lip(X, \mathbb{R})$ with a Lipschitz norm $||\cdot||_{Lip}$ with base point $x_0 \in X$. For each $x \in X$, let $\hat{x} : Lip(X, \mathbb{R}) \to \mathbb{R}$ be the evaluation map, i.e., $\hat{x}(f) = f(x)$. Then $x \mapsto \hat{x}$ maps X isometrically into the continuous dual of $\langle Lip(X, \mathbb{R}), ||\cdot||_{Lip} \rangle$ equipped the usual operator norm.*

Proof. We are going to show that $\phi(x) = \hat{x}$ is an isometry of $\langle X, d \rangle$ into $\langle \text{Lip}(X, \mathbb{R})^*, ||\cdot||_{\text{op}} \rangle$. First, if $x \in X$ and $f \in \text{Lip}(X, \mathbb{R})$ with $||f||_{\text{Lip}} \leq 1$, the triangle inequality gives

$$|\hat{x}(f)| = |f(x)| \leq |f(x_0)| + L(f)d(x, x_0) \leq 1 + d(x, x_0),$$

which shows that $\hat{x} \in \text{Lip}(X, \mathbb{R})^*$.

We next show that if $\{x_1, x_2\} \subseteq X$, then $d(x_1, x_2) = ||\widehat{x_1} - \widehat{x_2}||_{\mathrm{op}}$. First, if $||f||_{\mathrm{Lip}} \leq 1$, we have

$$|\widehat{x_1}(f) - \widehat{x_2}(f)| = |f(x_1) - f(x_2)| \leq L(f)d(x_1, x_2) \leq d(x_1, x_2).$$

On the other hand, if $f(x) := d(x_1, x) - d(x_1, x_0)$, then $||f||_{\mathrm{Lip}} = 1$ while $|\widehat{x_1}(f) - \widehat{x_2}(f)| = d(x_1, x_2)$. This shows that ϕ is an isometry. $\qquad\square$

As anticipated earlier, the operator norm closure of $\{\hat{x} : x \in X\}$ in $\mathrm{Lip}(X, \mathbb{R})^*$ is another way of obtaining the completion of the underlying metric space.

We have now done much of the work required to prove the Arens-Eells Theorem which comes next.

Theorem 4.3. *Each metric space $\langle X, d \rangle$ can be isometrically embedded as a closed subset of a normed linear space.*

Proof. If $\langle X, d \rangle$ is complete, we can use the embedding of Example 2.10. Otherwise, choose a complete metric space Y that contains X isometrically; for notational simplicity, we denote the metric of Y by d as well. For each $y \in Y$, let $\hat{y} : \mathrm{Lip}(Y, \mathbb{R}) \to \mathbb{R}$ be the linear functional defined by $\hat{y}(f) = f(y)$. Fix $y_0 \in Y$ and let $|| \cdot ||_{\mathrm{Lip}}$ be the Lipschitz norm with base point y_0. As we have seen, $\phi(y) = \hat{y}$ is an isometry of $\langle Y, d \rangle$ into $\langle \mathrm{Lip}(Y, \mathbb{R})^*, || \cdot ||_{\mathrm{op}} \rangle$. We intend to show that the elements of $\phi(Y)$ are linearly independent.

Suppose $y, y_1, y_2, \ldots, y_n$ are distinct elements of Y. Put $A = \{y_1, y_2, \ldots, y_n\}$ and let $f = d(\cdot, A) \in \mathrm{Lip}(Y, \mathbb{R})$. As $\hat{y}(f) \neq 0$ while $\widehat{y_j}(f) = 0$ for $j = 1, 2, \ldots, n$, the function $\hat{y}$ cannot be a linear combination of $\{\widehat{y_j} : 1 \leq j \leq n\}$.

Since $\langle Y, d \rangle$ is complete and ϕ is an isometry, $\phi(Y)$ is closed in $\mathrm{Lip}(Y, \mathbb{R})^*$. By linear independence of the elements of $\phi(Y)$, $\phi(X) = \mathrm{span}(\phi(X)) \cap \phi(Y)$, and this shows that $\phi(X)$ is closed in $\mathrm{span}(\phi(X))$ equipped with the relative operator norm topology. Finally, as the operator norm topology on $\mathrm{span}(\phi(X))$ agrees with this relative topology, the desired normed linear space is $\langle \mathrm{span}(\phi(X)), || \cdot ||_{\mathrm{op}} \rangle$. $\qquad\square$

Following Torunczyk [146], the next embedding theorem of Klee [108] regarding extending a homeomorphism between two suitably-placed closed subsets within a product $Y \times W$ of normed linear spaces to the entire product, combined with the Arens-Eells embedding theorem easily give a transparent proof of Hausdorff's extension theorem for metrics defined on closed subsets of a metrizable space [91].

Theorem 4.4. *Let Y and W be normed linear spaces and let $K \subseteq Y \times \{0_W\}$ and $E \subseteq \{0_Y\} \times W$ be nonempty homeomorphic closed subsets. Then if $f : K \to E$ is a homeomorphism, there exists a homeomorphism $F : Y \times W \to Y \times W$ that extends f.*

Proof. Denote the projections of $Y \times W$ onto the coordinate spaces by π_Y and π_W, respectively. Define $\hat{f} : \pi_Y(K) \to W$ by $\hat{f}(y) = \pi_W(f(y, 0_W))$ and then by the Dugundji extension theorem, find $\bar{f} \in C(Y, W)$ extending $\hat{f}$. Next consider the bijection ϕ_1 on $Y \times W$ defined by $\phi_1(y, w) = (y, w + \bar{f}(y))$. As the coordinate functions are continuous, ϕ_1 is also. To see that ϕ_1^{-1} is continuous, suppose $\langle \phi(y_n, w_n) \rangle$ converges to $\phi(y, w)$. Clearly $\lim_{n \to \infty} y_n = y$, so by continuity, $\lim_{n \to \infty} \bar{f}(y_n) = \bar{f}(y)$, and it follows that $\lim_{n \to \infty} w_n = w$. Thus, ϕ_1 is a homeomorphism.

Similarly, setting $g = f^{-1}$ let $\hat{g} : \pi_W(E) \to Y$ be given by $\hat{g}(w) = \pi_Y(g(0_Y, w))$ and extend $\hat{g}$ continuously to $\bar{g} : W \to Y$. Then $\phi_2 : Y \times W \to Y \times W$ defined by

$$\phi_2(y, w) := (y + \bar{g}(w), w)$$

can be shown to be homeomorphism in exactly the same way. Now suppose $(y, 0_W) \in K$; we compute

$$\phi_2(f(y, 0_W)) = \phi_2(0_Y, \bar{f}(y)) = (\bar{g}(\bar{f}(y)), \bar{f}(y)) = (y, \bar{f}(y)) = \phi_1(y, 0_W).$$

We have shown that $\phi_2 \circ f = \phi_1$ on K so that $F = \phi_2^{-1} \circ \phi_1$ is the desired extension of f to $Y \times W$. $\qquad\square$

We now come to Hausdorff's extension result, which we will directly use.

Theorem 4.5. *Let $\langle X, \tau \rangle$ be a metrizable space and let A be a nonempty closed subset. Let $\rho : A \times A \to [0, \infty)$ be a metric compatible with relative topology on A. Then there exists $\widetilde{\rho} \in \boldsymbol{D}(X)$ that extends ρ.*

Proof. Let d be a compatible metric for $\langle X, \tau \rangle$. By the Arens-Eells theorem, we can isometrically embed $\langle X, d \rangle$ as a closed subset of a normed linear space $\langle Y, ||\cdot||_1 \rangle$, and $\langle A, \rho \rangle$ as a closed subset of a second normed linear space $\langle W, ||\cdot||_2 \rangle$. Equipping $Y \times W$ with the norm $||(y, w)|| := ||y||_1 + ||w||_2$, we can isometrically embed $\langle X, d \rangle$ into $Y \times \{0_W\}$ and $\langle A, \rho \rangle$ into $\{0_Y\} \times W$, both as closed subsets. Calling these embeddings ϕ and ψ respectively, $f : \phi(A) \to \psi(A)$ defined by

$$f(\phi(a)) = \psi(a) \quad (a \in A)$$

is a homeomorphism. By the Klee result Theorem 4.4, extend f to a homeomorphism F of $Y \times W$ onto itself. Then the restriction of the metric determined by $||\cdot||$ to $F(\phi(X))$ under the identification $x \leftrightarrow F(\phi(x))$ for $x \in X$ does the job, because the restriction of this metric to $F(\phi(A))$ under the identification agrees with ρ. $\qquad\square$

In our brief discussion of weak topologies, we remarked that a function from a topological space into a completely regular space $\langle Y, \sigma \rangle$ is continuous provided whenever it is followed by a member of $C(Y, \mathbb{R})$, the composition is continuous. Actually, if the target space is a metric space $\langle Y, \rho \rangle$, then the statement remains true if $C(Y, \mathbb{R})$ is replaced by $\mathrm{Lip}(Y, \mathbb{R})$ as the smaller class still separates points from closed sets (see, e.g., [155, pp. 56–57]). But this can be seen without such elaborate machinery. If $h : \langle X, \tau \rangle \to \langle Y, \rho \rangle$ failed to be continuous at $p \in X$, then $f \circ h$ would also fail to be continuous at p where $f = d(\cdot, h(p))$, a 1-Lipschitz function.

The next result in the same vein is due to Garrido and Jaramillo [83, Theorem 3.9]. Our proof is a streamlined version of theirs, but it relies on Lipschitz norms in the same elegant way.

Theorem 4.6. *Let $\langle X, d \rangle$ and $\langle Y, \rho \rangle$ be metric spaces and let $h \in Y^X$. Then $h \in \mathrm{Lip}(X, Y)$ if and only if whenever $f \in \mathrm{Lip}(Y, \mathbb{R})$, we have $f \circ h \in \mathrm{Lip}(X, \mathbb{R})$.*

Proof. Only sufficiency requires proof. Let $x_0 \in X$ be fixed, put $y_0 = h(x_0)$, and equip $\mathrm{Lip}(X, \mathbb{R})$ (resp. $\mathrm{Lip}(Y, \mathbb{R})$) with a Lipschitz norm with base point x_0 (resp. y_0). Define the linear transformation $T : \mathrm{Lip}(Y, \mathbb{R}) \to \mathrm{Lip}(X, \mathbb{R})$ by $T(f) = f \circ h$. It is easy to see that T is continuous from the classical closed graph theorem of functional analysis [89, p. 158], which we may apply since both the domain and target space are Banach spaces. Suppose $f, f_1, f_2, f_3, \ldots$ is a sequence in $\mathrm{Lip}(Y, \mathbb{R})$ such that $||f_n - f||_{\mathrm{Lip}} \to 0$ while $||T(f_n) - g||_{\mathrm{Lip}} \to 0$ where $g \in \mathrm{Lip}(X, \mathbb{R})$. Since convergence in a Lipschitz norm forces pointwise convergence, $\langle f_n \rangle$ converges pointwise to f and $\langle T(f_n) \rangle$ converges pointwise to g. This means that for each $x \in X$,

$$\lim_{n \to \infty} f_n(h(x)) = f(h(x)) = T(f)(x)$$

while

$$\lim_{n \to \infty} f_n(h(x)) = \lim_{n \to \infty} T(f_n)(x) = g(x),$$

from which $T(f) = g$ as required.

With $K := ||T||_{\mathrm{op}}$, we intend to show that h is K-Lipschitz. Fix $x_1 \neq x_2$ in X and put $y_1 = h(x_1)$ and $y_2 = h(x_2)$. Let $f \in \mathrm{Lip}(Y, \mathbb{R})$ be defined by

$$f(y) := \rho(y, y_2) - \rho(y_0, y_2).$$

Notice that $L(f) = 1$ and $f(y_0) = T(f)(x_0) = 0$. From this, $||f||_{\mathrm{Lip}} = \max\{|f(y_0)|, L(f)\} = 1$ and similarly $||T(f)||_{\mathrm{Lip}} = L(T(f))$. The upcoming inequality string shows that $\rho(h(x_1), h(x_2)) \leq K \cdot d(x_1, x_2)$.

$$\rho(h(x_1), h(x_2)) = \rho(y_1, y_2) = |f(y_1) - f(y_2)| = |f(h(x_1) - f(h(x_2)))| = |(T(f))(x_1) - (T(f))(x_2)|$$

$$\leq L(T(f)) \cdot d(x_1, x_2) = ||T(f)||_{\mathrm{Lip}} \cdot d(x_1, x_2)$$

$$\leq K \cdot ||f||_{\mathrm{Lip}} \cdot d(x_1, x_2) = K \cdot d(x_1, x_2). \qquad \square$$

One wonders what other metric properties of a function $h : \langle X, d \rangle \to \langle Y, \rho \rangle$ hold provided whenever it is followed by a real-valued Lipschitz function f, the composition $f \circ h$ has the property. To close this section, we establish this behavior for boundedness, Cauchy continuity and uniform continuity [37].

Theorem 4.7. *Let $\langle X, d \rangle$ and $\langle Y, \rho \rangle$ be metric spaces and let $h \in Y^X$. Then h is a bounded function if and only if whenever $f \in Lip(Y, \mathbb{R})$, the composition $f \circ h$ is bounded.*

Proof. Suppose h is bounded. Since Lipschitz functions map bounded subsets to bounded subsets, whenever $f \in \mathrm{Lip}(Y, \mathbb{R})$, f maps the bounded set $h(X)$ to a bounded subset of the line. Thus, $f \circ h$ is a bounded function. Conversely, if h fails to be bounded, then taking $y_0 \in Y$, we see that $f \circ h$ is unbounded where $f = d(\cdot, y_0)$. $\qquad \square$

The remaining two assertions depend on the Efremovič lemma [77, 129]. As we shall apply this later as well, we provide a complete proof. Qualitatively, the Efremovič lemma says that if for two infinite sequences $\langle x_n \rangle$ and $\langle w_n \rangle$ in $\langle X, d \rangle$ we have inf $\{d(x_n, w_n) : n \in \mathbb{N}\} > 0$, then we can find an infinite subset $\mathbb{N}_1$ of $\mathbb{N}$ for which there is a positive gap between $\{x_n : n \in \mathbb{N}_1\}$ and $\{w_n : n \in \mathbb{N}_1\}$. Our proof follows [17, p. 92].

Theorem 4.8. *Let $\langle x_n \rangle$ and $\langle w_n \rangle$ be sequences in $\langle X, d \rangle$ such that $\forall n \in \mathbb{N}, d(x_n, w_n) > \varepsilon$. Then there is an infinite subset $\mathbb{N}_1$ of $\mathbb{N}$ such that $D_d(\{x_n : n \in \mathbb{N}_1\}, \{w_n : n \in \mathbb{N}_1\}) \geq \frac{\varepsilon}{4}$.*

Proof. Define for each $n \in \mathbb{N}$ subsets B_n and C_n of $\mathbb{N}$ as follows:

$$B_n := \{m \in \mathbb{N} : d(x_n, w_m) \leq \frac{\varepsilon}{4}\}, \quad C_n := \{m \in \mathbb{N} : d(w_n, x_m) \leq \frac{\varepsilon}{4}\}.$$

Suppose $\{m_1, m_2\} \subseteq B_n$. Since $d(w_{m_1}, w_{m_2}) \leq \varepsilon/2$, we get $d(x_{m_1}, w_{m_2}) \geq \varepsilon/2$, for if $d(x_{m_1}, w_{m_2}) < \varepsilon/2$, we would have

$$d(w_{m_1}, x_{m_1}) \leq d(w_{m_1}, w_{m_2}) + d(w_{m_2}, x_{m_1}) < \frac{\varepsilon}{2} + \frac{\varepsilon}{2} = \varepsilon,$$

and this violates the basic assumption for our paired sequences. Thus if B_n is infinite for some n, we can let $\mathbb{N}_1 = B_n$ and we are done; note that the achieved gap is at least $\varepsilon/2$ when this occurs. The case when C_n is infinite for some n is handled in the same way.

So we are left with the possibility that both B_n and C_n are finite for each $n \in \mathbb{N}$. We will construct our infinite set of integers $\mathbb{N}_1$ recursively. Start with $n_1 = 1$ and let $n_2 > 1$ be the first integer past $B_1 \cup C_1$. We have $d(x_1, w_{n_2}) > \varepsilon/4$ because $n_2 \notin B_1$ and $d(w_1, x_{n_2}) > \varepsilon/4$ because $n_2 \notin C_1$. From this, we have $D_d(\{x_{n_1}, x_{n_2}\}, \{w_{n_1}, w_{n_2}\}) > \varepsilon/4$. Now let $n_3 > n_2$ be the first integer past $B_{n_2} \cup C_{n_2}$. Then n_3 is not in $B_{n_1} \cup C_{n_1}$ either so that both $d(x_i, w_{n_3}) > \varepsilon/4$ for $i = 1, 2$ and $d(w_i, x_{n_3}) > \varepsilon/4$ for $i = 1, 2$. We conclude that

$$D_d(\{x_{n_1}, x_{n_2}, x_{n_3}\}, \{w_{n_1}, w_{n_2}, w_{n_3}\}) > \varepsilon/4.$$

Continuing in this way indefinitely yields $n_1 < n_2 < n_3 < \cdots$ that comprise $\mathbb{N}_1$ with the desired property. □

Theorem 4.9. *Let $\langle X, d \rangle$ and $\langle Y, \rho \rangle$ be metric spaces and let $h \in Y^X$. Then $h \in CC(X, Y)$ if and only if whenever $f \in Lip(Y, \mathbb{R})$, we have $f \circ h \in CC(X, \mathbb{R})$.*

Proof. Necessity is obvious. For sufficiency, we prove the contrapositive. Suppose $h \notin CC(X, Y)$; by definition, we can find a Cauchy sequence $\langle x_n \rangle$ in X that is not mapped by h into a Cauchy sequence in Y. As a result, for some $\varepsilon > 0$, we can find $n_1 < n_2 < n_3 < \cdots$ such for each $k \in \mathbb{N}$ we have $\rho(h(x_{n_{2k-1}}), h(x_{n_{2k}})) > \varepsilon$. For each $k \in \mathbb{N}$, put $w_k = h(x_{n_{2k-1}})$ and put $z_k = h(x_{n_{2k}})$. In view of the Efremovic lemma, there exists an infinite set of positive integers $\mathbb{N}_1$ such that

$$D_\rho(\{w_k : k \in \mathbb{N}_1\}, \{z_k : k \in \mathbb{N}_1\}) \geq \frac{\varepsilon}{4}.$$

Now put $f = \rho(\cdot, \{w_k : k \in \mathbb{N}_1\}) \in Lip(Y, \mathbb{R})$. By construction, $f \circ h$ fails to be take $\langle x_n \rangle$ to a Cauchy sequence of reals. □

The proof of the next companion theorem is similar and is left the reader (see [37, Theorem 3.3]).

Theorem 4.10. *Let $\langle X, d \rangle$ and $\langle Y, \rho \rangle$ be metric spaces and let $h \in Y^X$. Then $h \in UC(X, Y)$ if and only if whenever $f \in Lip(Y, \mathbb{R})$, we have $f \circ h \in UC(X, \mathbb{R})$.*

5. Nets and Uniformities

A relation $\preceq$ on a set Λ is said to *direct* Λ if it is reflexive, transitive, and whenever $\lambda_1, \lambda_2 \in \Lambda$, there exists $\lambda_3 \in \Lambda$ with $\lambda_1 \preceq \lambda_3$ and $\lambda_2 \preceq \lambda_3$. A subset Λ_0 of Λ is called *residual* if for some $\lambda_1 \in \Lambda, \lambda_1 \preceq \lambda \in \Lambda \Rightarrow \lambda \in \Lambda_0$, whereas Λ_0 is called *cofinal* if for each $\lambda_1 \in \Lambda, \{\lambda \in \Lambda : \lambda_1 \preceq \lambda\} \cap \Lambda_0 \neq \emptyset$. Given a family of directed sets $\{\langle \Lambda_i, \preceq_i \rangle : i \in I\}$, we direct $\prod_{i \in I} \Lambda_i$ coordinatewise, that is, $f \preceq g$ if and only if for each $i \in I, f(i) \preceq_i g(i)$.

By a *net* in a set X, we mean a function $s : \Lambda \to X$ where Λ is directed. Of course, a sequence is a net defined on $\mathbb{N}$ directed by the usual $\leq$. As with sequences, we can designate a net using subscripts: $\langle s_\lambda \rangle_{\lambda \in \Lambda}$ where s_λ means $s(\lambda)$. When the directed set is understood, we use the more compact notation $\langle s_\lambda \rangle$ for the net. By a *subnet* of such a net s defined on a second directed set $\langle \Sigma, \preceq' \rangle$, we mean a composition of the form $s \circ \phi : \Sigma \to X$ where $\phi : \Sigma \to \Lambda$ such that for each $\lambda \in \Lambda$, there exists $\sigma_\lambda \in \Sigma$ such that $\sigma \succeq' \sigma_\lambda \Rightarrow \phi(\sigma) \succeq \lambda$ [107, p. 70]. A stronger requirement preferred by some authors is that ϕ be monotone as well.

When X is equipped with a topology τ, we say that a net $s : \Lambda \to X$ *converges* to a point p if each neighborhood of p contains s_λ residually ($=$ for a residual set of indices). In this case, we call p a *limit point* of the net. We say the net has p as a *cluster point* if each neighborhood of p contains s_λ cofinally ($=$ for a cofinal set of indices). It is not hard to show that p is a cluster point of $\langle s_\lambda \rangle$ if and only if the net has a subnet convergent to p.

Topological notions in a general topological space $\langle X, \tau \rangle$ can be described in terms of nets, e.g.,

- The topology τ on X is finer than σ if and only if each τ-convergent net is σ-convergent to the same limits.

- τ is Hausdorff iff each convergent net in $\langle X, \tau \rangle$ can have at most one limit.

- $A \subseteq X$ is compact if and only if each net in A has a cluster point in A.

- $A \subseteq X$ is closed if and only if whenever a net in A converges to a point $p \in X$, then $p \in A$.

By a *convergence structure* on a set X, or more briefly a *convergence* on X, we mean a rule that assigns to each net $\langle s_\lambda \rangle$ in X a possibly empty subset of X consisting of those points to which the net converges. If p is such a point, we write $p \in \lim_\lambda s_\lambda$. A basic question relative to this framework is the following: exactly what conditions must one put on a convergence so that there is a topology τ on X so that a net $\langle s_\lambda \rangle$ converges to $p \in X$ with respect to the convergence rule if and only it is τ-convergent to p? In this case, we say that the convergence stucture is *topological*. It turns out the existence of such a topology occurs if and only if the rule satisfies the following conditions [107, 109]:

(1) Each constant net in X converges to the common value of the net;

(2) If a net $\langle s_\lambda \rangle$ converges to p, then so does each subnet;

(3) If $\langle s_\lambda \rangle$ is a net in X such that each subnet has in turn a subnet convergent to p, then $\langle s_\lambda \rangle$ converges to p;

(4) Let I be a directed set and for each $i \in I$, let Λ_i be a directed set. Suppose for each $i \in I$ and $\lambda_i \in \Lambda_i, s(i, \lambda_i) \in X$ and that for each fixed $i \in I$, $x_i \in \lim_{\lambda_i} s(i, \lambda_i)$. If $p \in \lim_i x_i$, then the net $(i, f) \mapsto s(i, f(i))$ defined on $I \times \prod_{i \in I} \Lambda_i$ converges to p as well.

The fourth condition in this list is called the *iterated limit condition*.

One of the practical benefits of working in a metric space is that it is possible to compare nearness for pairs of points in the space in terms of the distance functional. It is possible to do so somewhat more generally when the underlying set is equipped with a uniform structure. A uniformity is actually a family of subsets of $X \times X$, each containing the diagonal, satisfying certain properties. Elements of the uniformity are called *entourages*. It is usually more convenient to work with a base for the uniformity, from which the whole uniformity is recovered by taking supersets [155, p. 239].

Definition 5.1. Let X be a nonempty set. A family $\mathscr{D}$ of subsets of $X \times X$ is said to form a *uniform base* on X if it satisfies the following properties:

(1) each $D \in \mathscr{D}$ contains the diagonal of the product $\{(x, x) : x \in X\}$;

(2) for each $D \in \mathscr{D}$ there exists $E \in \mathscr{D}$ with $E^{-1} \subseteq D$;

(3) for each $D \in \mathscr{D}$ there exists $E \in \mathscr{D}$ with $E \circ E \subseteq D$;

(4) each $D_1, D_2 \in \mathscr{D}$ there exists $E \in \mathscr{D}$ with $E \subseteq D_1 \cap D_2$.

The uniform base $\mathscr{D}$ and the induced uniformity $\uparrow \mathscr{D}$ are called *separating* if for each $(x, w) \in X \times X$ with $x \neq w$, there exists $D \in \mathscr{D}$ such that $(x, w) \notin D$. Usually, in our description of a uniform base, we will have $D = D^{-1}$ so that condition (2) is already satisfied. In any case, each uniformity can be shown to have a symmetric base. Two uniformities are the same provided we can find bases for them whose members are mutually included in one another.

Example 5.2. Given a set equipped with a metric d, a base for the *metric uniformity* associated with d consists of all entourages of the form $D_\varepsilon = \{(x, w) : d(x, w) < \varepsilon\}$ where $\varepsilon > 0$. This uniformity is clearly separating. Two metrics give rise to the same metric uniformity if and only if they are uniformly equivalent.

Note that this uniformity has a countable base, namely $\{D_{1/n} : n \in \mathbb{N}\}$. It can be shown that a separating uniformity is a metric uniformity if and only if the uniformity has a countable base [155, p. 258].

Example 5.3. Given a family of real-valued functions $\{f_i : i \in I\}$ on X the *weak uniformity* determined by the family on X has a base consisting of all entourages of the form

$$D_{I_0, \varepsilon} := \{(x, w) : \forall i \in I_0, |f_i(x) - f_i(w)| < \varepsilon\}$$

where I_0 runs over the finite subsets of I and $\varepsilon > 0$. The uniformity is separating provided the family separates points of X. If additionally, I is countable, then from our comments in the last example, the uniformity is a metric uniformity.

Given a uniformity with base $\mathscr{D}$, for each $x \in X$ and $D \in \mathscr{D}$, put $D[x] = \{w \in X : (x, w) \in D\}$. Then $\{D[x] : D \in \mathscr{D}\}$ forms a local base at x for a topology on X that does not depend on the particular base for the uniformity chosen, and which is Hausdorff if and only if the uniformity is separating. The topology determined by the weak uniformity induced by a family of real-valued functions is the weak topology determined by the family of functions. Each completely regular topology arises in this way and conversely [155, p. 256]. For this reason as well as others, many analysts eschew topological spaces that fail to be completely regular as well as Hausdorff.

6. Some Basic Bornologies

Let us now denote the metrically bounded subsets of $\langle X, d \rangle$ by $\mathscr{B}_d(X)$. Clearly $\mathscr{B}_d(X)$ contains the singletons, is stable under finite unions, and is an hereditary family.

Definition 6.1. By a *bornology* $\mathscr{B}$ on a nonempty set X, we mean a family of subsets that contains the singletons, is stable under finite unions, and whenever $B \in \mathscr{B}$ and $B_0 \subseteq B$, then $B_0 \in \mathscr{B}$.

The largest bornology on X - called the *trivial bornology* - is $\mathscr{P}(X)$ and the smallest is $\mathscr{F}(X)$. We say the bornology $\mathscr{B}_1$ is *finer than* $\mathscr{B}_2$ (and $\mathscr{B}_2$ is *coarser than* $\mathscr{B}_1$) if $\mathscr{B}_2 \subseteq \mathscr{B}_1$. We denote the family of bornologies on X by $\mathfrak{B}_X$.

Note the wording "contains the singletons" can be replaced by "is a cover of X" without changing meaning in the definition. A family of subsets that is merely stable under finite unions and is hereditary is called an *ideal of subsets*. Thus a bornology is a family that is at once an ideal and a cover of X.

Definition 6.2. A subset A of a metric space $\langle X, d \rangle$ is called *totally bounded* if for each $\varepsilon > 0$ there exists $F_\varepsilon \in \mathscr{F}(X)$ with $A \subseteq S_d(F_\varepsilon, \varepsilon)$.

If $A \subseteq X$ is totally bounded and nonempty, then the for each $\varepsilon > 0$ the finite subset F_ε can be chosen inside A. We denote the totally bounded subsets of X by $\mathscr{T}\mathscr{B}_d(X)$. Evidently the totally bounded subsets form a bornology that is coarser than the bornology of metrically bounded subsets because a totally bounded set is a subset of a finite union of balls, and each ball belongs to $\mathscr{B}_d(X)$ which is stable under finite unions and is hereditary. Any infinite subset of a metric space equipped with the zero-one metric is bounded but not totally bounded. By a well-known theorem of Riesz, the closed unit ball of an infinite-dimensional normed linear space while bounded fails to be totally bounded (see, e.g., [89, p. 26]). It is a consequence of the Urysohn metrization theorem that a metrizable space has a compatible metric d for which $X \in \mathscr{T}\mathscr{B}_d(X)$ if and only if X is separable [155, p. 166].

Example 6.3. Let X be a nonempty set, let $\langle Y, \rho \rangle$ be a metric space and let $f : X \to Y$. Then $\{A \subseteq X : f(A)$ is metrically bounded$\}$ is a bornology on X. We denote this bornology by $\mathscr{B}(f)$. When the bornology is trivial, this means that f is a bounded function on X.

A Hausdorff space equipped with a bornology $\mathscr{B}$ is called a *bornological universe* [3, 23, 102, 152]. Thus a bornological universe is a triple $\langle X, \tau, \mathscr{B} \rangle$.

Definition 6.4. Let $\langle X, \tau \rangle$ be a Hausdorff space. Then $\{A \subseteq X : \mathrm{cl}(A)$ is compact$\}$ is a bornology on X which we call the *compact bornology*.

We denote the compact bornology by $\mathscr{K}(X)$. Its members are called *relatively compact sets*. That the relatively compact sets are stable under finite unions follows from the formula $\mathrm{cl}(A \cup B) = \mathrm{cl}(A) \cup \mathrm{cl}(B)$. Since compact subsets of a Hausdorff space are closed, A is relatively compact if and only A is a subset of some compact subset, which is an easier criterion to verify as we now illustrate.

Proposition 6.5. *Let $\langle X, \tau \rangle$ and $\langle Y, \sigma \rangle$ be Hausdorff spaces and suppose $f : X \to Y$ is continuous. Then f maps members of $\mathscr{K}(X)$ to members of $\mathscr{K}(Y)$.*

Proof. Let $A \in \mathscr{K}(X)$ be arbitrary. If $A = \emptyset$, then $f(A) = \emptyset \in \mathscr{K}(Y)$. Otherwise, by continuity $f(\mathrm{cl}(A))$ is compact and contains $f(A)$. Thus by our alternate criterion, $f(A)$ belongs to $\mathscr{K}(Y)$. $\qquad\square$

We now record some basic properties of the compact bornology for a metric space.

Proposition 6.6. *Let $\langle X, d \rangle$ be a metric space. Then $A \subseteq X$ is relatively compact if and only if each sequence in X has a subsequence convergent to a point of X.*

Proof. Suppose $A \neq \emptyset$ is relatively compact. If $\langle a_n \rangle$ is a sequence in A, then since the sequence lies in $\mathrm{cl}(A)$ and compactness agrees with sequential compactness for subsets of a metric space, $\langle a_n \rangle$ has a subsequence convergent to a point of $\mathrm{cl}(A)$. Conversely, suppose $A \notin \mathscr{K}(X)$. This means that there is a sequence $\langle x_n \rangle$ in $\mathrm{cl}(A)$ that has no convergent subsequence to any point of X. For each $n \in \mathbb{N}$, we can find $a_n \in A$ with $d(x_n, a_n) < 1/n$, and so $\langle a_n \rangle$ has no convergent subsequence either. $\qquad\square$

Proposition 6.7. *Let $\langle X, d \rangle$ be a metric space and let $\langle Y, \rho \rangle$ be a second metric space. Let $f : X \to Y$. The following conditions are equivalent.*

(1) *f belongs to $C(X, Y)$;*

(2) *the restriction of f to each nonempty relatively compact subset is uniformly continuous;*

(3) *the restriction of f to each nonempty relatively compact subset is continuous.*

Proof. For (1) $\Rightarrow$ (2), in the context of metrizable spaces, a globally continuous function is uniformly continuous when restricted to each compact subset and thus to any subset of a compact set. The implication (2) $\Rightarrow$ (3) is trivial. For (3) $\Rightarrow$ (1), let $p \in X$ be arbitrary and let $\langle x_n \rangle$ be a sequence convergent to p. Then $A = \{p, x_1, x_2, x_3, \ldots\}$ is compact, and as $f|_A$ is continuous by (3), we have $\lim_{n \to \infty} f(x_n) = f(p)$. We have shown that $f \in C(X, Y)$. $\qquad\square$

It is not true that (3) $\Rightarrow$ (1) holds in a general Hausdorff space even when $Y = \mathbb{R}$, but it does hold in Hausdorff spaces that are *compactly generated*, i.e., those in which a subset is closed iff its intersection with each compact subset is closed. This important class of spaces includes the first countable spaces (thus all metric spaces) and the locally compact spaces [155, p. 285].

We call a property of subsets of a metric space *intrinsic* if a subset A has it in one metric space, then it has it in each metric space in which A can be isometrically embedded. Metric boundedness, total boundedness and finiteness are all intrinsic properties, while relative compactness fails to be intrinsic. We are also interested in properties of subsets of a particular metrizable space that are invariant under a change of metric compatible with the topology. Having compact closure obviously is such a property. Having completely metrizable closure is another such property. This is because this property only depends on the relative topology of the closure. It is well-known that a subset A of a metrizable space is completely metrizable if and only if it can be topologically embedded in a completely metrizable space where it resides as a countable intersection of open subsets, i.e., as a so-called G_δ-subset. If this holds, then A must be a G_δ-subset of *any* metrizable space in which it can be topologically embedded [155, pp. 179–181].

Now consider within a metrizable space $\langle X, \tau \rangle$ the family

$$\mathscr{B} := \{A \in \mathscr{P}_0(X) : \mathrm{cl}(A) \text{ in its relative topology is completely metrizable}\} \cup \{\emptyset\}.$$

Evidently, $\mathscr{B}$ is hereditary and contains the singletons. Suppose $\{A_1, A_2, \ldots, A_n\} \subseteq \mathscr{B}$ where each A_i is nonempty, and let d_i be a compatible complete metric for $\mathrm{cl}(A_i)$. By the Hausdorff extension theorem, each d_i can be extended to a compatible metric for the entire space, so we can find a compatible metric ρ_i for $\langle X, \tau \rangle$ whose restriction to $\mathrm{cl}(A_i) \times \mathrm{cl}(A_i)$ is d_i for $i = 1, 2, \ldots, n$. Then the trace of $\max_{i \leq n} \rho_i$ on $\cup_{i=1}^n \mathrm{cl}(A_i) \times \cup_{i=1}^n \mathrm{cl}(A_i)$ is a compatible complete metric for $\mathrm{cl}(\cup_{i=1}^n A_i)$. This shows that $\mathscr{B}$ is a bornology.

Given a particular compatible metric d, a coarser related bornology is

$$\{A \subseteq X : \langle \mathrm{cl}(A), d \rangle \text{ is a complete metric space}\}.$$

By the Hausdorff extension theorem, the union of all such bornologies as d runs over $\mathbf{D}(X)$ produces the bornology $\mathscr{B}$ just described.

For a property P_d of subsets of a metrizable space $\langle X, \tau \rangle$ - which depends on the particular compatible metric chosen - such that $\{A \subseteq X : P_d(A)\}$ forms a bornology for each $d \in \mathbf{D}(X)$, we are of course interested in $\cap_{d \in \mathbf{D}(X)} \{A \subseteq X : P_d(A)\}$. Two important cases yield $\mathscr{K}(X)$ for the intersection.

Theorem 6.8. *Suppose A is a nonempty subset of a metrizable space $\langle X, \tau \rangle$. Then $A \in \mathscr{K}(X)$ if and only if for each $\rho \in \mathbf{D}(X)$, we have $A \in \mathscr{B}_\rho(X)$.*

Proof. Each relatively compact set is metrically bounded with respect to each compatible metric because compact subsets are metrically bounded. Suppose $A \subseteq X$ fails to be relatively compact. Choose a sequence $\langle a_n \rangle$ in A with distinct terms that has no cluster point. Then $a_n \mapsto n$ is a continuous real-valued function on the closed set $\{a_n : n \in \mathbb{N}\}$ equipped with its relative topology, so by the Tietze extension theorem, this assignment has a continuous extension $f : X \to \mathbb{R}$. If d is any compatible metric on X and $\rho(x, w) := d(x, w) + |f(x) - f(w)|$, then ρ is a compatible metric such that $A \notin \mathscr{B}_\rho(X)$. $\qquad\square$

Theorem 6.9. *Suppose A is a nonempty subset of a metrizable space $\langle X, \tau \rangle$. Then $A \in \mathscr{K}(X)$ if and only if for each $\rho \in \mathbf{D}(X)$, the trace of ρ on $\mathrm{cl}(A) \times \mathrm{cl}(A)$ is a complete metric.*

Proof. If A is relatively compact and nonempty, then $\langle \mathrm{cl}(A), \rho \rangle$ is a complete metric space for each $\rho \in \mathbf{D}(X)$ because each (Cauchy) sequence in $\mathrm{cl}(A)$ has a cluster point in $\mathrm{cl}(A)$. Conversely, suppose that $A \notin \mathscr{K}(X)$ and again let $\langle a_n \rangle$ be a sequence in A with distinct terms that has no cluster point. Define a metric d on the closed set of terms by

$$d(a_k, a_n) = \begin{cases} \sum_{j=k+1}^n 2^{-j} & \text{if } k < n \\ 0 & \text{if } k = n \end{cases}.$$

By construction, $\langle a_n \rangle$ is a Cauchy sequence, and the metric is compatible with the relative topology on $\{a_n : n \in \mathbb{N}\}$ which is discrete. By the Hausdorff extension theorem, we can find

$\rho \in \mathbf{D}(X)$ whose trace on $\{a_n : n \in \mathbb{N}\} \times \{a_n : n \in \mathbb{N}\}$ is d. The restriction of ρ to $\mathrm{cl}(A) \times \mathrm{cl}(A)$ fails to be complete. $\qquad\square$

In the sequel, we will encounter various families of bornologies on a metrizable space indexed by $\mathbf{D}(X)$ each containing $\mathscr{K}(X)$ and each member of which is either bounded or has complete closure for each choice of compatible metric. The totally bounded subsets as determined by compatible metrics d is such a family. In view of our last two results, the intersection of such a family as d runs over $\mathbf{D}(X)$ must be $\mathscr{K}(X)$.

Definition 6.10. A subset A of a Hausdorff space $\langle X, \tau \rangle$ is called *separable* if it has a countable dense subset.

Separability of a subset is an intrinsic property of metric spaces and separability of a subset of a particular metrizable space does not depend on the compatible metric chosen. Again in the metric context, $\{A \subseteq X : A \text{ is separable}\}$ is another bornology on X finer than $\mathscr{K}(X)$. However, this family may not form a bornology in a Hausdorff space, as separability is not in general an hereditary property. Consider the Sorgenfrey plane where $\mathbb{R}^2$ is equipped with the topology having as a base all sets of the form $[\alpha, \beta) \times [\delta, \gamma)$ where $\alpha < \beta$ and $\delta < \gamma$. While $\mathbb{Q} \times \mathbb{Q}$ is dense in the Sorgenfrey plane, an appropriately chosen line through the origin will have a discrete relative topology and thus does not have a countable dense subset.

Definition 6.11. A subset A of a Hausdorff space $\langle X, \tau \rangle$ is called *nowhere dense* provided $\mathrm{int}(\mathrm{cl}(A)) = \emptyset$.

To say that A is nowhere dense is the same as saying that $X \setminus A$ contains a dense open subset. As the intersection of two dense open subsets is of the same form, the family of nowhere dense subsets is stable under finite unions, and is clearly hereditary. Now $\{x\}$ is nowhere dense if and only if x is a limit point of X. Thus, the family of nowhere dense subsets forms a bornology if and only if $X = X'$.

Certainly, the most celebrated result around this notion is Baire's Theorem: in a complete metric space, the complement of a countable union of nowhere dense sets is dense [111, p. 414]. From Baire's Theorem, it follows that the rationals $\mathbb{Q}$ cannot be a G_δ subset of $\mathbb{R}$ equipped with the usual complete metric d, else $\mathbb{R} \setminus \mathbb{Q}$ would be a countable union of nowhere dense subsets and we could then express $\mathbb{R}$ as a countable union of nowhere dense subsets. From this, we immediately see that the property of having a completely metrizable closure is not an intrinsic property of subsets of metric spaces: $\mathbb{Q}$ as a subset of $\langle \mathbb{R}, d \rangle$ has completely metrizable closure, while $\mathbb{Q}$ as a subset of $\langle \mathbb{Q}, d \rangle$ does not.

Definition 6.12. Let $\langle X, \tau \rangle$ be a Hausdorff space. Then

$$\{A \subseteq X : \text{each base for } \tau \text{ has a finite subfamily that covers } A\}$$

is a bornology on X called the *Hindman bornology* [96].

We will call elements of the Hindman bornology *Hindman bounded*. In the context of metric spaces, as the family of all open balls of radius at most ε forms a base for the topology whatever $\varepsilon > 0$ may be, each Hindman bounded set must be totally bounded.

Proposition 6.13. *Let $\langle X, \tau \rangle$ be a Hausdorff space.*

(1) *The compact bornology on X is always coarser than the Hindman bornology.*

(2) *If $\langle X, \tau \rangle$ is regular, then the two bornologies agree.*

(3) *There exists a Hausdorff space for which the Hindman bornology is strictly finer than the compact bornology.*

(4) *A subset A of X is compact if and only if it is closed and Hindman bounded.*

Proof. For (1), suppose $\mathrm{cl}(A)$ is compact. Given a base for τ, it is an open cover of $\mathrm{cl}(A)$, so we can extract from it a finite subcover of $\mathrm{cl}(A)$ which also serves to cover A.

Moving to statement (2), by (1), we need only show that, with regularity, each Hindman bounded set has compact closure. We use regularity in this form: each neighborhood of a point $p \in X$ contains a closed neighborhood of p. So let A be Hindman bounded and let $\mathscr{U}$ be an open cover of $\mathrm{cl}(A)$. Define subfamilies $\mathscr{V}, \mathscr{W}$ of τ by

$$\mathscr{V} := \{V \in \tau : \exists U \in \mathscr{U} \text{ with } \mathrm{cl}(V) \subseteq U\}, \quad \mathscr{W} := \{V \in \tau : V \cap \mathrm{cl}(A) = \emptyset\}.$$

We claim that $\mathscr{V} \cup \mathscr{W}$ is a base for the topology. To see this, let $x \in W \in \tau$. If $x \in \mathrm{cl}(A)$, pick $U \in \mathscr{U}$ with $x \in U$. By regularity, there exists $V \in \tau$ with $x \in V \subseteq \mathrm{cl}(V) \subseteq U \cap W \subseteq U$. Of course, $x \in V \subseteq W$ as well. Otherwise, $x \in W \backslash \mathrm{cl}(A) \subseteq W$ and $W \backslash \mathrm{cl}(A) \in \mathscr{W}$ and the claim is established.

By Hindman boundedness, there is a finite subfamily $\mathscr{G}$ of the base that covers $\mathrm{cl}(A)$, and without loss of generality, $\mathscr{G} \subseteq \mathscr{V}$. For each $G \in \mathscr{G}$, pick $U_G \in \mathscr{U}$ with $G \subseteq U_G$; then $\{U_G : G \in \mathscr{G}\}$ is a finite subcover of the initial open cover of $\mathrm{cl}(A)$.

To illustrate statement (3), we will equip $[0,1]$ with a topology τ finer than the usual (relative) topology τ_d; of course, the finer topology is Hausdorff as well. Specifically τ is the the topology generated by $\tau_d \cup \{[0,1] \backslash \mathbb{Q}\}$. We will produce a τ-dense subset T that is Hindman bounded but whose closure, namely $[0,1]$, fails to be compact.

To see that $[0,1]$ is noncompact with respect to τ, let $\langle q_n \rangle$ be a sequence in $\mathbb{Q} \cap [0,1]$ with distinct terms convergent in the usual sense to an irrational number a. For each n choose $\varepsilon_n < |q_n - a|$ such that $\{S_d(q_n, \varepsilon_n) : n \in \mathbb{N}\}$ is a pairwise disjoint collection of balls. Notice that $[0,1] \backslash \mathbb{Q}$ is a τ-neighborhood of each irrational number in $[0,1]$ disjoint from $\{q_n : n \in \mathbb{N}\}$, and whenever $q \in \mathbb{Q} \cap [0,1]$ with $q \notin \{q_n : n \in \mathbb{N}\}$ we have $d(q, \{q_n : n \in \mathbb{N}\}) > 0$. As a result, the set $\{q_n : n \in \mathbb{N}\}$ is τ-closed so that

$$\{[0,1] \backslash \{q_n : n \in \mathbb{N}\}\} \cup \{S_d(q_n, \varepsilon_n) : n \in \mathbb{N}\}$$

is an open cover of $[0,1]$ with no finite subcover.

We finish this argument by showing that the τ-dense subset $T = [0,1] \backslash \mathbb{Q}$ is Hindman bounded. An arbitrary base for the topology can be decomposed into two parts: $\{U_i : i \in I\} \cup \{V_j \cap T : j \in J\}$ where $\{U_i : i \in I\} \subseteq \tau_d$ and $\{V_j : j \in J\} \subseteq \tau_d$. Since each base for τ is a cover of $[0,1]$, by the τ_d-compactness of $[0,1]$, we can find finite subsets I_0 of I and J_0 of J such that $[0,1] \subseteq \bigcup_{i \in I_0} U_i \cup \bigcup_{j \in J_0} V_j$. As a result,

$$T = T \cap \left(\bigcup_{i \in I_0} U_i \cup \bigcup_{j \in J_0} V_j\right) \subseteq \bigcup_{i \in I_0} U_i \cup \bigcup_{j \in J_0} V_j \cap T,$$

and this shows that T is Hindman bounded.

For (4), compactness of a subset of a Hausdorff space ensures that the subset is closed, and it is Hindman bounded by (1). Conversely, suppose A is closed and Hindman bounded. Let $\mathscr{V}$ be an open cover of A. Put

$$\mathscr{U} = \{U \in \tau : \text{ either } U \cap A = \emptyset \text{ or } \exists V \in \mathscr{V} \text{ with } U \subseteq V\}.$$

We claim that $\mathscr{U}$ is a base for τ. To see this, suppose $x \in W \in \tau$. If $x \notin A$, then $x \in A^c \cap W \subseteq W$ while $A^c \cap W \in \mathscr{U}$. If $x \in A$, then for some $V \in \mathscr{V}$ we have $x \in V$, and so $x \in W \cap V \subseteq W$ while $W \cap V \in \mathscr{U}$. By the definition of Hindman boundedness, we can find a finite subfamily $\{U_1, U_2, \ldots, U_n\}$ of $\mathscr{U}$ that covers A, and without loss of generality may assume $\forall i \leq n, U_i \cap A \neq \emptyset$. Choosing $V_i \in \mathscr{V}$ with $U_i \subseteq V_i$ for $i \leq n$, we have our finite subcover of A. $\qquad\square$

7. Total Boundedness Revisited and Bourbaki Boundedness

We first collect many basic facts about the totally bounded subsets in one place.

Proposition 7.1. *Let $\langle X, d \rangle$ be a metric space.*

(1) *The totally bounded subsets of X form a bornology contained in $\mathscr{B}_d(X)$.*

(2) *If $A \in \mathscr{T}\mathscr{B}_d(X)$, then for each $\varepsilon > 0$, we can find a finite subset A_ε of A with $A \subseteq S_d(A_\varepsilon, \varepsilon)$.*

(3) *Each subset of X with compact closure is totally bounded.*

(4) *$A \subseteq X$ is totally bounded if and only if each sequence in A has a Cauchy subsequence.*

(5) *The closure of each totally bounded set A is totally bounded.*

(6) *$A \subseteq X$ is compact if and only if A is totally bounded and d restricted to $A \times A$ is complete.*

(7) *$\mathscr{K}(X)$ agrees with $\mathscr{T}\mathscr{B}_d(X)$ if and only if the metric d is complete.*

(8) *$\mathscr{B}_d(X)$ agrees with $\mathscr{T}\mathscr{B}_d(X)$ if and only if the completion of $\langle X, d \rangle$ is a boundedly compact metric space.*

(9) *If d and ρ are uniformly equivalent metrics on X, then $\mathscr{T}\mathscr{B}_d(X) = \mathscr{T}\mathscr{B}_\rho(X)$.*

Proof. We have already noted (1). For (2), let $\varepsilon > 0$ and choose a finite subset E of X with $A \subseteq S_d(E, \frac{1}{2}\varepsilon)$. Without loss of generality, we may assume $\forall e \in E$ that $S_d(e, \frac{1}{2}\varepsilon) \cap A \neq \emptyset$. Then choosing $a_e \in A \cap S_d(e, \frac{1}{2}\varepsilon)$ we have

$$A \subseteq S_d(E, \frac{1}{2}\varepsilon) \subseteq S_d(\{a_e : e \in E\}, \varepsilon).$$

Moving to statement (3), if $\mathrm{cl}(A)$ is compact, then for each $\varepsilon > 0$ the open cover $\{S_d(x, \varepsilon) : x \in \mathrm{cl}\,(A)\}$ of A has a finite subcover. Thus A is totally bounded.

For (4), suppose A fails to be totally bounded. Then for some $\varepsilon > 0$, whenever $F \subseteq A$ is finite, we have $A \not\subseteq S_d(F, \varepsilon)$. This allows us to inductively find $a_1, a_2, a_3, \ldots$ in A such that for each $n \geq 1, a_{n+1} \notin \cup_{j=1}^n S_d(a_j, \varepsilon)$. Clearly, the sequence $\langle a_n \rangle$ has no Cauchy subsequence. Conversely, suppose A is totally bounded and $\langle a_n \rangle$ is a sequence in A. Since each subset of A is totally bounded, $\{a_n : n \in \mathbb{N}\}$ is contained in a finite union of balls of radius 1, so we can find an infinite subset $\mathbb{N}_1$ of $\mathbb{N}$ such that $\{a_n : n \in \mathbb{N}_1\}$ is contained in one of them. Since $\{a_n : n \in \mathbb{N}_1\}$ is contained in a finite union of balls of radius $\frac{1}{2}$, we can find an infinite subset $\mathbb{N}_2$ of $\mathbb{N}_1$ such that $\{a_n : n \in \mathbb{N}_2\}$ is contained in one of these smaller balls. Continuing, we produce a decreasing sequence of infinite sets of positive integers $\langle \mathbb{N}_j \rangle$ such that for each $j \in \mathbb{N}, \{a_n : n \in \mathbb{N}_j\}$ is contained in some ball of radius $\frac{1}{j}$. Let n_1 be the first integer in $\mathbb{N}_1$ and having chosen $n_1 < n_2 < n_3 < \cdots < n_j$ where $n_i \in \mathbb{N}_i$ for $i = 1, 2, \ldots, j$ let n_{j+1} be the smallest member of $\mathbb{N}_{j+1}$ exceeding n_j. Then $\langle a_{n_j} \rangle$ is a Cauchy subsequence of $\langle a_n \rangle$.

For statement (5), again let $\varepsilon > 0$; choosing any $\delta \in (0, \varepsilon)$ and then $F \in \mathscr{F}(X)$ with $A \subseteq S_d(F, \delta)$, we have $\mathrm{cl}(A) \subseteq S_d(F, \varepsilon)$.

Necessity in (6) is obvious. For sufficiency, let $\langle a_n \rangle$ be a sequence in A. By (2) $\langle a_n \rangle$ has a Cauchy subsequence, which by completeness of the trace of d on $A \times A$, must converge to a point of A. Thus, $\langle A, d \rangle$ is sequentially compact, which is equivalent to compactness in the context of metric spaces.

For (7), suppose the metric d is complete. Since by (3) each compact set is totally bounded, it suffices to show that each totally bounded set A has compact closure. But the trace of d on any closed subspace is clearly complete, so in particular, this is true for $\mathrm{cl}(A)$. Compactness of $\mathrm{cl}(A)$ now follows from statements (5) and (6). Conversely, suppose d fails to be complete. Let $\langle x_n \rangle$ be a d-Cauchy sequence in X that fails to converge. Easily, there can be no cluster point so that $\{x_n : n \in \mathbb{N}\}$ is a closed set. The set of terms is a closed totally bounded set that fails to be compact.

For (8), first suppose that closed and bounded subsets of the completion $\langle \hat{X}, \hat{d} \rangle$ are compact. Let A be a nonempty bounded subset of X; as A is $\hat{d}$-bounded viewed as a subset of the completion, its $\hat{d}$-closure is compact. By (6), A is $\hat{d}$-totally bounded, and by (2), it is d-totally bounded. Conversely, assume $\mathscr{B}_d(X) = \mathscr{T}\mathscr{B}_d(X)$ and let $E \subseteq \hat{X}$ be nonempty, $\hat{d}$-bounded and $\hat{d}$-closed. Let $\langle e_n \rangle$ be a sequence in E. By the density of X in $\hat{X}$ choose for each $n \in \mathbb{N}$ a point $x_n \in X$ with $\hat{d}(x_n, e_n) < \frac{1}{n}$. Now $\{x_n : n \in \mathbb{N}\}$ is a d-bounded subset of X so it must be d-totally bounded. By (4), $\langle x_n \rangle$ has a d-Cauchy subsequence, and the corresponding subsequence of $\langle e_n \rangle$ is $\hat{d}$-Cauchy and is thus $\hat{d}$-convergent to a point of E because we are assuming that E is $\hat{d}$-closed.

We leave the proof of statement (9) as an easy exercise. $\qquad \square$

By either property (2) or (4) above, if B is totally bounded in some metric space in which it is isometrically embedded, then it is totally bounded in all metric spaces in which it is isometrically embedded. That is, total boundedness is an intrinsic property of metric spaces. If B is a subset of a metrizable space, to be totally bounded with respect to each compatible metric means that $A \in \mathscr{K}(X)$ because totally bounded sets are metrically bounded (see Theorem 6.8).

We now give a variant of Proposition 6.7 perhaps due to Snipes [140].

Proposition 7.2. *Let $\langle X, d \rangle$ and $\langle Y, \rho \rangle$ be metric spaces and let $f : X \to Y$. The following conditions are equivalent:*

(1) *$f \in CC(X, Y)$;*

(2) *the restriction of f to each nonempty totally bounded subset of X is uniformly continuous;*

(3) *the restriction of f to each nonempty totally bounded subset of X is Cauchy continuous.*

Proof. For (1) $\Rightarrow$ (2), we prove the contrapositive. Suppose uniform continuity of f fails on $A \in \mathscr{T}\mathscr{B}_d(X)$. Then for some $\varepsilon > 0$ and each $n \in \mathbb{N}$ we can find $\{a_n, x_n\} \subseteq A$ with $d(a_n, x_n) < \frac{1}{n}$ but $\rho(f(a_n), f(x_n)) \geq \varepsilon$. By passing to a subsequence, we may assume $\langle a_n \rangle$ is a Cauchy sequence, so that $a_1, x_1, a_2, x_2, \ldots$ is Cauchy as well. Clearly the image sequence is not Cauchy, a contradiction. The implication (2) $\Rightarrow$ (3) is trivial, while (3) $\Rightarrow$ (1) uses the fact that the set of terms of a Cauchy sequence is a totally bounded set. $\qquad \square$

The next result characterizes membership to the bornology

$$\{B \subseteq X : \langle \mathrm{cl}(B), d \rangle \text{ is a complete metric space}\} \cup \{\emptyset\}$$

within a metric space $\langle X, d \rangle$ in terms of total boundedness.

Theorem 7.3. *Let $\langle X, d \rangle$ be a metric space and let B be a nonempty subset. The following conditions are equivalent:*

(1) *$\langle cl(B), d \rangle$ is a complete metric space;*

(2) *whenever $C \in \mathscr{C}_0(X) \cap \downarrow \{B\}$ and $A \in \mathscr{C}_0(X) \cap \mathscr{T}\mathscr{B}_d(X)$, we have $C \cap A = \emptyset \Rightarrow D_d(C, A) > 0$.*

Proof. $(1) \Rightarrow (2)$. Suppose (1) holds while (2) fails. Let $C \subseteq B$ be nonempty and closed and let A be nonempty, closed, totally bounded, and disjoint from C while $D_d(C, A) = 0$. Then we can find a sequence $\langle a_n \rangle$ in A with $\lim_{n \to \infty} d(a_n, C) = 0$. By passing to a subsequence, we can assume $\langle a_n \rangle$ is Cauchy and $d(a_n, C) < \frac{1}{n}$. Choose $c_n \in C$ with $d(a_n, c_n) < \frac{1}{n}$; by the completeness of $\langle cl(B), d \rangle$, the Cauchy sequence $\langle c_n \rangle$ converges. By construction, the limit lies in $A \cap C$, which is a contradiction.

$(2) \Rightarrow (1)$. Suppose condition (1) fails. Let $\langle x_n \rangle$ be a Cauchy sequence with distinct terms in $cl(B)$ that fails to cluster. Then we can find a sequence of positive scalars $\langle \alpha_n \rangle$ convergent to zero such that $\{S_d(x_n, \alpha_n) : n \in \mathbb{N}\}$ is a pairwise disjoint family of balls. For each $n \in \mathbb{N}$, pick $b_n \in B \cap S_d(x_n, \alpha_n)$; clearly, $\langle b_n \rangle$ is another Cauchy sequence with distinct terms that fails to cluster. With $C = \{b_{2n-1} : n \in \mathbb{N}\}$ and $A = \{b_{2n} : n \in \mathbb{N}\}$, condition (2) fails. $\square$

Condition (2) of the last result leads to the following conjecture: a nonempty subset B of $\langle X, d \rangle$ is relatively compact if and only if each closed subset of B fails to be near each nonempty closed subset disjoint from it. While relatively compact sets do have this property, it can hold more generally. The subsets with this property form a bornology that we will characterize in various ways in a subsequent section.

Given a subset A of $\langle X, d \rangle$ and $\varepsilon > 0$, we define $S_d^n(A, \varepsilon)$ for $n = 0, 1, 2, \ldots$ recursively by $S_d^0(A, \varepsilon) = A$ and $S_d^{n+1}(A, \varepsilon) = S_d(S_d^n(A, \varepsilon), \varepsilon)$. Observe that $S_d^n(A, \varepsilon) \subseteq S_d^{n+1}(A, \varepsilon)$ for all $n \geq 0$. We will write $S_d^n(x, \varepsilon)$ for $S_d^n(\{x\}, \varepsilon)$.

By an ε-*chain of length n from x to w* in $\langle X, d \rangle$ we mean a finite sequence of points (not necessarily distinct) $x_0, x_1, x_2, x_3, \ldots, x_n$ in X such that $x = x_0, w = x_n$ and for each $j \in \{1, 2, \ldots, n\}, d(x_{j-1}, x_j) < \varepsilon$. Clearly, there is an ε-chain of length n from x to w if and only if $w \in S_d^n(\{x\}, \varepsilon)$. The relation $\simeq_\varepsilon$ on X defined by $x \simeq_\varepsilon w$ if there is an ε-chain of some finite length form x to w is an equivalence relation whose equivalence classes are called ε-*step territories* [1, 114, 130]. The equivalence classes are closed subsets of X and the gap between any different two of them is at least ε.

Definition 7.4. A subset E of $\langle X, d \rangle$ is called *Bourbaki bounded* [34, 56, 86, 152] if for each $\varepsilon > 0$, there exist $F \in \mathscr{F}(X)$ and $n \in \mathbb{N}$ such that $E \subseteq S_d^n(F, \varepsilon)$.

As we noted earlier, the closed unit ball in any infinite dimensional normed linear space $\langle X, ||\cdot|| \rangle$ fails to be totally bounded. However, it obviously fulfills Definition 7.4 where for each $\varepsilon > 0$ we can take for F a single point, say the origin of the space $\{0_X\}$. With this choice for F, given $\varepsilon > 0$, we can choose any integer n exceeding $1/\varepsilon$. Perhaps surprisingly, using any other finite subset of X, we can't do better in terms of approximation of all points in the closed unit ball.

Let F be a nonempty finite subset of X and let S be the linear span of F. By finite dimensionality of S, the subspace is closed and each point of X has a nearest point in S (which may not be unique). Since $x \mapsto d(x, S)$ is positively homogeneous, we can find a point x_0 in X with

$d(x_0, S) = 1$. Letting $s_0 \in S$ be a nearest point to x_0 in S, we see that 0_X is nearest $x_0 - s_0$ in S and $x_0 - s_0$ lies on the surface of the closed unit ball. In particular, for each $w \in F$, we have

$$||x_0 - s_0 - w|| \geq ||x_0 - s_0|| = 1.$$

We denote the family of Bourbaki bounded subsets of a metric space $\langle X, d \rangle$ by $\mathscr{BB}_d(X)$. As with the totally bounded subsets, these are left unchanged by replacing d with a uniformly equivalent metric. Two sets of technical necessary and sufficient conditions that a metrizable space $\langle X, \tau \rangle$ have a compatible metric d for which $X \in \mathscr{BB}_d(X)$ have been identified by Aggarwal, Hasra and Kundu [1].

Proposition 7.5. *The Bourbaki bounded subsets of a metric space $\langle X, d \rangle$ form a bornology lying between $\mathscr{TB}_d(X)$ and $\mathscr{B}_d(X)$.*

Proof. Clearly, $\mathscr{BB}_d(X)$ is an hereditary family of subsets that contains $\mathscr{F}(X)$. Suppose A and B are Bourbaki bounded and $\varepsilon > 0$. Choose finite sets F_A and F_B and positive integers n and j such that $A \subseteq S_d^n(F_A, \varepsilon)$ and $B \subseteq S_d^j(F_B, \varepsilon)$. We obtain $A \cup B \subseteq S_d^{n+j}(F_A \cup F_B, \varepsilon)$, and this shows stability under finite unions.

If E is totally bounded, then for each $\varepsilon > 0$ there exists a finite set F with $E \subseteq S_d^1(F, \varepsilon)$. On the other hand, if E is Bourbaki bounded, then $E \subseteq S_d^n(F, 1)$ for some finite F and n gives $E \subseteq S_d(F, n)$. Thus, E is a subset of a finite union of balls and is metrically bounded. $\square$

Example 7.6. In any infinite dimensional normed linear space, each bounded set is Bourbaki bounded because each ball is Bourbaki bounded, while any ball in the space fails to be totally bounded. Each infinite subset of a set equipped with the zero-one metric while bounded fails to be Bourbaki bounded.

Bourbaki boundedness is not an intrinsic property of metric spaces. For example, if X is an infinite set equipped with the zero-one metric, then X is not a Bourbaki subset of itself. However, we can isometrically embed X into a Banach space in which, as a bounded subset, it must be Bourbaki bounded. It is easy to see that a nonempty subset A of a metric space $\langle X, d \rangle$ is Bourbaki bounded in each metric space in which it is isometrically embedded if and only if $\langle A, d \rangle$ is Bourbaki bounded.

The proof of the next result is left to the reader as a routine exercise.

Proposition 7.7. *Let $\langle X, d \rangle$ and $\langle Y, \rho \rangle$ be metric spaces and suppose $f \in UC(X, Y)$. If A is a totally bounded (resp. Bourbaki bounded) subset of X, then $f(A)$ is a totally bounded (resp. Bourbaki bounded) subset of Y.*

We have shown that a subset A of a metric space is totally bounded provided each sequence in A has a Cauchy subsequence. Our next goal is to provide a parallel characterization of Bourbaki boundedness of a subset as discovered by Garrido and Meroño [86].

Definition 7.8. A sequence $\langle x_n \rangle$ in a metric space $\langle X, d \rangle$ is called *Bourbaki-Cauchy* provided for each $\varepsilon > 0$, there exist $\{m, n_0\} \subseteq \mathbb{N}$ and such that $\forall n \geq n_0$, $\forall k \geq n_0$ we have $x_n \in S_d^m(x_k, \varepsilon)$.

Notice that if $\langle x_n \rangle$ is Bourbaki-Cauchy sequence, then $\{x_n : n \in \mathbb{N}\} \in \mathscr{BB}_d(X)$. A Cauchy sequence is a Bourbaki-Cauchy sequence where we can choose $m = 1$ for each $\varepsilon > 0$ in the above definition.

Proposition 7.9. *A subset A of a metric space $\langle X, d \rangle$ is Bourbaki bounded if and only if each sequence in A has a Bourbaki-Cauchy subsequence.*

Proof. Assume first that A is Bourbaki bounded and $\langle a_n \rangle$ is a sequence in A. As the set of terms is Bourbaki bounded, for $\varepsilon = 1$ we can find a nonempty finite subset F of X and $m_1 \in \mathbb{N}$ such that $\{a_n : n \in \mathbb{N}\} \subseteq S_d^{m_1}(F, 1)$. Thus, we can find an infinite subset $\mathbb{N}_1$ of $\mathbb{N}$ and $p_1 \in F$ such that $\{a_n : n \in \mathbb{N}_1\} \subseteq S_d^{m_1}(p_1, 1)$. Letting ε successively run over $\frac{1}{2}, \frac{1}{3}, \frac{1}{4}, \ldots$ we can produce a decreasing sequence $\langle \mathbb{N}_k \rangle$ of infinite subsets of $\mathbb{N}$, a sequence of points $\langle p_k \rangle$ in X, and a sequence $\langle m_k \rangle$ in $\mathbb{N}$ such that for each $k \in \mathbb{N}$,

$$\{a_n : n \in \mathbb{N}_k\} \subseteq S_d^{m_k}\left(p_k, \frac{1}{k}\right)$$

Taking n_k to be the k-th member of $\mathbb{N}_k$ with respect to the natural order it inherits from the positive integers, it is easy to verify that $\langle a_{n_k} \rangle$ is a Bourbaki-Cauchy subsequence of $\langle a_n \rangle$.

Conversely, suppose A fails to be Bourbaki bounded. Then for some $\varepsilon > 0$ and each finite subset F of X and each $m \in \mathbb{N}$, $S_d^m(F, \varepsilon)$ fails to contain A. Fix $a_0 \in A$, and for each $n \in \mathbb{N}$, recursively choose $a_n \in A$ such that $a_n \notin S_d^n(\{a_0, \ldots, a_{n-1}\}, \varepsilon)$. We claim $\langle a_n \rangle$ so constructed cannot have a Bourbaki-Cauchy subsequence. Otherwise, for this ε and some positive integer m, there exists an infinite subset $\mathbb{N}_\varepsilon$ of $\mathbb{N}$ such that whenever $\{n, k\} \subseteq \mathbb{N}_\varepsilon$ we have $a_n \in S_d^m(a_k, \varepsilon)$. Take $n > k \geq m$ in $\mathbb{N}_\varepsilon$. We compute

$$a_n \in S_d^m(a_k, \varepsilon) \subseteq S_d^n(a_k, \varepsilon) \subseteq S_d^n(\{a_0, \ldots, a_{n-1}\}, \varepsilon),$$

and this is a contradiction.   □

Functions that map Bourbaki-Cauchy sequences to Bourbaki-Cauchy sequences, often called *Bourbaki-Cauchy regular* functions [1], need not be continuous. For example, on the metric subspace of the real line $X = \{0, 1, 1/2, 1/3, 1/4, \ldots\}$, consider $f : X \to \ell_2$ mapping 0 to the origin and $1/n$ to e_n where e_n is the sequence whose nth term is 1 and whose other terms are zero. Beyond the Cauchy continuous functions, in this monograph, we will not be concerned with functions between metric spaces that preserve sequences of various types.

Definition 7.10. We call a metric space $\langle X, d \rangle$ *Bourbaki complete* provided each Bourbaki-Cauchy sequence in X clusters.

Since each Cauchy sequence is Bourbaki-Cauchy, and each Cauchy sequence that clusters is convergent, each Bourbaki complete metric space is complete in the usual sense.

Proposition 7.11. *A metric space $\langle X, d \rangle$ is Bourbaki complete if and only if $\mathscr{K}(X) = \mathscr{BB}_d(X)$.*

Proof. For necessity, we need only show that if $\langle X, d \rangle$ is Bourbaki complete, then $\mathscr{BB}_d(X) \subseteq \mathscr{K}(X)$. Let A be nonempty and Bourbaki bounded and let $\langle a_n \rangle$ be a sequence in A. Then $\langle a_n \rangle$ has a Bourbaki-Cauchy subsequence which clusters by Bourbaki completeness. Thus $\langle a_n \rangle$ itself clusters so that A is relatively compact. For sufficiency, suppose the space is not Bourbaki complete. Let $\langle x_n \rangle$ be a Bourbaki-Cauchy sequence in X that fails to cluster. Then its set of terms is a Bourbaki bounded subset that is not relatively compact.   □

There are many characterizations of complete metrizability of metrizable spaces. One is this: a metrizable space $\langle X, \tau \rangle$ is completely metrizable if and only if it is a G_δ subset of its Stone-Čech compactification βX [155, p. 180]. Along these lines, Garrido and Meroño [86, Theorem 22] have given this characterization of Bourbaki completely metrizable spaces.

Theorem 7.12. *Let* $\langle X, d \rangle$ *be a metrizable space. Then the space has a compatible Bourbaki complete metric if and only if* X *can be expressed as a countable intersection of open paracompact subspaces of* βX.

We close this section by describing when $\mathscr{T}\mathscr{B}_d(X) = \mathscr{B}\mathscr{B}_d(X)$ (see, e.g., [4, Theorem 2.7] and [88, Proposition 36]). In view of Proposition 7.11, the next result should come as no surprise.

Proposition 7.13. *Let* $\langle X, d \rangle$ *be a metric space and let* $\langle \hat{X}, \hat{d} \rangle$ *be its completion. The following conditions are equivalent.*

(1) $\langle \hat{X}, \hat{d} \rangle$ *is Bourbaki complete;*

(2) $\mathscr{T}\mathscr{B}_d(X) = \mathscr{B}\mathscr{B}_d(X)$;

(3) *each Bourbaki-Cauchy sequence in* X *has a Cauchy subsequence.*

Proof. (1) $\Rightarrow$ (2). We need only show that each nonempty Bourbaki bounded subset B is totally bounded. Let $\langle b_n \rangle$ be a sequence in B which we can assume by Proposition 7.9 is Bourbaki-Cauchy. As $\langle b_n \rangle$ is Bourbaki-Cauchy with respect to $\hat{d}$, by condition (1) we can find a subsequence $\langle b_{n_k} \rangle$ and $\hat{p} \in \hat{X}$ such that $\lim_{k \to \infty} \hat{d}(b_{n_k}, \hat{p}) = 0$. It follows that $\langle b_{n_k} \rangle$ is d-Cauchy, and by Proposition 7.1, B is totally bounded.

(2) $\Rightarrow$ (3). Suppose $\langle x_n \rangle$ is a Bourbaki-Cauchy sequence in X. Its range is Bourbaki bounded so that by condition (2) it is totally bounded. Thus, $\langle x_n \rangle$ has a Cauchy subsequence.

(3) $\Rightarrow$ (1). Suppose $\langle \hat{x}_n \rangle$ is a Bourbaki-Cauchy sequence in the completion. For each $n \in \mathbb{N}$, pick $x_n \in X$ with $\hat{d}(x_n, \hat{x}_n) < \frac{1}{n}$. It is routine (but tedious) to show that $\langle x_n \rangle$ is a Bourbaki-Cauchy sequence in X. By condition (3), $\langle x_n \rangle$ has a Cauchy subsequence, and the corresponding subsequence of $\langle \hat{x}_n \rangle$ is $\hat{d}$-Cauchy and hence convergent in the completion. Thus, the completion is Bourbaki complete. $\qquad\square$

8. Locally Lipschitz Functions

We first identify some classes of Lipschitz-type functions, arranged in ascending order in terms of levels of strength. These will appear later in the monograph with respect to basic uniform density results.

Definition 8.1. Let $f : \langle X, d \rangle \to \langle Y, \rho \rangle$.

(1) f is called *locally Lipschitz* provided for each $x \in X$, there exists $\delta_x > 0$ such that the restriction of f to $S_d(x, \delta_x)$ is Lipschitz (where the local Lipschitz constant and δ_x may depend on x);

(2) f is called *Cauchy-Lipschitz* [36] if f is Lipschitz when restricted to the range of each Cauchy sequence;

(3) f is called *uniformly locally Lipschitz* [34] if f is locally Lipschitz and δ_x can be chosen independent of x;

(4) f is called *Lipschitz in the small* [34, 84, 114, 116] if it is uniformly locally Lipschitz and the local Lipschitz constant can also be chosen independent of x, i.e., there exists $\delta > 0$ and $\lambda > 0$ such that $d(x, w) < \delta \Rightarrow \rho(f(x), f(w)) \leq \lambda d(x, w)$.

Each of these classes of functions, like the Lipschitz functions, is stable under composition. In the case of real-valued functions, like $\mathrm{Lip}(X, \mathbb{R})$, each forms a vector lattice because the absolute value of each function in the class is again in the class. The vector lattice of locally Lipschitz real-valued functions, like $C(X, \mathbb{R})$, is in general stable under pointwise product and under taking reciprocals of nonvanishing members.

Proposition 8.2. *Let $\langle X, d \rangle$ be a metric space.*

(1) *whenever f and g are locally Lipschitz real-valued functions on X, then fg is locally Lipschitz;*

(2) *whenever f is a locally Lipschitz real-valued function on X such that $\forall x \in X, f(x) \neq 0$, then $\frac{1}{f}$ is locally Lipschitz.*

Proof. By Theorem 1.2, we only need verify statement (2). Let $p \in X$ be arbitrary and choose $\delta > 0$ and $\lambda > 0$ such that f is λ-Lipschitz on $S_d(p, \delta)$. By the triangle inequality, λ serves as a Lipschitz constant for $|f|$ restricted to the ball as well, and so we can find $\delta_0 \in (0, \delta)$ such that whenever $d(x, p) < \delta_0$, we have

$$|f(x)| \geq |f(p)| - \lambda d(x, p) \geq \frac{1}{2}|f(p)|.$$

As a result, if $x_1, x_2 \in S_d(p, \delta_0)$, we obtain

$$\left| \frac{1}{f(x_1)} - \frac{1}{f(x_2)} \right| = \frac{|f(x_2) - f(x_1)|}{|f(x_1)f(x_2)|} \leq \frac{4\lambda}{|f(p)|^2} d(x_1, x_2),$$

and so the reciprocal of f is Lipschitz in a neighborhood of p. $\qquad\square$

It is routine to show that the class of Cauchy-Lipschitz real-valued functions and the class of uniformly locally Lipschitz real-valued functions are always stable under pointwise product. On the other hand, the class of Lipschitz in the small real-valued functions is stable under pointwise product if and only if $UC(X, \mathbb{R})$ is stable under pointwise product [33]; additional more tangible characterizations of such spaces will be given later in this monograph. None of these three classes is in general stable under taking reciprocals of nonvanishing members: consider $f(x) = x$ on $(0, \infty)$.

Paralleling the Tietze extension theorem, we state without proof the following extension result of Czipszer and Gehér [71] (see also [65, Theorem 4.17]). The proof is quite involved.

Theorem 8.3. *Let $\langle X, d \rangle$ be a metric space and let A be a nonempty closed subset of X. Then each locally Lipschitz real-valued function on A has a locally Lipschitz extension to X.*

Example 8.4. $f(x) = \sqrt{x}$ is locally Lipschitz on $(0, \infty)$ but has no locally Lipschitz extension to $[0, \infty)$.

Obviously, each Lipschitz in the small function as we have defined it is λ-Lipschitz on each open ball of radius $\delta/2$. To see that the definitions are made in ascending order of strength, we only need to show that each Cauchy-Lipschitz function is locally Lipschitz and each uniformly locally Lipschitz function is Cauchy-Lipschitz. The first claim follows from a sequential characterization of the locally Lipschitz property.

Proposition 8.5. *Let $\langle X, d \rangle$ and $\langle Y, \rho \rangle$ be metric spaces. Then $f : X \to Y$ is locally Lipschitz if and only if the restriction of f to the range of each convergent sequence in X is Lipschitz.*

Proof. For sufficiency, suppose for some $p \in X$, f fails to be Lipschitz in each ball with center p. Then for each $n \in \mathbb{N}$, there exist $x_n \neq w_n$ in $S_d(p, 1/n)$ such that $\rho(f(x_n), f(w_n)) > nd(x_n, w_n)$. Clearly, the sequence $x_1, w_1, x_2, w_2, \ldots$ converges to p but the restriction of f to its set of terms is not Lipschitz. For necessity, suppose f is locally Lipschitz and let $\langle x_n \rangle$ be a sequence in X convergent to p. Choose $\delta > 0$ such that f restricted to $S_d(p, \delta)$ is Lipschitz. Denote the set of terms inside the ball by E_1 and the finite set of terms outside by E_2. Since $\{p\} \cup E_1$ is compact, we have $D_d(E_1, E_2) > 0$. Since $f(\{x_n : n \in \mathbb{N}\})$ is a bounded set and $f|_{E_1}$ is Lipschitz, it is clear that the restriction to $E_1 \cup E_2$ is Lipschitz. $\qquad\square$

Proposition 8.6. *Let $\langle X, d \rangle$ and $\langle Y, \rho \rangle$ be metric spaces.*

(1) *Each Cauchy-Lipschitz function from X to Y is locally Lipschitz;*

(2) *Each uniformly locally Lipschitz function from X to Y is Cauchy-Lipschitz;*

(3) *Each Lipschitz in the small function is uniformly locally Lipschitz.*

Proof. Statement (1) is immediate from Proposition 8.5. For statement (2), choose $\delta > 0$ such that f is Lipschitz on each ball of radius δ. Evidently, such a function is bounded on each totally bounded subset of X and thus on the range of each Cauchy sequence (cf. [34, Theorem 3.2]). Now let $\langle x_n \rangle$ be a Cauchy sequence in X. To show that f is Lipschitz on the range, we may assume without loss of generality that the sequence has distinct terms. By the definition of Cauchy sequence there exists a finite subset $\mathbb{N}_0$ of $\mathbb{N}$ and $p \in X$ such that $d(x_n, p) > \frac{\delta}{2}$ if and only if $n \in \mathbb{N}_0$. If $\mathbb{N}_0 = \emptyset$, then $\{x_n : n \in \mathbb{N}\}$ is contained in $S_d(p, \delta)$ and the restriction of f to the this ball is already Lipschitz. Otherwise, $\mathbb{N}_0$ is nonempty, and we let $\lambda > 0$ be a Lipschitz constant for f restricted to $\{x \in X : d(p, x) \leq \frac{\delta}{2}\}$.

We consider three mutually exclusive and exhaustive conditions on distinct indices n and k to show that $\sup_{n \neq k} \rho(f(x_n), f(x_k))/d(x_n, x_k) < \infty$: (a) both n and k belong to $\mathbb{N}_0$; (b) neither n nor k belongs to $\mathbb{N}_0$; (c) exactly one of n and k belongs to $\mathbb{N}_0$, say n.

Choose $\alpha > 0$ such that $\operatorname{diam}_\rho \{f(x_n) : n \in \mathbb{N}\} < \alpha$. In case (a) put $\beta := \inf \{d(x_n, x_k) : n \in \mathbb{N}_0, k \in \mathbb{N}_0, n \neq k\} > 0$, a positive number since in case (a), $\mathbb{N}_0$ is a finite set that contains at least two points. We compute

$$\frac{\rho(f(x_n), f(x_k))}{d(x_n, x_k)} \leq \frac{\alpha}{\beta}.$$

In case (b) we have $\rho(f(x_n), f(x_k))/d(x_n, x_k) \leq \lambda$. In case (c), we first specify a scalar γ by the formula $\gamma := \inf_{n \in \mathbb{N}_0} d(x_n, p) - \frac{\delta}{2} > 0$. Then

$$\frac{\rho(f(x_n), f(x_k))}{d(x_n, x_k)} \leq \frac{\alpha}{\gamma}.$$

These three estimates together show that $\sup_{n \neq k} \rho(f(x_n), f(x_k))/d(x_n, x_k) < \infty$ where n, k range over $\mathbb{N}$. Finally, statement (3) is obvious. $\qquad \square$

Example 8.7. $f : (0, \infty) \to \mathbb{R}$ defined by $f(\alpha) = \frac{1}{\alpha}$ is locally Lipschitz because it has locally bounded derivative. But it is not Lipschitz when restricted to the set of terms of the Cauchy sequence $\langle \frac{1}{n} \rangle$.

Example 8.8. We equip $\mathbb{N} \times \mathbb{N}$ with a metric determined by distances between distinct points as follows:

- $d((n_1, k_1), (n_2, k_2)) = 17$ if $n_1 \neq n_2$;

- $d((n, k_1), (n, k_2)) = \frac{1}{n}$ if $k_1 \neq k_2$;

Notice that a sequence in $\mathbb{N} \times \mathbb{N}$ is d-Cauchy if and only it is eventually constant. With this in mind, we can easily define a real-valued function on $\mathbb{N} \times \mathbb{N}$ that is Cauchy-Lipschitz but not uniformly locally Lipschitz: put $f(n, k) = k$ for all n and k.

Example 8.9. By the mean-value theorem, $f : \mathbb{R} \to \mathbb{R}$ defined by $f(x) = x^2$ is uniformly locally Lipschitz but it is not Lipschitz in the small.

Each Lipschitz function is obviously Lipschitz in the small.

Example 8.10. If $X = \cup_{n=1}^{\infty} [n - \frac{1}{4}, n + \frac{1}{4}]$ equipped with the usual metric of the line, then the locally constant function defined by $f(x) = n^2$ if $n - \frac{1}{4} \leq x \leq n + \frac{1}{4}$ is Lipschitz in the small but fails to be Lipschitz on X.

There is a simple characterization of those Lipschitz in the small functions that are already Lipschitz as discovered by Beer, García-Lirola and Garrido [33].

Theorem 8.11. *Let $f : \langle X, d \rangle \to \langle Y, \rho \rangle$ be Lipschitz in the small. Then f is Lipschitz if and only if its modulus of continuity function m_f has a majorant of the form $t \mapsto \alpha t + \beta$, that is, an affine majorant.*

Proof. If f is λ-Lipschitz, then as we noted earlier, $t \mapsto \lambda t$ majorizes m_f. Conversely, if $t \mapsto \alpha t + \beta$ majorizes m_f then $\beta \geq 0$ and $\alpha \geq 0$ because m_f has no negative values. If f is Lipschitz in the small with distance control δ and uniform local Lipschitz constant λ, then $t \mapsto \lambda t$ majorizes m_f on $(0, \delta)$. By increasing λ if necessary we may assume $\lambda > \alpha$ and $\lambda \delta > \alpha \delta + \beta$. As a result, $t \mapsto \lambda t$ majorizes $t \mapsto \alpha t + \beta$ on $[\delta, \infty)$ and so $t \mapsto \lambda t$ majorizes the modulus of continuity function on $[0, \infty)$. $\square$

The easy proof of the following result is left to the reader.

Proposition 8.12. *Let $f : \langle X, d \rangle \to \langle Y, \rho \rangle$. If f is locally Lipschitz, then f is continuous; if f is Cauchy-Lipschitz (and thus if it is uniformly locally Lipschitz), then f is Cauchy continuous; if f is Lipschitz in the small, then f is uniformly continuous.*

Theorem 8.13. *Let $\langle X, d \rangle$ be a metric space. The following conditions are equivalent.*

(1) *the metric d is complete;*

(2) *each locally Lipschitz function on X with values in an arbitrary metric space $\langle Y, \rho \rangle$ is Cauchy-Lipschitz;*

(3) *each real-valued locally Lipschitz function is Cauchy-Lipschitz;*

(4) *the reciprocal of each nonvanishing real-valued Cauchy-Lipschitz function is Cauchy-Lipschitz;*

(5) *the reciprocal of each nonvanishing member of $Lip(X, \mathbb{R})$ is Cauchy-Lipschitz.*

Proof. The implication (1) $\Rightarrow$ (2) is a consequence of Proposition 8.5, and the implication (2) $\Rightarrow$ (3) is trivial. The implication (3) $\Rightarrow$ (4) follows from the fact that the nonvanishing real-valued locally Lipschitz functions are stable under reciprocation, and (4) $\Rightarrow$ (5) is trivial.

For (5) $\Rightarrow$ (1), suppose the metric is not complete, and let $\langle x_n \rangle$ be a nonconvergent Cauchy sequence in X. Let $\langle \hat{X}, \hat{d} \rangle$ be the completion of the initial metric space, and let $\hat{p}$ be the point in the completion to which the sequence is $\hat{d}$-convergent. Define $f : X \to \mathbb{R}$ by $f(x) = 1/\hat{d}(x, \hat{p})$. Clearly, f is the reciprocal of a nonvanishing 1-Lipschitz function on X. As f fails to bounded on the range of the sequence $\langle x_n \rangle$, it cannot be Cauchy-Lipschitz. $\square$

Each continuous function on $\langle X, d \rangle$ is locally Lipschitz if and only if $X' = \emptyset$ (if $x_0 \in X'$, consider $x \mapsto \sqrt{d(x, x_0)}$). Metric spaces on which each Lipschitz in the small function is already Lipschitz are of particular interest, for it is exactly in such spaces that $Lip(X, \mathbb{R})$ is uniformly dense in $UC(X, \mathbb{R})$ [84]. We will prove this later in the monograph, as well as present an internal description of such spaces [114]. We will also consider other pairwise coincidences within our four classes of locally Lipschitz functions.

The remainder of this section is concerned with the local Lipschitzian behavior of a real-valued convex function defined on an open convex subset C of a normed linear space $\langle X, || \cdot || \rangle$. We will use two easily verified facts about such functions.

- if $f : C \to \mathbb{R}$ is convex and $\sum_{i=1}^{n} \alpha_i x_i$ is a convex combination of points of C, then $f(\sum_{i=1}^{n} \alpha_i x_i) \leq \sum_{i=1}^{n} \alpha_i f(x_i)$;

- if $f : C \to \mathbb{R}$ is convex and if x_2 is a convex combination of distinct points x_1 and x_3 in C with positive coefficients and $\frac{f(x_2)-f(x_1)}{||x_2-x_1||} > 0$, then

$$\frac{f(x_3) - f(x_2)}{||x_3 - x_2||} \geq \frac{f(x_2) - f(x_1)}{||x_2 - x_1||}.$$

Proposition 8.14. *Let C be an open convex subset of a normed linear space $\langle X, || \cdot || \rangle$ and let f be a convex function on C. Suppose f is bounded above in some ball with center $c_0 \in C$. Then f is locally bounded on C, i.e., $\forall x \in C \, \exists \varepsilon_x > 0$ such that f is bounded on $x + \varepsilon_x U_X$.*

Proof. By translation, we may assume that $c_0 = 0_X$. Suppose $f(x) \leq \mu$ for each $x \in \delta U_X \subseteq C$. Let x be an arbitrary element of the ball; as $f(0_X) \leq \frac{1}{2}f(x) + \frac{1}{2}f(-x)$ and $-f(-x) \geq -\mu$, we get

$$f(x) \geq 2f(0_X) - f(-x) \geq 2f(0_X) - \mu,$$

and so f is bounded below in the ball as well.

Now let $x \in C$ be arbitrary. We can find $c \in C$ and $\alpha \in (0,1)$ such that $x = \alpha c + (1-\alpha)0_X = \alpha c$. Then $\alpha c + (1-\alpha)\delta U_X \subseteq C$ and for each w in this ball, we have $f(w) \leq \alpha f(c) + (1-\alpha)\mu$. By the initial argument, f is bounded on this ball with center x. □

In $\mathbb{R}^n$, a convex function f defined on an open convex subset C is bounded above in a neighborhood of each point $x \in C$, because there is a polytope with $n+1$ vertices $\{c_1, c_2, \ldots, c_{n+1}\}$ in C containing x in its interior and clearly $f(c) \leq \max\{f(c_1), f(c_2), \ldots, f(c_{n+1})\}$ for each c in the polytope. On the other hand, if $\langle X, || \cdot || \rangle$ is infinite dimensional, then this is not always true. If $\langle x_n \rangle$ is a sequence of linearly independent norm-one vectors in X, we can find a linear functional f on X such that $f(x_n) = n$ for each $n \in \mathbb{N}$. Clearly, f is a convex function on X that fails to be bounded above in a neighborhood of any point of X.

Theorem 8.15. *Let C be an open convex subset of a normed linear space $\langle X, || \cdot || \rangle$ and let f be a convex function on C that is bounded above in some ball contained in C. Then f is locally Lipschitz on C.*

Proof. Fix $x_0 \in C$. By the last proposition, we can find $\varepsilon > 0$ and $\mu > 0$ such that $x \in x_0 + 2\varepsilon U_X \subseteq C \Rightarrow |f(x)| \leq \mu$. We claim that $2\mu/\varepsilon$ is a Lipschitz constant for f restricted to $x_0 + \varepsilon U_X$. If this fails, then we can find distinct x_1 and x_2 in $x_0 + \varepsilon U_X$ such that

$$\frac{f(x_2) - f(x_1)}{||x_2 - x_1||} > \frac{2\mu}{\varepsilon}.$$

Choose $\alpha > 0$ such that with $x_3 = x_2 + \alpha(x_2 - x_1)$, we have $||x_3 - x_2|| = \varepsilon$. Then $x_3 \in x_0 + 2\varepsilon U_X$ and by the convexity of f,

$$\frac{f(x_3) - f(x_2)}{||x_3 - x_2||} \geq \frac{f(x_2) - f(x_1)}{||x_2 - x_1||} > \frac{2\mu}{\varepsilon},$$

and this yields $|f(x_3) - f(x_2)| > 2\mu$, a contradiction to the definition of μ. □

Example 8.16. A bounded convex function defined on a convex subset C that fails to be open need not be continuous on C, even for $X = \mathbb{R}$: let $C = [0, \infty)$ and let f be the characteristic function of $\{0\}$.

For our final result of this section, we will use Baire's Theorem.

Theorem 8.17. *Let C be a nonempty open convex subset of a Banach space $\langle X, || \cdot || \rangle$ and let f be a convex lower semicontinuous real-valued function on C. Then f is locally Lipschitz on C.*

Proof. By lower semicontinuity, for each $n \in \mathbb{N}$, $\{c \in C : f(c) \leq n\}$ is a relatively closed subset of C. In the case that $C = X$, put $A_n := \{c \in C : f(c) \leq n\}$; otherwise, put

$$A_n := \{c \in C : d(c, X \backslash C) \geq \frac{1}{n} \text{ and } f(c) \leq n\}.$$

In either case, each A_n is a relatively closed subset of C and $\cup_{n=1}^{\infty} A_n = C$. By Baire's Theorem, some A_n must have nonempty interior; applying Theorem 8.15, f is locally Lipschitz on C. $\quad\square$

9. Common Sets of Boundedness for Classes of Continuous Functions

The discussion below follows [34]. It is very well-known that a subset A of a metric space $\langle X, d \rangle$ has compact closure if and only if each continuous function on X with values in a second metric space maps A to a metrically bounded set. One direction is trivial, as the continuous image of a compact set is compact and each compact set is bounded. Conversely, if A fails to have compact closure, then we can find a sequence $\langle a_n \rangle$ in A that fails to cluster, so that the function $a_n \mapsto n$ is continuous on the closed set $\{a_n : n \in \mathbb{N}\}$. By the Tietze extension theorem, there is a continuous real-valued function on X that is unbounded on $\{a_n : n \in \mathbb{N}\}$.

Given a function $f : \langle X, d \rangle \to \langle Y, \rho \rangle$, we have agreed to write $\mathscr{B}(f)$ for the bornology of subsets B of X such that $f(B) \in \mathscr{B}_d(Y)$. In this section, we identify $\cap_{f \in \mathscr{C}} \mathscr{B}(f)$ where $\mathscr{C}$ is a prescribed family of continuous functions. We shall see that $\mathscr{C}$ can be reduced in several important cases to a certain family of Lipschitz-type functions within $\mathscr{C}$. We begin with the class of Lipschitz functions itself.

Theorem 9.1. *Let $\langle X, d \rangle$ be a metric space and let B be a nonempty subset. The following conditions are equivalent:*

(1) *$B \in \mathscr{B}_d(X)$;*

(2) *whenever $f : \langle X, d \rangle \to \langle Y, \rho \rangle$ is Lipschitz, then $f(B) \in \mathscr{B}_\rho(Y)$;*

(3) *whenever f is a real-valued Lipschitz function on $\langle X, d \rangle$, then $f(B)$ is metrically bounded.*

Proof. (1) $\Rightarrow$ (2). Suppose f is λ-Lipschitz; then if B is metrically bounded, we compute $\mathrm{diam}_\rho f(B) \leq \lambda \cdot \mathrm{diam}_d(B)$ and so $f(B) \in \mathscr{B}_\rho(Y)$. The implication (2) $\Rightarrow$ (3) is trivial. For (3) $\Rightarrow$ (1), if B fails to be metrically bounded, then with $p \in X$ and $f = d(p, \cdot)$, a 1-Lipschitz function, $f(B)$ fails to be metrically bounded. $\qquad\square$

In the case that X is a real normed linear space, $\mathrm{Lip}(X, \mathbb{R})$ can be replaced by X^* and still produce $\mathscr{B}_d(X)$ for the sets of common boundedness. This statement is just a rephrasing of a particular case of the Uniform Boundedness Principle from functional analysis [143, Theorem 9.2], whose proof of course uses the Baire category theorem.

Our next result refines the observation of the first paragraph of this section.

Theorem 9.2. *Let $\langle X, d \rangle$ be a metric space and let B be a nonempty subset. The following conditions are equivalent:*

(1) *$\mathrm{cl}(B)$ is compact;*

(2) *whenever $\langle Y, \rho \rangle$ is a metric space and $f : X \to Y$ is continuous, then $f(B) \in \mathscr{B}_\rho(Y)$;*

(3) *whenever $\langle Y, \rho \rangle$ is a metric space and $f : X \to Y$ is locally Lipschitz, then $f(B) \in \mathscr{B}_\rho(Y)$;*

(4) *whenever $f : X \to \mathbb{R}$ is locally Lipschitz, then $f(B)$ is a bounded set of real numbers.*

Proof. The implication (1) $\Rightarrow$ (2) follows from the fact that $f(\mathrm{cl}(B))$ must be compact if f is continuous, while (2) $\Rightarrow$ (3) and (3) $\Rightarrow$ (4) are trivial.

For (4) $\Rightarrow$ (1), we prove the contrapositive. Suppose $\mathrm{cl}(B)$ fails to be compact; then we can find a sequence $\langle b_n \rangle$ in B with distinct terms that has no cluster point in X. For each $n \in \mathbb{N}$, put $\mu_n = d(b_n, \{b_j : j \neq n\}) > 0$ and then put $\varepsilon_n := \min\{\frac{1}{n}, \frac{1}{3}\mu_n\}$. The family of open balls $\{S_d(b_n, \varepsilon_n) : n \in \mathbb{N}\}$ is a pairwise disjoint family, as whenever $n \neq j, \varepsilon_n + \varepsilon_j < \max\{\mu_n, \mu_j\}$. For each $n \in \mathbb{N}$, let $g_n : X \to \mathbb{R}$ be the Lipschitz function defined by

$$g_n(x) = n - \frac{n}{\varepsilon_n} d(x, b_n).$$

Notice that $g_n(x) > 0$ if and only if $d(x, b_n) < \varepsilon_n$.

We are now ready to describe our globally defined badly behaved locally Lipschitz function f:

$$f(x) = \begin{cases} g_n(x) & \text{if } x \in S_d(b_n, \varepsilon_n) \\ 0 & \text{otherwise.} \end{cases}$$

Since $f(b_n) = n, f(B)$ is unbounded. To see that f is locally Lipschitz, let $x_0 \in X$ be arbitrary. Since $\varepsilon_n \leq \frac{1}{n}$ for each n and x_0 is not a cluster point of $\langle b_n \rangle$, there exists $\delta > 0$ such that $S_d(x_0, \delta)$ either fails to hit any $S_d(b_n, \varepsilon_n)$ or it hits $S_d(b_n, \varepsilon_n)$ for at most finitely many n, say $n_1, n_2, \ldots, n_k$. In the first case, f restricted to $S_d(x_0, \delta)$ is the zero function, while in the second, whenever $d(x, x_0) < \delta$, we have $f(x) = \max\{0, g_{n_1}(x), g_{n_2}(x), \ldots, g_{n_k}(x)\}$. Either way, f restricted to $S_d(x_0, \delta)$ is Lipschitz. $\qquad\square$

We next give a parallel result for Cauchy continuous functions with respect to both the uniformly locally Lipschitz functions and the Cauchy-Lipschitz functions.

Theorem 9.3. *Let $\langle X, d \rangle$ be a metric space and let B be a nonempty subset. The following conditions are equivalent:*

(1) *B is totally bounded;*

(2) *whenever $\langle Y, \rho \rangle$ is a metric space and $f : X \to Y$ maps Cauchy sequences to Cauchy sequences, then $f(B) \in \mathscr{B}_\rho(Y)$;*

(3) *whenever $\langle Y, \rho \rangle$ is a metric space and $f : X \to Y$ is Cauchy-Lipschitz, then $f(B) \in \mathscr{B}_\rho(Y)$;*

(4) *whenever $\langle Y, \rho \rangle$ is a metric space and $f : X \to Y$ is uniformly locally Lipschitz, then $f(B) \in \mathscr{B}_\rho(Y)$;*

(5) *whenever $f : X \to \mathbb{R}$ is uniformly locally Lipschitz, then $f(B)$ is a bounded set of real numbers.*

Proof. (1) $\Rightarrow$ (2). Suppose f maps Cauchy sequences to Cauchy sequences. Using the sequential characterization of total boundedness, it is easy to see that $f(B)$ is actually totally bounded, else one could find a sequence $\langle b_n \rangle$ in B such that $\inf\{\rho(f(b_j), f(b_n)) : j \neq n\} > 0$ while $\langle b_n \rangle$ has a Cauchy subsequence.

The implications (2) $\Rightarrow$ (3), (3) $\Rightarrow$ (4), and (4) $\Rightarrow$ (5) are evident. For (5) $\Rightarrow$ (1), suppose B fails to be totally bounded. Then for some $\varepsilon > 0$, the set B fails to be contained in the ε-enlargement of any finite subset of X. Inductively we can construct a sequence $\langle b_n \rangle$ in B such that for each $n, b_{n+1} \notin S_d(\{b_1, b_2, \ldots, b_n\}, \varepsilon)$. The family of balls $\{S_d(b_n, \frac{1}{4}\varepsilon) : n \in \mathbb{N}\}$ is uniformly discrete: $\forall x \in X$, $S_d(x, \frac{1}{4}\varepsilon)$ intersects at most one of the balls $S_d(b_n, \frac{1}{4}\varepsilon)$. As a result, $f : X \mapsto \mathbb{R}$ defined by

$$f(x) = \begin{cases} n - \frac{4n}{\varepsilon}d(x, b_n) & \text{if } x \in S_d(b_n, \frac{1}{4}\varepsilon) \\ 0 & \text{otherwise.} \end{cases}$$

is a uniformly locally Lipschitz function that is unbounded on B. $\square$

The reader should now be able to anticipate the next result. It is due to Marino, Lewicki and Pietramala [119] and rediscovered by Beer and Garrido [34]. The proof given below is an adaptation of techniques due to Atsuji [7, pp. 14–15].

Theorem 9.4. *Let $\langle X, d \rangle$ be a metric space and let B be a nonempty subset. The following conditions are equivalent:*

(1) *B is Bourbaki bounded;*

(2) *whenever $\langle Y, \rho \rangle$ is a metric space and $f : X \to Y$ is uniformly continuous, then $f(B) \in \mathscr{B}_\rho(Y)$;*

(3) *whenever $\langle Y, \rho \rangle$ is a metric space and $f : X \to Y$ is Lipschitz in the small, then $f(B) \in \mathscr{B}_\rho(Y)$;*

(4) *whenever $f : X \to \mathbb{R}$ is Lipschitz in the small, then $f(B)$ is a bounded set of real numbers.*

Proof. For the string (1) $\Rightarrow$ (2) $\Rightarrow$ (3) $\Rightarrow$ (4) $\Rightarrow$ (1), only (1) $\Rightarrow$ (2) and (4) $\Rightarrow$ (1) require proof. For (1) $\Rightarrow$ (2), choose $\delta > 0$ such that

$$\forall x \in X \; \forall w \in X, \; d(x, w) < \delta \Rightarrow \rho(f(x), f(w)) < 1.$$

By Bourbaki boundedness, choose $\{x_1, x_2, \ldots, x_k\} \subseteq X$ and $n \in \mathbb{N}$ such that $B \subseteq \cup_{i=1}^k S_d^n(x_i, \delta)$; then $f(B) \subseteq S_\rho(\{f(x_1), f(x_2), \ldots, f(x_k)\}, n)$ and so $f(B) \in \mathscr{B}_\rho(Y)$.

(4) $\Rightarrow$ (1). Suppose B fails to be Bourbaki bounded. Choose $\varepsilon > 0$ such that whenever F is a finite subset of X and $n \in \mathbb{N}$ we have $B \nsubseteq S_d^n(F, \varepsilon)$.

We now denote the equivalence class of $x \in X$ with respect to the equivalence relation $\simeq_\varepsilon$ by $x/ \simeq_\varepsilon$. This is the ε-step territory to which x belongs. We have $x/ \simeq_\varepsilon = \cup_{n=0}^\infty S_d^n(x, \varepsilon)$. We consider two mutually exclusive and exhaustive cases for the structure of B:

(a) for some $b \in B$, $\forall n \in \mathbb{N} \; \exists j \in \mathbb{N}$ such that $B \cap S_d^n(b, \varepsilon)$ is a proper subset of $B \cap S_d^{n+j}(b, \varepsilon)$;

(b) for all $b \in B$, $\exists n \in \mathbb{N} \; \forall j \in \mathbb{N}$, $B \cap S_d^n(b, \varepsilon) = B \cap S_d^{n+j}(b, \varepsilon)$, that is, $B \cap S_d^n(b, \varepsilon) = B \cap b/ \simeq_\varepsilon$.

In case (a), for $x \simeq_\varepsilon b$, let $n(x)$ be the smallest n such that $x \in S_d^n(b, \varepsilon)$. We define our real-valued function f by

$$f(x) = \begin{cases} (n(x) - 1)\varepsilon + d(x, S_d^{n(x)-1}(b, \varepsilon)) & \text{if } x \neq b \text{ and } x \simeq_\varepsilon b \\ 0 & \text{otherwise} \end{cases}.$$

By assumption (a), f is unbounded on B. We intend to show that if $x \neq w$ and $d(x, w) < \varepsilon$ then $|f(x) - f(w)| \leq 2d(x, w)$. Now if either x or w is not related to b, then the same is true for the other because $d(x, w) < \varepsilon$. Thus $|f(x) - f(w)| = 0 < 2d(x, w)$. We now pass to the situation where both $x \simeq_\varepsilon b$ and $w \simeq_\varepsilon b$.

Without loss of generality we may assume $n(x) \geq n(w)$. If $n(w) = 0$, that is, $w = b$, then $0 < d(x, w) < \varepsilon$ implies $n(x) = 1$ and

$$|f(x) - f(w)| = f(x) = (1 - 1)\varepsilon + d(x, S_d^0(b, \varepsilon)) = d(x, b) = d(x, w).$$

Otherwise, $n(w) \geq 1$ and either $n(x) = n(w)$ or $n(x) = n(w) + 1$. If $n(x) = n(w)$ we easily compute

$$|f(x) - f(w)| = |d(x, S_d^{n(x)-1}(b, \varepsilon)) - d(w, S_d^{n(x)-1}(b, \varepsilon))| \leq d(x, w).$$

We are left with the possibility that $n(x) = n(w) + 1$. This gives $f(x) > f(w)$, because

$$(n(x) - 1)\varepsilon - (n(w) - 1)\varepsilon = \varepsilon$$

and

$$d(x, S_d^{n(x)-1}(b, \varepsilon)) - d(w, S_d^{n(w)-1}(b, \varepsilon)) > 0 - \varepsilon.$$

We now claim that $d(w, S_d^{n(w)-1}(b, \varepsilon)) \geq \varepsilon - d(x, w)$. If the inequality fails, then

$$d(x, S_d^{n(w)-1}(b, \varepsilon)) \leq d(x, w) + d(w, S_d^{n(w)-1}(b, \varepsilon)) < d(x, w) + \varepsilon - d(x, w),$$

making $n(x) \leq n(w) - 1 + 1$ which is impossible. The claim established, we get

$$|f(x) - f(w)| = f(x) - f(w) = \varepsilon + d(x, S_d^{n(x)-1}(b, \varepsilon)) - d(w, S_d^{n(w)-1}(b, \varepsilon))$$
$$\leq \varepsilon + d(x, w) - (\varepsilon - d(x, w)) \leq 2d(x, w).$$

Case (b) is easier. Let $b_1 \in B$ be arbitrary. Choose n_1 such that $B \cap S_d^{n_1}(b_1, \varepsilon) = B \cap b_1 / \simeq_\varepsilon$. Since B is not Bourbaki bounded, there exists $b_2 \in B$ with $b_2 \notin S_d^{n_1}(b_1, \varepsilon)$. By the choice of $n_1, b_1 / \simeq_\varepsilon \neq b_2 / \simeq_\varepsilon$. Choose $n_2 > n_1$ such that $B \cap S_d^{n_2}(b_2, \varepsilon) = B \cap b_2 / \simeq_\varepsilon$. Since $B \not\subseteq \cup_{j=1}^2 S_d^{n_j}(b_j, \varepsilon)$, we can find $b_3 \in B \backslash (b_1 / \simeq_\varepsilon \cup b_2 / \simeq_\varepsilon)$. Continuing, we can find a sequence $\langle b_j \rangle$ with distinct terms in B such that whenever $j \neq k$, we have $b_j / \simeq_\varepsilon \neq b_k / \simeq_\varepsilon$. We now define $f : X \to \mathbb{R}$ by

$$f(x) = \begin{cases} j & \text{if } x \simeq_\varepsilon b_j \text{ for some } j \\ 0 & \text{otherwise} \end{cases}$$

Since $f(b_j) = j$, f is unbounded on B; further, for each $x \in X$, f is constant on $S_d(x, \varepsilon)$ and so f is Lipschitz in the small. $\qquad \square$

10. Hejcman's Theorem and its Analog for Totally Bounded Subsets

One of the most attractive results with respect to the large structure of metric spaces is a characterization of the bornology $\mathscr{B}\mathscr{B}_d(X)$ due to Hejcman [93, Theorem 1.12]. Recall that a subset of $\langle X, d \rangle$ is relatively compact if and only if it is bounded with respect to each equivalent metric. Hejcman proved that a subset belongs to $\mathscr{B}\mathscr{B}_d(X)$ if and only if it is bounded with respect to each uniformly equivalent metric. In our proof, we call on a metric for an ε-step territory of $\langle X, d \rangle$ first explicitly considered by O'Farrell [130].

We call an ε-chain $x_0, x_1, x_2, \ldots, x_n$ in our metric space $\langle X, d \rangle$ *irreducible* provided for each $j \in \{1, 2, 3, \ldots, n-1\}$ we have

$$d(x_{j-1}, x_j) + d(x_j, x_{j+1}) \geq \varepsilon.$$

If $x \simeq_\varepsilon w$, then we can always join x to w by an irreducible ε-chain. For an irreducible ε-chain, we cannot have both $d(x_{j-1}, x_j) < \frac{\varepsilon}{2}$ and $d(x_j, x_{j+1}) < \frac{\varepsilon}{2}$ for any such j, and this implies that $d(x_{j-1}, x_j) \geq \frac{\varepsilon}{2}$ for at least $\frac{n-1}{2}$ values of j in $\{1, 2, 3, \ldots, n\}$.

Now let ε be positive and let E be an ε-step territory of $\langle X, d \rangle$. We define O'Farrell's metric on E by the formula

$$d_\varepsilon(x, w) := \inf \left\{ \sum_{j=1}^n d(x_{j-1}, x_j) : x_0, x_1, \ldots, x_n \text{ is an } \varepsilon - \text{chain from } x \text{ to } w \right\}.$$

In terms of estimating $d_\varepsilon(x, w)$ where $x \simeq_\varepsilon w$, we can restrict our estimators to irreducible ε-chains from x to w. We list some properties of O'Farrell's metric that should be obvious from our above comments.

(a) $d \leq d_\varepsilon$ on E;

(b) if either $d(x, w) < \varepsilon$ or $d_\varepsilon(x, w) < \varepsilon$ for $\{x.w\} \subseteq E$, then $d_\varepsilon(x, w) = d(x, w)$;

(c) $I_E : \langle E, d \rangle \to \langle E, d_\varepsilon \rangle$ is uniformly continuous in both directions;

(d) if $f : \langle E, d_\varepsilon \rangle \to \langle Y, \rho \rangle$ is Lipschitz, then f is Lipschitz in the small on $\langle E, d \rangle$.

We now come to Hejcman's Theorem.

Theorem 10.1. *Let $\langle X, d \rangle$ be a metric space. Then*

$$\mathscr{B}\mathscr{B}_d(X) = \bigcap \{\mathscr{B}_\rho(X) : \rho \in \boldsymbol{D}(X) \text{ is uniformly equivalent to } d\}.$$

Proof. Each member of $\mathscr{B}\mathscr{B}_d(X)$ is Bourbaki bounded with respect to each uniformly equivalent metric and thus bounded with respect to each uniformly equivalent metric. For the reverse inclusion, we show that if $A \notin \mathscr{B}\mathscr{B}_d(X)$, then we can find a uniformly equivalent metric ρ such that $A \notin \mathscr{B}_\rho(X)$. We can find $\varepsilon < 1$ such that whenever F is finite and nonempty and $n \in \mathbb{N}$, A is not contained in $S_d^n(F, \varepsilon)$.

Let $\{E_i : i \in I\}$ be the family of ε-step territories determined by d where $E_i \neq E_j$ for $i \neq j$ so that $D_d(E_i, E_j) \geq \varepsilon$ for $i \neq j$. We next define a function $\phi : X \to \mathbb{N}$ to be used in describing our bad metric. If $\{i \in I : A \cap E_i \neq \emptyset\}$ is finite, we put $\phi(x) = 1$ for each all $x \in X$. Otherwise, let $\langle i_k \rangle$ be a sequence of distinct terms in I such that for each $k \in \mathbb{N}$, $E_{i_k} \cap A \neq \emptyset$ and define ϕ by

$$\phi(x) = \begin{cases} k & \text{if } x \in E_{i_k} \text{ for some } k \in \mathbb{N} \\ 1 & \text{otherwise.} \end{cases}$$

We are now ready to define a uniformly equivalent metric ρ on X with respect to which A is unbounded.

$$\rho(x, w) := \begin{cases} d_\varepsilon(x, w) & \text{if } x \simeq_\varepsilon w \\ \phi(x) + \phi(w) & \text{otherwise.} \end{cases}$$

We only verify that the triangle inequality is satisfied. This is clearly the case if the three points - call them x, w and z - all belong to the same ε-step territory because d_ε is a metric. Suppose $x \simeq_\varepsilon w$ while w and z lie in different ε-step territories. As x and z are in different ε-step territories and $\phi(x) = \phi(w)$,

$$\rho(x, z) = \phi(x) + \phi(z) \leq d_\varepsilon(x, w) + \phi(w) + \phi(z) = \rho(x, w) + \rho(w, z).$$

Finally, if all three points belong to distinct ε-step territories,

$$\rho(x, z) = \phi(x) + \phi(z) \leq \phi(x) + 2\phi(w) + \phi(z) = \rho(x, w) + \rho(w, z).$$

For uniform equivalence of the two metrics, if $\rho(x, w) < \varepsilon$, since $\varepsilon < 1$, we have $x \simeq_\varepsilon w$ and $d(x, w) = d_\varepsilon(x, w) = \rho(x, w)$. On the other hand, $d(x, w) < \varepsilon$, then $x \simeq_\varepsilon w$ and $\rho(x, w) = d_\varepsilon(x, w) = d(x, w)$ so that $\langle X, d \rangle$ and $\langle X, \rho \rangle$ are in fact uniformly locally isometric.

It remains to show that $A \notin \mathscr{B}_\rho(X)$. Now either A hits infinitely many E_i or not. In the first case, if $j \neq k$, choosing $a_j \in E_{i_j} \cap A$ and $a_k \in E_{i_k} \cap A$, we have $\rho(a_j, a_k) = k + j$ so that $\mathrm{diam}_\rho(A) = \infty$. Otherwise, let $i_1, i_2, i_3, \ldots, i_m$ be those members of I for which $E_i \cap A \neq \emptyset$. Since A is not Bourbaki bounded, for some $j \in \{1, 2, 3, \ldots, m\}$, $\forall n \in \mathbb{N}, \forall a \in A \cap E_{i_j} \exists \bar{a} \in A \cap E_{i_j}$ for which no irreducible ε-chain exists from a to $\bar{a}$ of length at most n. But in view of irreducibility we have

$$\rho(a, \bar{a}) = d_\varepsilon(a, \bar{a}) \geq \frac{(n+1) - 1}{2} \cdot \frac{\varepsilon}{2} = \frac{n\varepsilon}{4},$$

and from this we conclude $\mathrm{diam}_\rho(A) \geq \mathrm{diam}_\rho(A \cap E_{i_j}) = \infty$. $\quad\square$

It is possible that there is a uniformly equivalent metric ρ for which $\mathscr{B}_\rho(X)$ is smallest, i.e., such that $\mathscr{BB}_d(X) = \mathscr{B}_\rho(X)$. Hejcman [94] called metric spaces $\langle X, d \rangle$ with this property *B-simple*. This class was subsequently studied in [87]. We will internally characterize this family of spaces later in the monograph.

If we replace the class of metrics uniformly equivalent to d by the class of metrics that are Cauchy equivalent to d in the statement of Hejcman's theorem, then we must replace $\mathscr{BB}_d(X)$ by $\mathscr{TB}_d(X)$.

Theorem 10.2. *Let $\langle X, d \rangle$ be a metric space. Then*

$$\mathscr{TB}_d(X) = \bigcap \{\mathscr{B}_\rho(X) : \rho \in \boldsymbol{D}(X) \text{ is Cauchy equivalent to } d\}.$$

Proof. Suppose ρ is Cauchy equivalent to d. Since $I_X : \langle X, d \rangle \to \langle X, \rho \rangle$ is Cauchy continuous, by Theorem 9.3, $A \in \mathscr{TB}_d(X) \Rightarrow A = I_X(A) \in \mathscr{B}_\rho(X)$. Suppose $A \notin \mathscr{TB}_d(X)$. If $A \notin \mathscr{B}_d(X)$, then A is not in the displayed intersection. Otherwise, we can find a sequence $\langle a_n \rangle$ in A and $\alpha > \delta > 0$ such that for all $n \neq k \in \mathbb{N}$ we have $\alpha > d(a_n, a_k) > \delta$.

Define a metric d_0 on $\{a_n : n \in \mathbb{N}\}$ by $d_0(a_n, a_k) := |n - k|\alpha$. Notice that for all $(n, k) \in \mathbb{N} \times \mathbb{N}$, we have $d(a_n, a_k) \leq d_0(a_n, a_k)$, and $\{a_n : n \in \mathbb{N}\}$ is a uniformly discrete (closed) subset of the completion $\langle \hat{X}, \hat{d} \rangle$ of the original metric space. Since d_0 is compatible with the relative topology on $\{a_n : n \in \mathbb{N}\}$, by the Hausdorff extension theorem, there is a metric $\hat{d}_0$ on $\hat{X}$ compatible with $\tau_{\hat{d}}$ that extends d_0. Put $\hat{\mu} = \max\{\hat{d}, \hat{d}_0\}$; for all $(n, k) \in \mathbb{N} \times \mathbb{N}$, we have $\hat{\mu}(a_n, a_k) = d_0(a_n, a_k)$.

We claim that $\hat{d}$ and $\hat{\mu}$ are Cauchy equivalent. First, $I_{\hat{X}} : \langle \hat{X}, \hat{d} \rangle \to \langle \hat{X}, \hat{\mu} \rangle$ is Cauchy continuous because $I_{\hat{X}}$ is continuous and the domain is a complete metric space. Second, $I_{\hat{X}} : \langle \hat{X}, \hat{\mu} \rangle \to \langle \hat{X}, \hat{d} \rangle$ is Cauchy continuous because it is nonexpansive, and the claim is verified.

Now let μ denote the restriction of $\hat{\mu}$ to $X \times X$; clearly d and μ are Cauchy equivalent, and

$$\text{diam}_\mu(A) \geq \text{diam}_\mu(\{a_n : n \in \mathbb{N}\}) = \text{diam}_{d_0}(\{a_n : n \in \mathbb{N}\}) = \infty.$$

This proves that $A \notin \bigcap\{\mathscr{B}_\rho(X) : \rho \in \mathbf{D}(X) \text{ is Cauchy equivalent to } d\}$.   □

11. General Constructions

We now turn to considering general constructions involving bornologies.

Proposition 11.1. *Let* $\{\mathscr{B}_i : i \in I\}$ *be a family of bornologies on a set* X. *Then* $\cap_{i\in I}\mathscr{B}_i$ *is a bornology on* X.

Since $\mathscr{P}(X)$ is a bornology, given any family $\mathscr{A}$ of subsets of X, there is a smallest bornology containing the family, namely the intersection of all bornologies containing $\mathscr{A}$. This is called *the bornology generated by* $\mathscr{A}$ which we will denote by born($\mathscr{A}$). Constructively, we have

Proposition 11.2. *Let* $\mathscr{A}$ *be a family of subsets of a nonempty set* X. *Then*
$$born(\mathscr{A}) = \downarrow \sum(\mathscr{A} \cup \mathscr{F}(X)) = \sum \downarrow (\mathscr{A} \cup \mathscr{F}(X)).$$
If $\mathscr{A}$ *is already a cover, then* $born(\mathscr{A}) = \downarrow \sum(\mathscr{A})$. *If* $\mathscr{A}$ *is an hereditary family of subsets, then* $born(\mathscr{A}) = \sum(\mathscr{A} \cup \mathscr{F}(X))$.

Proof. Since any bornology containing $\mathscr{A}$ contains $\mathscr{A} \cup \mathscr{F}(X)$ and is stable under finite unions and under taking subsets, the smallest such bornology must contain $\downarrow \sum(\mathscr{A}\cup\mathscr{F}(X))$. It remains to show that this family is a bornology. By property (2) of Proposition 1.1, the family contains $\mathscr{A} \cup \mathscr{F}(X)$ and thus the singletons. Since $\downarrow$ is idempotent, the family is hereditary. Finally the family is stable under finite unions as the operators $\sum$ and $\downarrow$ commute and $\sum$ is idempotent.

For the second statement, use the fact that the operators $\downarrow$ and $\sum$ commute and $\downarrow (\mathscr{A}\cup\mathscr{F}(X)) = \downarrow \mathscr{A}$. The third statement is a consequence of $\sum(\mathscr{A} \cup \mathscr{F}(X))$ being already hereditary. $\square$

In particular, if $A \subseteq X$, the bornology born($\{A\}$) consists of all sets of the form $B \cup F$ where $B \subseteq A$ and F is finite. We call this the *principal bornology* $\mathscr{B}(A)$ determined by the subset A.

The next result is obvious.

Proposition 11.3. *If* $\mathscr{B}_1$ *and* $\mathscr{B}_2$ *are bornologies on a set* X, *then* $born(\mathscr{B}_1 \cup \mathscr{B}_2) = \{B_1 \cup B_2 : B_1 \in \mathscr{B}_1 \text{ and } B_2 \in \mathscr{B}_2\}$.

From the above discussion it is clear that $\mathfrak{B}_X$ is a lattice with meet and join
$$\mathscr{B}_1 \wedge \mathscr{B}_2 = \mathscr{B}_1 \cap \mathscr{B}_2 \quad \text{and} \quad \mathscr{B}_1 \vee \mathscr{B}_2 = \{B_1 \cup B_2 : B_1 \in \mathscr{B}_1 \text{ and } B_2 \in \mathscr{B}_2\}.$$

In fact, $\mathfrak{B}_X$ is a complete lattice.

Definition 11.4. Let $\mathscr{B}$ be a bornology on a set X and let A be a nonempty subset. Then
$$\mathscr{B}_A := \{B \cap A : B \in \mathscr{B}\}$$

is a bornology on A called the *relative bornology* determined by $\mathscr{B}$. Further, if X is equipped with a topology τ, then the triple $\langle A, \tau_A, \mathscr{B}_A \rangle$ is called a *subuniverse*.

The next proposition explains how $\mathscr{T}\mathscr{B}_d(X)$ arises as a relative bornology.

Proposition 11.5. *Let $\langle X, d\rangle$ be a metric space. Then $\mathscr{T}\mathscr{B}_d(X)$ is the relative bornology induced by $\mathscr{K}(\hat{X})$ on the completion $\langle \hat{X}, \hat{d}\rangle$ of $\langle X, d\rangle$.*

Proof. Let B be a totally bounded subset of $\langle X, d\rangle$. Then B is a totally bounded subset of the completion so that $\mathrm{cl}_{\hat{X}}(B)$ being complete and totally bounded is compact. Thus $B \in \mathscr{K}(\hat{X})$ and $B = B \cap X$. Conversely, if $C \subseteq \hat{X}$ is relatively compact, then $C \cap X$ is $\hat{d}$-totally bounded and thus is d-totally bounded as totally boundedness is an intrinsic property of subsets. $\square$

Definition 11.6. Let $\mathscr{B}$ be a bornology on X. Then a subfamily $\mathscr{B}_0$ is called a *base* for $\mathscr{B}$ if $\mathscr{B} = \downarrow \mathscr{B}_0$.

Obviously, a base for a bornology generates the bornology.

Example 11.7. The compact bornology on a Hausdorff space X has a closed base, namely the family of compact subsets of X.

Example 11.8. If a bornological universe has a closed (resp. open) base for its bornology, then the same is true for each subuniverse.

Example 11.9. A countable open base for the metrically bounded subsets of a metric space $\langle X, d\rangle$ is $\{S_d(p, n) : n \in \mathbb{N}\}$ where p is a fixed point of X, while $\{\{x \in X : d(p, x) \le n\} : n \in \mathbb{N}\}$ is a countable closed base.

Example 11.10. On $\mathbb{R}$ equipped with the Euclidean metric, $\mathscr{F}(\mathbb{R})$ has a closed base (in fact, each member is closed) but not an open base. A bornology on $\mathbb{R}$ having an open base but not a closed base consists of all sets of the form $E \cup F$ where $E \subseteq \cup_{n \in \mathbb{Z}}(n, n+1)$ and $F \in \mathscr{F}(\mathbb{R})$.

The next result is obvious.

Proposition 11.11. *A bornology $\mathscr{B}$ on a set X has a finite base iff $\mathscr{B} = \mathscr{P}(X)$.*

Proposition 11.12. *Let $\mathscr{A}$ be a family of subsets of a nonempty set X. Then $\mathscr{A}$ is a base for some bornology on X if and only if $\mathscr{A}$ is a cover of X that is directed by inclusion.*

Proof. Suppose $\mathscr{A}$ is a base for a bornology $\mathscr{B}$. Since each singleton $\{x\}$ belongs to $\mathscr{B}$, there exists $A_x \in \mathscr{A}$ with $\{x\} \subseteq A_x$. Thus $\mathscr{A}$ is a cover of X. To see that the cover is directed by inclusion, suppose $A_1 \in \mathscr{A}$ and $A_2 \in \mathscr{A}$; since $A_i \in \mathscr{B}$ for $i = 1, 2$, we have $A_1 \cup A_2 \in \mathscr{B}$ which, by the definition of base, means that it is a subset of some $A_3 \in \mathscr{A}$. In particular $A_i \subseteq A_3$ for $i = 1, 2$. Conversely, if $\mathscr{A}$ is a cover directed by inclusion, it is routine to verify that $\downarrow \mathscr{A}$ is a bornology. $\square$

Definition 11.13. Let $\mathscr{B}$ be a bornology on a Hausdorff space $\langle X, \tau\rangle$. By the *closure* of $\mathscr{B}$, we mean the finer bornology $\overline{\mathscr{B}}$ having as a base $\{\mathrm{cl}(B) : B \in \mathscr{B}\}$.

That this family of sets is directed by inclusion is a consequence of the formula $\mathrm{cl}(B_1 \cup B_2) = \mathrm{cl}(B_1) \cup \mathrm{cl}(B_2)$.

Proposition 11.14. *Let $\mathscr{B}$ be a bornology on a Hausdorff space $\langle X, \tau \rangle$. The following statements are equivalent:*

(1) *$\mathscr{B}$ has a closed base;*

(2) *$\forall B \in \mathscr{B}$, $cl(B) \in \mathscr{B}$;*

(3) *$\mathscr{B} = \overline{\mathscr{B}}$.*

Proof. Suppose $\mathscr{B}$ has a closed base. Let $B \in \mathscr{B}$ be arbitrary and choose a closed $A \in \mathscr{B}$ with $B \subseteq A$. Then $cl(B) \subseteq A$ and as $\mathscr{B} = \downarrow \mathscr{B}$ we have $cl(B) \in \mathscr{B}$. If (2) holds, then $\mathscr{B}$ contains a base for the formally finer bornology $\overline{\mathscr{B}}$, so the two coincide. Finally, (3) $\Rightarrow$ (1) is trivial. $\square$

Example 11.15. As the closure of a totally bounded set (resp. Bourbaki bounded set) is totally bounded (resp. Bourbaki bounded), both $\mathscr{T}\mathscr{B}_d(X)$ and $\mathscr{B}\mathscr{B}_d(X)$ have closed bases.

Suppose $\mathscr{B}$ is a nontrivial bornology on a Hausdorff space $\langle X, \tau \rangle$ with a closed base. Generalizing the standard construction of the Alexandroff one-point compactification of the space when the bornology is $\mathscr{K}(X) \neq \mathscr{P}(X)$ (see, e.g., [107, 155]), we can form a one-point extension of the space by adjoining an ideal point p to X and taking for the open subsets of $X \cup \{p\}$ the family

$$\tau \cup \{(X \cup \{p\}) \backslash B : B \in \mathscr{B} \text{ and } B \text{ is closed}\}.$$

Following [64], we denote this extension by $o(\mathscr{B}, p)$. If $\mathscr{B}_0$ is any closed base for the bornology, then $\{(X \cup \{p\}) \backslash B : B \in \mathscr{B}_0\}$ is a local base for the extension topology at the ideal point. Obviously, $o(\mathscr{B}, p)$ satisfies the T_1-separation property, that is, its singleton subsets are closed. If $x \in X$, then $(X \cup \{p\}) \backslash \{x\}$ is open because $\{x\} \in \mathscr{B} \cap \mathscr{C}(X)$. On the other hand, $\{p\}$ is closed in the extension because its complement is $X \in \tau$.

Proposition 11.16. *Let $\langle X, \tau \rangle$ be a Hausdorff space and let $\langle X \cup \{p\}, \sigma \rangle$ be an extension in which p is a limit point of the space and $\{p\}$ is closed. Then there is a nontrivial bornology $\mathscr{B}$ on X with closed base such that $\langle X \cup \{p\}, \sigma \rangle = o(\mathscr{B}, p)$.*

Proof. Let $\mathscr{A}$ denote the family of complements of the neighborhoods of $\{p\}$ in the extension and put $\mathscr{B} := \downarrow \mathscr{A}$. $\square$

Topological properties of such extensions have been well-studied [19, 71, 64, 151]. We devote an entire subsequent section to selectively considering some of them.

The following discussion involves bornologies induced by a function. Let $\mathscr{A}$ be a bornology on X and let $\mathscr{B}$ be a bornology on Y. Suppose $f : X \to Y$ is a function. Then $\{A \subseteq X : f(A) \in \mathscr{B}\}$ is called *the bornology on X induced by f and $\mathscr{B}$*. A base for this bornology is $\{f^{-1}(B) : B \in \mathscr{B}\}$. It is not hard to show $\{f^{-1}(B) : B \in \mathscr{B}\}$ gives the entire bornology if and only if f is one-to-one. Dually, $\{f(A) : A \in \mathscr{A}\}$ is a bornology on Y if and only if f is onto.

We next turn to products. We first give a construction analogous to the product topology.

Definition 11.17. Let $\{X_i : i \in I\}$ be a family of nonempty sets and for each $i \in I$, let $\mathscr{B}_i$ be a bornology on X_i. Then the *product bornology* $\mathscr{B}_{\text{prod}}$ on $\prod_{i \in I} X_i$ is generated by

$$\{\pi_i^{-1}(B_i) : B_i \in \mathscr{B}_i, i \in I\}$$

where for each $i \in I, \pi_i$ is the natural projection map of the product onto X_i.

Note that if any of the factor bornologies is trivial, then the same is true for the product bornology! A base for the product bornology consists of all sets of the form $\cup_{j=1}^{n} \pi_{i_j}^{-1}(B_{i_j})$ where $\{i_1, i_2, \ldots, i_n\} \subseteq I$ and $B_{i_j} \in \mathscr{B}_{i_j}$. If each X_i carries a topology τ_i, we call the triple $\langle \prod_{i \in I} X_i, \tau_{\text{prod}}, \mathscr{B}_{\text{prod}} \rangle$ a *product universe*.

The proof of the next proposition is left to the reader.

Proposition 11.18. *Let $\{\langle X_i, \tau_i, \mathscr{B}_i \rangle : i \in I\}$ be a family of bornological universes. If each universe has an open (resp. closed) base for its bornology, then so does $\mathscr{B}_{\text{prod}}$.*

Definition 11.19. Let $\{X_i : i \in I\}$ be a family of sets and for each $i \in I$, let $\mathscr{B}_i$ be a bornology on X_i. Then the *box bornology* $\mathscr{B}_{\text{box}}$ on $\prod_{i \in I} X_i$ consists of all $E \subseteq \prod_{i \in I} X_i$ such that $\forall i \in I, \pi_i(E) \in \mathscr{B}_i$.

Evidently, the box bornology is the intersection of the bornologies induced on $\prod_{i \in I} X_i$ by the projection maps, and it is thus coarser than the product bornology. A base for the box bornology consists of all sets of the form $\prod_{i \in I} B_i$ where the B_i run over $\mathscr{B}_i$. In the case that we have a finite product of metric spaces each equipped with its bornology of metrically bounded sets, then the box bornology on the product coincides with the metric bornology of the box metric on the product.

If we pair the product topology with the box bornology and the bornology of each factor has a closed base, then the same is true of the box bornology. However, the bornology of each factor can have an open base while the box bornology need not have one.

Example 11.20. Consider $\mathbb{R}$ equipped with the standard metric bornology induced by the Euclidean metric. Then a base for the product bornology on $\mathbb{R}^2$ consists of all sets of the form

$$[B_1, B_2] := \{(\alpha, \beta) : \alpha \in B_1 \text{ or } \beta \in B_2\}$$

where B_i is a bounded set of reals for $i = 1, 2$. Each looks like a cross whose arms have bounded width. On the other hand, elements of the box bornology consist of the metrically bounded subsets of $\mathbb{R}^2$ with respect to the Euclidean metric on the plane.

12. Properties of Bornologies

Definition 12.1. Let $\langle X, \tau, \mathscr{B} \rangle$ be a bornological universe. The bornology is called *local* if each $x \in X$ has a neighborhood in $\mathscr{B}$.

Clearly each bornology with an open base is local.

Example 12.2. On $\mathbb{R}^2$ equipped with the usual topology, consider the bornology $\mathscr{B}$ with countable closed base consisting of all sets of the form

$$B_n := \{(\alpha, 0) : \alpha \in \mathbb{R}\} \cup ([-n, n] \times [-n, n]) \quad for \ n = 1, 2, 3, \ldots.$$

As each point of the plane lies in $(-n, n) \times (-n, n)$ for some $n \in \mathbb{N}$, the bornology is local. Clearly it has no open base. Notice that the bornology lies between the usual metric bornology for $\mathbb{R}^2$ and the product bornology, both of which have open bases.

Example 12.3. Since the closure of each totally bounded set is totally bounded, $\mathscr{T}\mathscr{B}_d(X)$ has a closed base. The bornology can fail be local and thus not have an open base, e.g., in an infinite dimensional normed linear space. The situation is the same for the Bourbaki bornology, for if $E \subseteq S_d^n(F, \varepsilon)$ where F is finite and $n \in \mathbb{N}$, then $\mathrm{cl}(E) \subseteq S_d^{n+1}(F, \varepsilon)$. For an example of a point in a space no neighborhood of which is Bourbaki bounded, in ℓ_2 with standard orthonormal base $\{e_n : n \in \mathbb{N}\}$ and origin 0_X, consider the constellation $X = \{0_X\} \cup \{\frac{1}{k} e_n : n \in \mathbb{N}, k \in \mathbb{N}\}$ as a metric subspace, where 0_X has no Bourbaki bounded neighborhood.

We now address the question: when is a product universe local?

Proposition 12.4. *Let* $\{\langle X_i, \tau_i, \mathscr{B}_i \rangle : i \in I\}$ *be a family of bornological universes. Then the product universe they determine is local if and only if for each* $x \in \prod_{i \in I} X_i$ *there exist* $i \in I$ *and* $V_i \in \tau_i \cap \mathscr{B}_i$ *with* $x_i \in V_i$.

Proof. Sufficiency is trivial as $x \in \pi_i^{-1}(V_i)$. For necessity, we may assume that each factor bornology is nontrivial. Suppose

$$x \in \cap_{j=1}^n \pi_{i_j}^{-1}(V_{i_j}) \in \mathscr{B}_{\mathrm{prod}}$$

where $\{i_1, i_2, \ldots, i_n\}$ are distinct members of I and $V_{i_j} \in \tau_{i_j}$ for $j \leq n$. By the definition of the product bornology, we can find distinct $\{k_1, k_2, \ldots, k_m\} \subseteq I$ with

$$\cap_{j=1}^n \pi_{i_j}^{-1}(V_{i_j}) \subseteq \cup_{l=1}^m \pi_{k_l}^{-1}(B_{k_l})$$

where $B_{k_l} \in \mathscr{B}_{k_l}$ for $l \leq m$. It is not hard to show that there must exist j and l such that $i_j = k_l$ and $V_{i_j} \subseteq B_{i_j}$. From this we have

$$x_{i_j} \in V_{i_j} \in \tau_{i_j} \cap \mathscr{B}_{i_j}$$

as required. $\qquad\square$

From the last proposition, if each factor universe has a local bornology, then the product universe will be local. Unless all but finitely many of our factor bornologies are trivial, if we equip our product with the product topology and the box bornology, locality of each factor bornology will not guarantee locality of $\mathscr{B}_{\text{box}}$.

The next theorem appears in an article of Vipera [151, p. 821].

Theorem 12.5. *Let $\mathscr{B}$ be a bornology with a closed base on a Hausdorff space $\langle X, \tau \rangle$. Then the one-point extension topology $o(\mathscr{B}, p)$ is Hausdorff if and only if $\mathscr{B}$ is local.*

Proof. First, suppose that $o(\mathscr{B}, p)$ is Hausdorff. Let $x \in X$ be arbitrary; since x and p have disjoint neighborhoods, we can find $V \in \tau$ and $B \in \mathscr{B} \cap \mathscr{C}(X)$ such that $x \in V$ and $V \cap X \backslash B = \emptyset$. This means that $x \in V \subseteq B$ and so $\mathscr{B}$ is local because a bornology is hereditary. Conversely, suppose $\mathscr{B}$ is local. Without this assumption, distinct points of X have disjoint neighborhoods in the extension as $\tau \subseteq \tau(\mathscr{B}, p)$. On the other, hand if $x \in X$ choose $V \in \tau \cap \mathscr{B}$ with $x \in V$. Then V and $(X \backslash \text{cl}(V)) \cup \{p\}$ are disjoint neighborhoods of x and p, respectively, in the extension. $\qquad\square$

Definition 12.6. Let $\mathscr{B}$ be a bornology on a metric space $\langle X, d \rangle$. We say $\mathscr{B}$ is *stable under small enlargements* provided for each $A \in \mathscr{B}$ there exists $\varepsilon > 0$ such that $S_d(A, \varepsilon) \in \mathscr{B}$.

Clearly if $\mathscr{B}$ is stable under small enlargements, then $\mathscr{B}$ has an open base.

Example 12.7. By the Tietze extension theorem there is a continuous function f from $(0, \infty)$ to $\mathbb{R}$ such that $f(1/n) = n$ when n is even and $f(1/n) = 0$ when n is odd. Then $\mathscr{B}_f = \{B \subseteq (0, \infty) : f(B) \text{ is bounded}\}$ has an open base but is not stable under small enlargements.

Here are some easily established facts:

- each local bornology on a Hausdorff space must contain $\mathscr{K}(X)$;
- the metric bornology for any metric space is stable under small enlargements;
- the compact bornology in a metric space is stable under small enlargements if and only if it is local;
- the bornology of totally bounded subsets in a normed linear space is local if and only if the space is finite dimensional.

We next introduce a property of bornologies that is weaker than stability under small enlargements and that has a number of applications [29, 46]. Given a subset E of a metric space $\langle X, d \rangle$, we say that a superset A *shields E from closed sets* if each neighborhood of A contains an enlargement of E. Put differently, A shields E from closed sets if for each nonempty closed set C disjoint from A, $\exists \varepsilon > 0$ such that $S_d(E, \varepsilon) \cap C = \emptyset$. If this occurs, we say A is a *shield* for E.

Obviously, X is a shield for each of its subsets, while each compact set shields itself from closed sets (in particular $\emptyset$ shields itself from closed sets). Also, any superset of a shield is a shield as well.

The proof of our first elementary lemma is left to the reader.

Lemma 12.8. *Suppose A is a shield for E in a metric space $\langle X, d \rangle$. Then $cl(E) \subseteq A$ and A is a shield for $cl(E)$.*

Proposition 12.9. *A nonempty closed subset E of $\langle X, d \rangle$ is shielded from closed sets by a superset A if and only if whenever $\langle x_n \rangle$ is a sequence in $X \backslash A$ with $\lim_{n \to \infty} d(x_n, E) = 0$, then $\langle x_n \rangle$ has a cluster point (in E).*

Proof. Suppose E is a nonempty closed set and A is a shield for E. Let $\langle x_n \rangle$ be a sequence in $X \backslash A$ with $\lim_{n \to \infty} d(x_n, E) = 0$. If $\langle x_n \rangle$ fails to cluster, then $\{x_n : n \in \mathbb{N}\}$ a nonempty closed subset of $X \backslash A$ that hits each enlargement of E, a contradiction. Conversely, suppose the superset A fails to be a shield for $E \neq \emptyset$. There exists a nonempty closed set C disjoint from A such that $\forall n \in \mathbb{N}, S_d(E, \frac{1}{n}) \cap C \neq \emptyset$. For each $n \in \mathbb{N}$, pick x_n in the intersection. Clearly, $\langle x_n \rangle$ can't cluster, else the cluster point would lie in $C \cap E \subseteq C \cap A$ which is impossible. □

Definition 12.10. Let $\mathscr{A}$ be a family of subsets of a metric space $\langle X, d \rangle$. We say that $\mathscr{A}$ is *shielded from closed sets* if each member of $\mathscr{A}$ has a shield in $\mathscr{A}$.

Clearly,

- the relatively compact subsets are shielded from closed sets, for if $E \in \mathscr{K}(X)$, then $cl(E)$ is a shield for E relative to $\mathscr{K}(X)$;

- if $\mathscr{A}$ contains an enlargement of each of its members, then $\mathscr{A}$ is shielded from closed sets;

- if a bornology $\mathscr{B}$ has a base $\mathscr{B}_0$ such that each member of $\mathscr{B}_0$ has a shield in $\mathscr{B}_0$, then $\mathscr{B}$ is shielded from closed sets;

- if a bornology $\mathscr{B}$ is shielded from closed sets, then for each base $\mathscr{B}_0$, each member of $\mathscr{B}_0$ has a shield in $\mathscr{B}_0$;

- Given a bornology $\mathscr{B}$, $\{B \in \mathscr{B} : B$ has a shield in $\mathscr{B}\}$ is a bornology.

Example 12.11. Let $X = \mathbb{R}^2 \backslash \{(0, \beta) : \beta > 0\}$, and let $\mathscr{B}$ be the bornology on X having closed base $\{B_n : n \in \mathbb{N}\}$ where for each $n, B_n = \{(\alpha, \beta) : \beta \leq n|\alpha|\}$. Notice B_{n+1} is a shield for B_n as each closed set disjoint from B_{n+1} misses some neighborhood of $(0,0)$. We remark that this bornology fails to be local, as it contains no neighborhood of $(0,0)$.

The next result is an immediate consequence of Lemma 12.8.

Proposition 12.12. *Let $\mathscr{B}$ be a bornology on $\langle X, d \rangle$ that is shielded from closed sets. Then $\mathscr{B}$ has a closed base.*

Example 12.13. On $\mathbb{R}$ equipped with the usual metric, $\mathscr{B}$ consisting of all sets of the form $E \cup F$ where $E \subseteq \mathbb{N}$ and $F \in \mathscr{F}(\mathbb{R})$ has a closed base but is not shielded from closed sets; in particular, $\mathbb{N}$ has no shield in $\mathscr{B}$.

Example 12.14. Let $\langle X, || \cdot || \rangle$ be an infinite dimensional normed linear space. As each finite dimensional linear subspace of X is closed [89, p. 25], $\{B : B \subseteq X$ and B lies in a finite dimensional linear subspace$\}$ is a bornology with closed base, but it also fails to be shielded from closed sets. To see this, pick $p \in X$ with $||p|| > 0$ and let $E = \{np : n \in \mathbb{N}\}$. Then the closed set E lies in the bornology, but if $\{x_n : n \in \mathbb{N}\}$ is a sequence of linearly independent norm one vectors, then eventually $\langle \frac{1}{n} x_n + np \rangle$ lies outside of any finite dimensional subspace containing E while $\lim_{n \to \infty} d(\frac{1}{n} x_n + np, E) = 0$. Apply Proposition 12.9.

We shall next show that a local bornology on a metric space that is shielded from closed sets must already be stable under small enlargements. This remarkable fact was established in [29]. We provide a more economical proof.

Theorem 12.15. *Let $\mathscr{B}$ be a local bornology in a metric space $\langle X, d \rangle$ that is shielded from closed sets. Then $\mathscr{B}$ is stable under small enlargements.*

Proof. We know that $\mathscr{B}$ has a closed base, so we need only show that each nonempty closed subset B of $\mathscr{B}$ has an enlargement in $\mathscr{B}$. Let $A \in \mathscr{B}$ be a shield for B. If A already contains an enlargement of B, we are done. Otherwise, for each $n \in \mathbb{N}$ the set $E_n := A^c \cap S_d(B, \frac{1}{n})$ is nonempty and by Proposition 12.9, $K := \cap_{n=1}^{\infty} \mathrm{cl}(E_n)$ is a nonempty subset of B. We claim K is compact. Let $\langle x_n \rangle$ be a sequence in K. Choose for each $n \in \mathbf{N}, e_n \in E_n$ with $d(e_n, x_n) < \frac{1}{n}$. Again by Proposition 12.9, $\langle e_n \rangle$ has a subsequence convergent to a point of K, and the corresponding subsequence of $\langle x_n \rangle$ converges to the same point.

Since the bornology is local and K is compact, there exists $V \in \tau_d \cap \mathscr{B}$ with $K \subseteq V$. An easy argument gives $E_k \subseteq V$ for some large $k \in \mathbb{N}$. By construction, all the points of $S_d(B, \frac{1}{k})$ either lie in V or in A. As a result $V \cup A$ belongs to $\mathscr{B}$ and $V \cup A$ contains an enlargement of B. $\square$

If $\langle X, d \rangle$ is a complete metric space, then $\mathscr{T}\mathscr{B}_d(X)$ is shielded from closed sets as $\mathscr{T}\mathscr{B}_d(X) = \mathscr{K}(X)$. However, completeness of d is not necessary, as if X is itself totally bounded, then $\mathscr{T}\mathscr{B}_d(X)$ is shielded from closed sets. That $\mathscr{T}\mathscr{B}_d(X)$ need not be shielded from closed sets is a consequence of the next proposition.

Proposition 12.16. *Let $\langle X, || \cdot || \rangle$ be a normed linear space. Then $\mathscr{T}\mathscr{B}_d(X)$ is shielded from closed sets if and only if $\langle X, || \cdot || \rangle$ is a Banach space.*

Proof. Sufficiency follows the last remarks. For necessity, suppose the norm fails to be complete. The space is then infinite dimensional, as all norms on a finite dimensional space are equivalent, and thus the metrics they determine are uniformly equivalent and so are all complete [89, p. 24]. Take a Cauchy sequence $\langle x_n \rangle$ in X which fails to converge. As the sequence can have no cluster point in X, $B := \{x_n : n \in \mathbb{N}\}$ is a closed totally bounded set. We claim that if B_0 is a totally bounded superset of B, then B_0 cannot be a shield for B. To see this, let $\{\delta_n : n \in \mathbb{N}\}$ be positive scalars such that for each $n, \delta_n < \frac{1}{n}$ and the family of balls $\{S_{\delta_n}(x_n) : n \in \mathbb{N}\}$ is pairwise disjoint. Since balls in an infinite dimensional space are not totally bounded [89, p. 26], we can choose for each n, $c_n \in S_{\delta_n}(x_n) \backslash B_0$. Then with $C := \{c_n : n \in \mathbb{N}\}$, we get a closed set disjoint from B_0 with $D_d(C, B) = 0$ as required. $\square$

As the next example shows, the bornology $\mathscr{B}\mathscr{B}_d(X)$ need not be shielded from closed sets in a complete metric space [37].

Example 12.17. Let B be the closed unit ball in ℓ_2 and let $\{e_n : n \in \mathbb{N}\}$ be the standard orthonormal base. Let X be this complete metric subspace of ℓ_2:

$$X := B \cup \{e_n + \frac{1}{(n+1)k} e_j : (n, k, j) \in \mathbb{N}^3\}.$$

It is easy to see that for each $n \in \mathbb{N}$, $E_n := \{e_n + \frac{1}{(n+1)k} e_j : (k, j) \in \mathbb{N}^2\}$ is not Bourbaki bounded. Since the family of Bourbaki bounded subsets is hereditary, if B_1 is Bourbaki bounded superset of B in X, then for each $n \in \mathbb{N}$, there exists $x_n \in E_n$ with $x_n \notin B_1$. As $d(x_n, e_n) < \frac{1}{n}$, $\{x_n : n \in \mathbb{N}\}$ is a closed subset of $X \backslash B_1$ that is near B.

We will provide tangible necessary and sufficient conditions for $\mathscr{TB}_d(X)$ and $\mathscr{BB}_d(X)$ to be shielded from closed sets in Section 22.

Definition 12.18. Let X be a set equipped with a bornology $\mathscr{B}$. A net $\langle s_\lambda \rangle_{\lambda \in \Lambda}$ in X is said to *converge to* ∞ with respect to $\mathscr{B}$ if for each $B \in \mathscr{B}, X \backslash B$ contains s_λ residually.

Proposition 12.19. *Let X be a set equipped with a bornologies $\mathscr{B}_1$ and $\mathscr{B}_2$. Then $\mathscr{B}_1 \supseteq \mathscr{B}_2$ iff each net in X convergent to ∞ with respect to $\mathscr{B}_1$ also converges to ∞ with respect to $\mathscr{B}_2$.*

Proof. Necessity is obvious. For sufficiency, suppose $\mathscr{B}_1$ is not finer than $\mathscr{B}_2$. Then there exist $B_2 \in \mathscr{B}_2$ such for each $B \in \mathscr{B}_1$, we can find $x_B \in B_2 \backslash B$. Then with $\mathscr{B}_1$ directed by inclusion, $B \mapsto x_B$ converges to ∞ with respect to $\mathscr{B}_1$ but not $\mathscr{B}_2$. $\square$

Example 12.20. Let $\mathscr{B}_1$ be the product bornology on $\mathbb{R}^2$ and let $\mathscr{B}_2$ be the coarser usual metric bornology. Each net $\mathscr{B}_1$-convergent to ∞ is $\mathscr{B}_2$-convergent, for if a point is outside a cross, it is outside the intersection of its arms. On the other hand, the sequence $\langle (0, n) \rangle$ converges to ∞ in the usual sense but not with respect to $\mathscr{B}_1$ as the vertical axis belongs to $\mathscr{B}_1$.

We now give definitions for functions between two sets X and Y equipped with bornologies $\mathscr{B}_1$ and $\mathscr{B}_2$, respectively, that parallel the definitions of openness and continuity for functions between topological spaces.

Definition 12.21. Let $\mathscr{B}_1$ be a bornology on X and $\mathscr{B}_2$ be a bornology on Y. Suppose $f : X \to Y$.

(1) we call f *bornological* if $\forall B_1 \in \mathscr{B}_1, f(B_1) \in \mathscr{B}_2$;

(2) we call f *coercive* if $\forall B_2 \in \mathscr{B}_2, f^{-1}(B_2) \in \mathscr{B}_1$.

Example 12.22. Let $X = Y = \mathbb{R}$, each equipped with the usual metric bornology. The squaring function $f : \mathbb{R} \to \mathbb{R}$ defined by $f(x) = x^2$ is both bornological and coercive. The function $g(x) = x\sin x$ while bornological fails to be coercive. The function h defined by

$$h(x) = \begin{cases} \ln |x| & \text{if } x \neq 0 \\ 0 & \text{if } x = 0 \end{cases}$$

is coercive but is not bornological. Finally, the function v below is neither bornological nor coercive:

$$h(x) = \begin{cases} 1/x & \text{if } x \neq 0 \\ 0 & \text{if } x = 0 \end{cases}$$

Example 12.23. If X and Y are normed linear spaces, a (continuous) linear operator $T : X \to Y$ is called *compact* if the image of the closed unit ball has compact closure in Y. By linearity this means that T is bornological where X is equipped with its metric bornology and Y is equipped with the compact bornology.

The term coercive is borrowed from optimization theory [53, 136]; it means that nets that go to infinity in the domain space are mapped to nets that do the same in the codomain, as made precise by the next statement. The simple proof is left to the reader.

Proposition 12.24. *Let $\mathscr{B}_1$ be a bornology on X and $\mathscr{B}_2$ be a bornology on Y. Then $f : X \to Y$ is coercive if and only if whenever $\langle x_\lambda \rangle$ converges to ∞ with respect to $\mathscr{B}_1$, then $\langle f(x_\lambda) \rangle$ converges to ∞ with respect to $\mathscr{B}_2$.*

The next three results are companion theorems; as the proofs are very similar, we just prove the second.

Theorem 12.25. *Let $\langle X, d \rangle$ and $\langle Y, \rho \rangle$ be metric spaces, both equipped with their respective compact bornologies $\mathscr{K}(X)$ and $\mathscr{K}(Y)$. Then $f \in Y^X$ is bornological if and only if for each convergent sequence $\langle x_n \rangle$ in X, $\langle f(x_n) \rangle$ has a convergent subsequence.*

Theorem 12.26. *Let $\langle X, d \rangle$ and $\langle Y, \rho \rangle$ be metric spaces, both equipped with their respective bornologies of totally bounded sets. Then $f \in Y^X$ is bornological if and only if each Cauchy sequence in X is mapped into a sequence having a Cauchy subsequence.*

Proof. Suppose f preserves total boundedness, and $\langle x_n \rangle$ is Cauchy; as $\{x_n : n \in \mathbb{N}\}$ is totally bounded, so is $\{f(x_n) : n \in \mathbb{N}\}$. By the sequential characterization of total boundedness, $\langle f(x_n) \rangle$ has a Cauchy subsequence. Conversely, suppose the image of each Cauchy sequence has a Cauchy subsequence and $B \subseteq X$ is totally bounded and nonempty. If $\langle y_n \rangle$ is a sequence in $f(B)$, choose $b_n \in B$ with $f(b_n) = y_n$ for each n. Since $B \in \mathscr{TB}_d(X)$, we can choose $n_1 < n_2 < n_3 < \cdots$ such that $\langle b_{n_k} \rangle$ is Cauchy. By assumption, $\langle f(b_{n_k}) \rangle$ has a Cauchy subsequence, but this is a subsequence of $\langle y_n \rangle$ as well. This shows that $f(B)$ is totally bounded. $\square$

Theorem 12.27. *Let $\langle X, d \rangle$ and $\langle Y, \rho \rangle$ be metric spaces, both equipped with their respective bornologies of Bourbaki bounded sets. Then $f \in Y^X$ is bornological if and only if each Bourbaki-Cauchy sequence in X is mapped into a sequence having a Bourbaki-Cauchy subsequence.*

Any bounded real-valued function maps relatively compact sets to relatively compact sets. Each Cauchy continuous function belongs to the class of functions that preserve total boundedness. On the other hand, membership to this class does not imply Cauchy continuity, even for continuous functions. A continuous function that preserves totally bounded sets that is not Cauchy continuous is $f(\frac{1}{n}) = (-1)^n$ defined on $\{\frac{1}{n} : n \in \mathbb{N}\}$ equipped with the usual metric of the line. Such a function cannot be constructed on a complete metric space, as we have the following improvement of Proposition 3.2; the details are left to the reader.

Proposition 12.28. *Let $\langle X, d \rangle$ be a metric space. The following conditions are equivalent:*

(1) *$\langle X, d \rangle$ is a complete metric space;*

(2) *each continuous function on X is a Cauchy continuous function;*

(3) *each continuous function on X preserves totally bounded sets.*

We give one last result giving necessary and sufficient conditions for a function to be bornological, where this time, domain and target space are equipped with different bornologies.

Theorem 12.29. *Let $\langle X, d \rangle$ be a metric space equipped with the bornology $\mathscr{K}(X)$ and let $\langle Y, \rho \rangle$ be a second metric space equipped with the bornology $\mathscr{B}_\rho(Y)$. Then $f \in Y^X$ is bornological if and only if f is locally bounded: each $x \in X$ has a neighborhood V_x such that $f(V_x) \in \mathscr{B}_\rho(Y)$.*

Proof. Suppose f is locally bounded and B has compact closure. We can find $x_1, x_2, \ldots, x_n$ in $\mathrm{cl}(B)$ such that $B \subseteq \cup_{i=1}^n V_{x_i}$ and $f(V_{x_i})$ is metrically bounded for each $i \leq n$. Thus $f(B) \subseteq \cup_{i=1}^n f(V_{x_i})$ is metrically bounded. Conversely, suppose $x_0 \in X$ has no neighborhood whose image is metrically bounded. By assumption we can find $x_1 \in S_d(x_0, 1)$ such that

$\rho(f(x_1), f(x_0)) > 1$. Continuing produce $x_1, x_2, x_3, \ldots$ in X such that for all $n \in \mathbb{N}, d(x_n, x_0) < \frac{1}{n}$ and $\rho(f(x_n), f(x_0)) > n$. Then f maps the relatively compact set $\{x_n : n \in \mathbb{N}\}$ to a metrically unbounded subset of Y. $\qquad\square$

Proposition 12.30. *Let X and Y be nonempty sets.*

(1) *If $\mathscr{B}$ is a bornology on Y and $f : X \to Y$, then $\{A \subseteq X : f(A) \in \mathscr{B}\}$, which we called the bornology on X induced by f and $\mathscr{B}$, is the finest bornology on X that makes f bornological and the coarsest bornology on X making f coercive.*

(2) *If $\mathscr{A}$ is a bornology on X and $f : X \to Y$ is onto, then $\{f(A) : A \in \mathscr{A}\}$ is the coarsest bornology on Y such that f is bornological.*

Proof. We just verify statement (1). Put $\mathscr{A} := \{A \subseteq X : f(A) \in \mathscr{B}\}$. By definition f is bornological with respect to $\mathscr{A}$ and $\mathscr{B}$, and there are no other subsets of X mapped to members of $\mathscr{B}$, so this is the largest possible bornology. It is also coercive for if $B \in \mathscr{B}$, then $f(f^{-1}(B)) \subseteq B \Rightarrow f(f^{-1}(B)) \in \downarrow \mathscr{B} = \mathscr{B}$. As noted earlier, $\{f^{-1}(B) : B \in \mathscr{B}\}$ is a base for $\mathscr{A}$ and must be contained in any bornology on X with respect to which f is coercive. $\qquad\square$

Example 12.31. When f is onto and $\mathscr{A}$ is a bornology on X, then f need not be coercive with respect to the bornology $\{f(A) : A \in \mathscr{A}\}$ on Y. Let $X = Y = \mathbb{R}$, and consider $f(x) = x\sin x$ where the domain is equipped with $\mathscr{F}(\mathbb{R})$.

Proposition 12.32. *Let $\{X_i : i \in I\}$ be a family of nonempty sets and let $\mathscr{B}_i$ be a bornology on X_i for each $i \in I$.*

(1) *The product bornology on $\prod_{i \in I} X_i$ is the coarsest bornology making each projection π_i coercive.*

(2) *The box bornology on $\prod_{i \in I} X_i$ is the finest bornology making each projection π_i bornological.*

Proof. We just prove statement (2). The finest possible bornology would consist of all $A \subseteq \prod_{i \in I} X_i$ such that $\forall i \in I, \pi_i(A) \in \mathscr{B}_i$. By definition the family of such subsets is exactly $\mathscr{B}_{\text{box}}$. $\qquad\square$

We collect some other easily verified facts about bornological and coercive functions.

- If $\mathscr{B}_1$ and $\mathscr{B}_2$ are bornologies on X, then $\mathscr{B}_1$ is coarser than (resp. finer than) $\mathscr{B}_2$ if and only if the identity map $I_X : \langle X, \mathscr{B}_1 \rangle \to \langle X, \mathscr{B}_2 \rangle$ is bornological (resp. coercive).

- If $\mathscr{B}$ is a bornology on a set X, and E is a nonempty subset equipped with the relative bornology, then inclusion map from E to X is both bornological and coercive.

- If $\langle X, \tau \rangle$ and $\langle Y, \sigma \rangle$ are Hausdorff spaces and $f \in C(X, Y)$, then f is bornological with respect to the compact bornologies on the domain and codomain.

- If X and Y are normed linear spaces equipped with their respective metric bornologies and $T : X \to Y$ is linear, then T is continuous if and only if T is bornological.

- If $\langle X, d \rangle$ and $\langle Y, \rho \rangle$ are metric spaces and $f : X \to Y$ is Lipschitz, then f is bornological with respect to $\mathscr{B}_d(X)$ and $\mathscr{B}_\rho(Y)$.

13. Approximation by Members of a Bornology

If a subset A of $\langle X, d \rangle$ is d-totally bounded, then $\forall \varepsilon > 0 \, \exists \, F \in \mathscr{F}(X)$ with $F \subseteq A \subseteq S_d(F, \varepsilon)$, that is, we can take the approximating subset F inside A. But if we replace $\mathscr{F}(X)$ by an arbitrary bornology $\mathscr{B}$, this is no longer the case: the condition (1) $\forall \varepsilon > 0 \, \exists \, B \in \mathscr{B}$ with $B \subseteq A \subseteq S_d(B, \varepsilon)$ may be properly stronger than (2) $\forall \varepsilon > 0 \, \exists \, B \in \mathscr{B}$ with $A \subseteq S_d(B, \varepsilon)$. Following Lechicki, Levi and Spakowski [113], we call the former condition $\mathscr{B}$-*total boundedness* and the latter condition *weak $\mathscr{B}$-total boundedness*, and consistent with their notation, we denote the $\mathscr{B}$-totally bounded sets by $\mathscr{B}_*$ and the weakly $\mathscr{B}$-totally bounded sets by $\mathscr{B}^*$.

Example 13.1. In the Euclidean plane with the usual metric d, let $\mathscr{B}$ be the bornology generated by the family of all vertical lines. Let $A = \{(x, 1/x) : 0 < x \leq 1\}$. Then $A \in \mathscr{B}^*$, for given $\varepsilon > 0$, choose $n \in \mathbb{N}$ with $\frac{1}{n} < \varepsilon$. Then A is contained in the ε-enlargement of $\{\frac{k}{n} : 1 \leq k \leq n\} \times \mathbb{R}$. However, $A \notin \mathscr{B}_*$ because if $B \in \mathscr{B}$ with $B \subseteq A$, then B is a finite set.

Proposition 13.2. *Let $\mathscr{B}$ be a bornology on a metric space $\langle X, d \rangle$. Then*

(1) $\mathscr{B} \subseteq \mathscr{B}_* \subseteq \mathscr{B}^*$;

(2) $\mathscr{B}^*$ *is a bornology;*

(3) $A \in \mathscr{B}^*$ *if and only if $\forall \varepsilon > 0$ there exists $B \in \mathscr{B}$ with $H_d(A, B) < \varepsilon$.*

Proof. Statement (1) is obvious. For statement (2), $\mathscr{B}^*$ contains the singletons because $\mathscr{B} \subseteq \mathscr{B}^*$. If $A_1 \subseteq S_d(B_1, \varepsilon)$ and $A_2 \subseteq S_d(B_2, \varepsilon)$ with $\{B_1, B_2\} \subseteq \mathscr{B}$, then $B_1 \cup B_2 \in \mathscr{B}$ and $A_1 \cup A_2 \subseteq S_d(B_1 \cup B_2, \varepsilon)$. This argument shows that $\mathscr{B}^*$ is stable under finite unions. That $\mathscr{B}^*$ is an hereditary family is equally simple. For statement (3), let $A \in \mathscr{B}^*$ and let $\varepsilon > 0$. Choose $B \in \mathscr{B}$ with $A \subseteq S_d(B, \varepsilon)$. Then with $B_0 = B \cap S_d(A, \varepsilon) \in \mathscr{B}$, we have $H_d(B_0, A) < \varepsilon$. $\quad\square$

In qualitative terms, the last statement says that the $\mathscr{B}$-weakly totally bounded subsets are the closure of the initial bornology with respect to the Hausdorff pseudometric topology on $\mathscr{P}(X)$. As taking the closure is idempotent, a family $\mathscr{A}$ of subsets of X forms the $\mathscr{B}$-weakly totally bounded subsets for some bornology $\mathscr{B}$ if and only if $\mathscr{A}$ is a bornology that is closed subset of $\mathscr{P}(X)$ with respect to the Hausdorff pseudometric topology. Further, different bornologies can induce the same totally bounded or weakly totally bounded subsets, e.g., $\mathscr{T}\mathscr{B}_d(X)$ is equally well induced by $\mathscr{K}(X)$.

The $\mathscr{B}$-totally bounded subsets while containing the initial bornology and being stable under finite unions, need not be an hereditary family. In the last example, $[0, 1] \times \mathbb{R}$ is $\mathscr{B}$-totally bounded whereas $\{(x, 1/x) : 0 < x \leq 1\}$ is not.

To further familiarize the reader with these two ideas, we revisit the principal bornology $\mathscr{B}(A)$ determined by a subset A of $\langle X, d \rangle$ consisting of all sets of the form $B \cup F$ where $B \subseteq A$ and F is finite. The weakly totally bounded sets and the totally bounded sets as determined by $\mathscr{B}(A)$ need not coincide nor are the $\mathscr{B}(A)$-totally bounded sets always an hereditary family.

The remaining results and examples of this section were established by Beer and Levi [44]; there is much more in this article that space limitations do not allow us to include.

Proposition 13.3. *Let A be a nonempty subset of a metric space $\langle X, d \rangle$. Then for the principal bornology $\mathscr{B}(A)$, we have*

$$\mathscr{B}(A)^* = \{C \subseteq X : \forall \, \varepsilon > 0 \quad C \backslash S_d(A, \varepsilon) \text{ is } d\text{-totally bounded }\}$$

Proof. Let $\varepsilon > 0$. If $C \subseteq X$ is such that $C \backslash S_d(A, \varepsilon)$ is d-totally bounded, then $\exists F \in \mathscr{F}(X)$ such that $C \backslash S_d(A, \varepsilon) \subseteq S_d(F, \varepsilon)$, whence

$$C \subseteq S_d(A, \varepsilon) \cup S_d(F, \varepsilon) = S_d(A \cup F, \varepsilon),$$

with $A \cup F \in \mathscr{B}(A)$. This shows $C \in \mathscr{B}(A)^*$.

Suppose on the other hand we have $C \in \mathscr{B}(A)^*$ and $\varepsilon > 0$. To show that $C \backslash S_d(A, \varepsilon) \in \mathscr{T}\mathscr{B}_d(X)$, we need only show that we can δ-approximate the set by a finite set for each $\delta \in (0, \varepsilon)$. For such a δ, by the nature of a principal bornology, we can choose $F_\delta \in \mathscr{F}(X)$ with

$$C \subseteq S_d(A \cup F_\delta, \delta) = S_d(A, \delta) \cup S_d(F_\delta, \delta).$$

Since $\delta < \varepsilon$, $C \backslash S_d(A, \varepsilon) \subseteq C \backslash S_d(A, \delta) \subseteq S_d(F_\delta, \delta)$ as required. $\qquad\square$

Proposition 13.4. *Let A be a nonempty subset of a metric space $\langle X, d \rangle$. Then for the principal bornology $\mathscr{B}(A)$, we have*

$$\mathscr{B}(A)_* = \{C \subseteq X : \forall \, \varepsilon > 0 \quad C \backslash S_d(C \cap A, \varepsilon) \text{ is } d\text{-totally bounded }\}$$

Proof. Let $\varepsilon > 0$ be arbitrary, and suppose $C \backslash S_d(C \cap A, \varepsilon) \in \mathscr{T}\mathscr{B}_d(X)$. As this set is a subset of C, we can find $F \in \mathscr{F}(X)$ such that $F \subseteq C$ and

$$C \backslash S_d(C \cap A, \varepsilon) \subseteq S_d(F, \varepsilon).$$

From this, we obtain the following inclusion string:

$$(C \cap A) \cup F \subseteq C \subseteq S_d((C \cap A) \cup F, \varepsilon).$$

As $\varepsilon > 0$ was arbitrary and $(C \cap A) \cup F \in \mathscr{B}(A)$, the inclusion $\supseteq$ for the two families is established.

For the inclusion $\subseteq$ for the families, suppose $C \in \mathscr{B}(A)_*$ and let $\varepsilon > 0$ be arbitrary. By definition, for each $\delta \in (0, \varepsilon)$, there exists $A_\delta \subseteq A$ and a finite set F_δ with

$$A_\delta \cup F_\delta \subseteq C \subseteq S_d(A_\delta \cup F_\delta, \delta).$$

From this it follows that

$$C \backslash S_d(A \cap C, \varepsilon) \subseteq C \backslash S_d(A_\delta, \delta) \subseteq S_d(F_\delta, \delta).$$

We have shown that $C \backslash S_d(A \cap C, \varepsilon)$ is d-totally bounded. $\qquad\square$

Example 13.5. Let ℓ_2 be the Hilbert space of square summable sequences with the usual orthonormal base $e_1, e_2, e_3, \ldots$ and closed unit ball U. Consider $C = U \cup \{\frac{n+1}{n} e_n : n \in \mathbb{N}\}$. Then $C \in \mathscr{B}(U)_*$ while its subset $\frac{1}{2} U \cup \{\frac{n+1}{n} e_n : n \in \mathbb{N}\}$ fails to be totally bounded with respect to the principal bornology because each member of the principal bornology can contain only finitely many points of the form $\frac{n+1}{n} e_n$. Again, the totally bounded sets and weakly totally bounded sets do not coincide.

In this section, we want to answer the following questions.

(1) When is a family of subsets $\mathscr{A}$ of X equal to $\mathscr{B}_*$ for some bornology $\mathscr{B}$?

(2) What is the structure of the bornologies $\mathscr{B}$ that induce a particular family of sets as their $\mathscr{B}_*$? Is there a largest and a smallest bornology among them, and if so, what do they look like?

(3) Under what conditions is $\mathscr{B}_*$ a bornology?

(4) Under what conditions does $\mathscr{B}_* = \mathscr{B}^*$?

It makes sense to consider the operators $\mathscr{A} \mapsto \mathscr{A}_*$ and $\mathscr{A} \mapsto \mathscr{A}^*$ on the power set of $\mathscr{P}(X)$, rather than restricting them to bornologies. For one thing, we want to apply them to families of $\mathscr{B}$-totally bounded subsets with respect to a bornology $\mathscr{B}$. Obviously, both operators are monotone on the power set of $\mathscr{P}(X)$ and we have $\mathscr{A} \subseteq \mathscr{A}_* \subseteq \mathscr{A}^*$ whenever $\mathscr{A} \subseteq \mathscr{P}(X)$.

Recall that an operator $A \mapsto \phi(A)$ on the power set of a set Y is called a *closure operator* provided it satisfies the following four properties (see, e.g., [74, p. 73] or [155, p. 25]) :

(a) whenever $A \subseteq Y$, $A \subseteq \phi(A)$;

(b) whenever $A \subseteq Y$, $\phi(\phi(A)) = \phi(A)$;

(c) $\phi(\emptyset) = \emptyset$;

(d) ϕ preserves finite unions.

The fixed points of a closure operator are of course the closed subsets of a topology on Y.

Proposition 13.6. *Let $\langle X, d \rangle$ be a metric space. Then the operators $\mathscr{A} \mapsto \mathscr{A}_*$ and $\mathscr{A} \mapsto \mathscr{A}^*$ on the power set of $\mathscr{P}(X)$ are closure operators.*

Proof. We just verify the conditions for $\mathscr{A} \mapsto \mathscr{A}_*$. We have already noted (a). For (b) the inclusion $\mathscr{A}_* \subseteq \mathscr{A}_{**}$ follows from $\mathscr{A} \subseteq \mathscr{A}_*$ and the monotonicity of the operator. For the reverse inclusion, let $E \in \mathscr{A}_{**}$ and let $\varepsilon > 0$ be arbitrary. Choose $D \in \mathscr{A}_*$ with $D \subseteq E \subseteq S_d(D, \frac{\varepsilon}{2})$ and choose $A \in \mathscr{A}$ with $A \subseteq D \subseteq S_d(A, \frac{\varepsilon}{2})$. This yields $A \subseteq E \subseteq S_d(A, \varepsilon)$ as required.

Condition (c) is immediate from the fact that nothing can be approximated from an empty family of sets, including $\emptyset$. For (d), let $\mathscr{A}$ and let $\mathscr{C}$ be families of subsets of X. To show $(\mathscr{A} \cup \mathscr{C})_* = \mathscr{A}_* \cup \mathscr{C}_*$, by the monotonicity of the operator, we need only check $(\mathscr{A} \cup \mathscr{C})_* \subseteq \mathscr{A}_* \cup \mathscr{C}_*$. Given $E \in (\mathscr{A} \cup \mathscr{C})_*$, for each positive integer n we can find $W_n \in \mathscr{A} \cup \mathscr{C}$ with $W_n \subseteq E \subseteq S_d(W_n, \frac{1}{n})$. But frequently either $W_n \in \mathscr{A}$ or $W_n \in \mathscr{C}$, establishing the required inclusion. □

We content ourselves with one corollary that says in particular that our two operators commute.

Corollary 13.7. *Let $\langle X, d \rangle$ be a metric space and let $\mathscr{A} \subseteq \mathscr{P}(X)$. Then*

$$(\mathscr{A}_*)^* = (\mathscr{A}^*)_* = \mathscr{A}^*.$$

Proof. By monotonicity and idempotency, $(\mathscr{A}_*)^* \subseteq (\mathscr{A}^*)^* = \mathscr{A}^*$. Similarly, $(\mathscr{A}^*)_* \subseteq (\mathscr{A}^*)^* = \mathscr{A}^*$. □

Our next result in qualitative terms says that the weakly totally bounded sets as determined by some bornology $\mathscr{B}$ are subsets of $\mathscr{B}$-totally bounded sets.

Theorem 13.8. *Let $\mathscr{B}$ be a bornology on a metric space $\langle X, d \rangle$. Then $\downarrow (\mathscr{B}_*) = \mathscr{B}^*$.*

Proof. Since $\downarrow (\mathscr{B}^*) = \mathscr{B}^*$, $\mathscr{B}_* \subseteq \mathscr{B}^*$ and $\downarrow$ is monotone, we have $\downarrow (\mathscr{B}_*) \subseteq \mathscr{B}^*$. For the reverse inclusion, suppose $A \in \mathscr{B}^*$. For each $n \in \mathbb{N}$ choose $B_n \in \mathscr{B}$ with $A \subseteq S_d(B_n, 1/n)$. Let $T_n = B_n \cap S_d(A, 1/n)$ so that $T_n \in \mathscr{B}$ and $H_d(A, T_n) \leq 1/n$. Now for each $n \in \mathbb{N}$ set $E_n = \cup_{j=1}^n T_j \in \mathscr{B}$. Let $\delta > 0$; we claim that whenever $\frac{1}{n} < \frac{\delta}{2}$, we have

$$E_n \subseteq A \cup \bigcup_{j=1}^{\infty} T_j \subseteq S_d(E_n, \delta).$$

The first inclusion is obvious. For the second, $H_d(A, T_n) \leq 1/n < \delta/2$ gives

$$A \subseteq S_d(T_n, \frac{\delta}{2}) \subseteq S_d(E_n, \delta).$$

If $j \leq n$ then $T_j \subseteq E_n \subseteq S_d(E_n, \delta)$. On the other hand, if $j > n$ then $H_d(A, T_j) < \delta/2$ and so

$$H_d(T_n, T_j) \leqq H_d(T_n, A) + H_d(A, T_j) < \delta.$$

Thus, in this case, $T_j \subseteq S_d(T_n, \delta) \subseteq S_d(E_n, \delta)$. These three estimates together establish the claim. Since $\delta > 0$ was arbitrary we may conclude that $A \cup (\cup_{j=1}^{\infty} T_j) \in \mathscr{B}_*$ and so $A \in \downarrow (\mathscr{B}_*)$. $\square$

With respect to the discussion in this section, a family of subsets $\mathscr{A}$ of $\langle X, d \rangle$ can be closed in two different senses: it can be a fixed point of the upper star operator or it can be a fixed point of the lower star operator. For bornologies, there is no difference, so that if we speak of a *closed bornology*, there is no ambiguity.

Corollary 13.9. *Let $\mathscr{B}$ be a bornology on a metric space $\langle X, d \rangle$. Then $\mathscr{B} = \mathscr{B}^*$ if and only if $\mathscr{B} = \mathscr{B}_*$.*

Proof. Suppose $\mathscr{B} = \mathscr{B}_*$ and let $A \in \mathscr{B}^*$. By the last result, $\exists C \in \mathscr{B}_*$ with $A \subseteq C \in \mathscr{B}$. But $\mathscr{B}$ is hereditary and so $A \in \mathscr{B}$. We always have $\mathscr{B} \subseteq \mathscr{B}^*$. Conversely, we always have $\mathscr{B} \subseteq \mathscr{B}_* \subseteq \mathscr{B}^*$, so that $\mathscr{B} = \mathscr{B}^*$ gives $\mathscr{B} = \mathscr{B}_*$. $\square$

It is easy to check that $\mathscr{B}\mathscr{B}_d(X)$ is a closed bornology. If a bornology $\mathscr{B}$ is closed, then it has a closed base, i.e., $\forall B \in \mathscr{B}$, $\text{cl}(B) \in \mathscr{B}$ because $H_d(B, \text{cl}(B)) = 0$. However, a bornology can have a closed base without being closed in the sense of this section, e.g., $\mathscr{F}(X)$ always has a closed base.

As we mentioned earlier, a family of subsets $\mathscr{A}$ of a metric space forms the weakly bounded subsets for some bornology if and only if $\mathscr{A}$ is itself a bornology closed with respect to the Hausdorff pseudometric topology; more formally, $\mathscr{A}$ is a bornology with $\mathscr{A} = \mathscr{A}^*$. We now characterize those families of subsets $\mathscr{A}$ of X that are the totally bounded subsets induced by some bornology, answering our first question posed above. The key idea here is that the *hereditary core* of $\mathscr{A}$, defined by

$$\text{hc}(\mathscr{A}) := \{ A \in \mathscr{A} : \mathscr{P}(A) \subseteq \mathscr{A} \}$$

be large enough.

Theorem 13.10. *Let $\mathscr{A}$ be a family of subsets of $\langle X, d \rangle$. Then $\mathscr{A} = \mathscr{B}_*$ for some bornology $\mathscr{B}$ on X if and only if each of the following conditions is satisfied:*

(1) $\mathscr{F}(X) \subseteq \mathscr{A}$;

(2) $\Sigma(\mathscr{A}) = \mathscr{A}$;

(3) $\mathscr{A} = (\mathrm{hc}(\mathscr{A}))_*$.

Proof. Suppose $\mathscr{A} = \mathscr{B}_*$ for some bornology $\mathscr{B}$. Condition (1) holds since always $\mathscr{F}(X) \subseteq \mathscr{B} \subseteq \mathscr{B}_*$. For (2), we need only consider unions of size two. Let $A_1 \in \mathscr{A}$ and $A_2 \in \mathscr{A}$ and let $\varepsilon > 0$ be arbitrary. Choose B_1 and $B_2 \in \mathscr{B}$ such that for $j = 1, 2$, $B_j \subseteq A_j \subseteq S_d(B_j, \varepsilon)$. This immediately yields

$$B_1 \cup B_2 \subseteq A_1 \cup A_2 \subseteq S_d(B_1 \cup B_2, \varepsilon).$$

Condition (3) holds as $\mathrm{hc}(\mathscr{A}) \subseteq \mathscr{A}$ implies

$$(\mathrm{hc}(\mathscr{A}))_* \subseteq \mathscr{A}_* = \mathscr{B}_{**} = \mathscr{B}_* = \mathscr{A}$$

on the one hand, while $\mathscr{B} \subseteq \mathscr{B}_* = \mathscr{A}$ implies $\mathscr{B} \subseteq \mathrm{hc}(\mathscr{A})$ so that $\mathscr{A} = \mathscr{B}_* \subseteq (\mathrm{hc}(\mathscr{A}))_*$ on the other.

Conversely, suppose conditions (1) through (3) are satisfied for $\mathscr{A}$. By condition (1) $\mathscr{F}(X) \subseteq \mathrm{hc}(\mathscr{A})$ as $\mathscr{F}(X)$ is hereditary. By definition $\mathrm{hc}(\mathscr{A})$ is hereditary. Also condition (2) implies that $\mathrm{hc}(\mathscr{A})$ is stable under finite unions. Altogether, we see that $\mathrm{hc}(\mathscr{A})$ is a bornology. Finally, by condition (3) we have $\mathscr{A} = \mathscr{B}_*$ with $\mathscr{B} = \mathrm{hc}(\mathscr{A})$. $\qquad\square$

Example 13.11. In the real line let $\mathscr{A} := \mathscr{F}(\mathbb{R}) \cup \{F \cup [-n, n] : F \in \mathscr{F}(\mathbb{R}) \text{ and } n \in \mathbb{N}\}$. While $\mathscr{A}$ satisfies conditions (1) and (2), it fails to satisfy (3) as $\mathrm{hc}(\mathscr{A}) = \mathscr{F}(\mathbb{R})$. It is left to the reader to construct families of subsets satisfying (a) conditions (1) and (3) but not (2), and (b) conditions (2) and (3) but not (1).

Notice that when $\mathscr{A}$ is a totally bounded family induced by some bornology, $\mathrm{hc}(\mathscr{A})$ is the largest bornology whose induced totally bounded sets equal $\mathscr{A}$, for if $\mathscr{B}'$ is a bornology with $\mathscr{B}'_* = \mathscr{A}$ and $B \in \mathscr{B}'$, then

$$\mathscr{P}(B) \subseteq \mathscr{B}' \subseteq \mathscr{A}$$

so that $B \in \mathrm{hc}(\mathscr{A})$ by definition.

Let $\mathscr{B}$ be a bornology on $\langle X, d \rangle$. Since $\mathscr{B}_*$ is closed under finite unions, minimally $\mathrm{hc}(\mathscr{B}_*)$ contains $\mathscr{T}\mathscr{B}_d(X) \vee \mathscr{B}$. To see that the containment can be proper, in ℓ_2 consider the bornology $\mathscr{B}$ of its countable subsets, where $\mathscr{B}_* = \mathscr{P}(\ell_2) = \mathrm{hc}(\mathscr{B}_*)$.

We next show that $\mathrm{hc}(\mathscr{B}_*)$ for the bornology $\mathscr{B}$ introduced in Example 13.1 is no larger than the minimal possible family.

Example 13.12. Consider $\mathscr{B}_*$ for the bornology $\mathscr{B}$ in the plane generated by the family of all vertical lines. We claim

$$\mathrm{hc}(\mathscr{B}_*) = \{E_1 \cup E_2 : E_1 \text{ is bounded and } E_2 \in \mathscr{B}\} = \mathscr{T}\mathscr{B}_d(\mathbb{R}^2) \vee \mathscr{B}.$$

As we have already noted, the inclusion $\supseteq$ always holds. For the reverse inclusion, suppose $A \subseteq \mathbb{R}^2$ cannot be written as such a union. Let L_0 be an arbitrary vertical line and pick $(x_1, y_1) \in A$ with $(x_1, y_1) \notin L_0 \cup [-1, 1] \times [-1, 1]$. Let L_1 be the line with equation $x = x_1$ and then pick $(x_2, y_2) \in A$ with $(x_2, y_2) \notin (L_0 \cup L_1)$ and $(x_2, y_2) \notin [-2, 2] \times [-2, 2]$. Continuing we produce a sequence of points $(x_1, y_1), (x_2, y_2), (x_3, y_3), \ldots$ in A such that no two lie on a common vertical line and $\forall n \in \mathbb{N}$, $\max \{|x_n|, |y_n|\} > n$. It follows that $\{(x_n, y_n) : n \in \mathbb{N}\} \notin \mathscr{B}_*$ and so $A \notin \mathrm{hc}(\mathscr{B}_*)$. Note however that $\mathrm{hc}(\mathscr{B}_*)$ is a proper subset of $\mathscr{B}_*$ as $[0, 1] \times \mathbb{R} \in \mathscr{B}_*$.

Our next theorem at once answers the third and fourth questions posed earlier in this section.

Theorem 13.13. *Let $\mathscr{B}$ be a bornology on a metric space $\langle X, d \rangle$. The following conditions are equivalent:*

(1) $\mathscr{B}_* = \mathscr{B}^*$;

(2) $\mathscr{B}_*$ *is a bornology;*

(3) $\mathscr{B}_*$ *is hereditary;*

(4) $hc(\mathscr{B}_*) = \mathscr{B}_*$;

(5) $hc(\mathscr{B}_*)_* = hc(\mathscr{B}_*)$;

(6) $hc(\mathscr{B}_*) = \mathscr{B}^*$.

Proof. $(1) \Rightarrow (2)$. Since $\mathscr{B}^*$ is always a bornology, equality of the totally bounded and weakly totally bounded sets as determined by $\mathscr{B}$ forces $\mathscr{B}_*$ to be a bornology.

$(2) \Rightarrow (3)$. This is trivial.

$(3) \Rightarrow (4)$. If $\mathscr{B}_*$ is hereditary, then by the definition of $hc(\mathscr{B}_*)$, we have $hc(\mathscr{B}_*) = \mathscr{B}_*$.

$(4) \Rightarrow (5)$. If condition (4) holds, by the idempotency of the lower star operator we obtain

$$hc(\mathscr{B}_*)_* = \mathscr{B}_{**} = \mathscr{B}_* = hc(\mathscr{B}_*).$$

$(5) \Rightarrow (6)$. In view of Corollary 13.9 and the monotonicity of the upper star operator, (5) yields

$$\mathscr{B}^* \subseteq hc(\mathscr{B}_*)^* = hc(\mathscr{B}_*) \subseteq \mathscr{B}_* \subseteq \mathscr{B}^*.$$

$(6) \Rightarrow (1)$. From condition (6), $\mathscr{B}^* = hc(\mathscr{B}_*) \subseteq \mathscr{B}_*$ while the inclusion $\mathscr{B}_* \subseteq \mathscr{B}^*$ always holds. $\qquad\square$

For completeness, we state without proof a result that gives necessary and sufficient conditions for the totally bounded sets to coincide with the weakly totally bounded subsets as determined by a principal bornology [44, Theorem 4.4].

Theorem 13.14. *Let A be a nonempty subset of $\langle X, d \rangle$. The following conditions are equivalent:*

(1) $\mathscr{B}(A)_* = \mathscr{B}(A)^*$;

(2) $\forall \lambda > 0 \; \exists \delta > 0 \; \exists F \in \mathscr{F}(X)$ *with* $S_d(A, \delta) \backslash A \subseteq S_d(F, \lambda)$;

(3) $\mathscr{B}(A)^* = \{E_1 \cup E_2 : E_1 \subseteq A \text{ and } E_2 \in \mathscr{T}\mathscr{B}_d(X)\}$;

(4) $\mathscr{B}(A)_* = \{E_1 \cup E_2 : E_1 \subseteq A \text{ and } E_2 \in \mathscr{T}\mathscr{B}_d(X)\}$.

Our next goal is to show that there is a smallest bornology that induces a given family of totally bounded sets. The following simple lemma is the key to identifying this smallest bornology.

Lemma 13.15. *Let $\mathscr{B}$ be a bornology on $\langle X, d \rangle$ and let $A \in \mathscr{B}_*$.*

(1) $\forall \varepsilon > 0$ *there exists a uniformly discrete element D of $\mathscr{B}$ with $D \subseteq A \subseteq S_d(D, \varepsilon)$;*

(2) *if A itself is uniformly discrete then $A \in \mathscr{B}$.*

Proof. For (1), choose $B \in \mathscr{B}$ with $B \subseteq A \subseteq S_d(B, \frac{\varepsilon}{2})$. If we partially order the $\frac{\varepsilon}{2}$-discrete subsets of B by inclusion, by Zorn's lemma, there exists a maximal $\frac{\varepsilon}{2}$-discrete subset D and it follows that $D \subseteq A \subseteq S_d(D, \varepsilon)$. For (2), if A is λ-discrete and $B \in \mathscr{B}$ satisfies $B \subseteq A \subseteq S_d(B, \lambda)$, then we must have $A = B$. $\qquad\square$

By the second statement of the last lemma, given a family $\mathscr{A}$ of subsets of X induced as the totally bounded subsets of some bornology, each bornology that so induces $\mathscr{A}$ must contain the uniformly discrete subsets of $\mathscr{A}$. From this observation we get

Theorem 13.16. *Let $\langle X, d \rangle$ be a metric space and let $\mathscr{B}$ be a bornology on X. Then there is a smallest bornology on X whose totally bounded sets coincide with $\mathscr{B}_*$, namely*

$$\mathscr{B}^{small} := \{E : E \text{ is a finite union of uniformly discrete subsets of } X \text{ which are in } \mathscr{B}_*\}.$$

Proof. $\mathscr{B}^{small}$ as described is evidently a bornology because (i) each finite subset belongs to $\mathscr{B}_*$ and is uniformly discrete, and (ii) the uniformly discrete subsets are an hereditary family. By Lemma 13.15 we have

- each bornology that induces $\mathscr{B}_*$ under the lower star operator must contain $\mathscr{B}^{small}$ as bornologies are stable under finite unions,

- $\mathscr{B}^{small}_* = \mathscr{B}_*$,

completing our proof. $\qquad\square$

Corollary 13.17. *Let $\mathscr{B}_1$ and $\mathscr{B}_2$ be two bornologies on $\langle X, d \rangle$. Then $(\mathscr{B}_1)_* = (\mathscr{B}_2)_*$ if and only if $\mathscr{B}_1$ and $\mathscr{B}_2$ have the same uniformly discrete subsets of X.*

As we have noted, each finite subset of a metric space is uniformly discrete. It turns out that there is a nice tangible way to visualize those infinite subsets expressible as a finite union of uniformly discrete subsets. We close this section with this description.

Proposition 13.18. *Let A be an infinite subset of $\langle X, d \rangle$. The following conditions are equivalent:*

(1) $A = \cup_{j=1}^n E_j$ *where each E_j is uniformly discrete;*

(2) $\exists\, \varepsilon > 0\ \exists\, n \in \mathbb{N}\ \forall\, x \in X,\ S_d(x, \varepsilon)$ *contains at most n points of A.*

Proof. (1) $\Rightarrow$ (2). Suppose $A = \cup_{j=1}^n E_j$ where each E_j is ε_j-discrete. Let $\varepsilon = \frac{1}{2}\min\{\varepsilon_j : j \leqslant n\}$. By the pigeon-hole principle, for each $x \in X$, $S_d(x, \varepsilon)$ can contain at most n points of A.

(2) $\Rightarrow$ (1). Suppose ε and n are as stated in condition (2). By Zorn's lemma A has a subset E_1 such that (1) for all $x \neq y \in E_1$, $d(x, y) \geq \varepsilon$, and (2) for all $z \in A\ \exists x \in E_1$ with $d(x, z) < \varepsilon$. By definition E_1 is ε-discrete and for each $z \in A \backslash E_1$, $S_d(z, \varepsilon) \cap (A \backslash E_1)$ contains at least one less point of A than does $S_d(z, \varepsilon) \cap A$, i.e., it contains at most n - 1 points of A. Applying the same construction to $A \backslash E_1$ we obtain an ε-discrete ε-net E_2 of $A \backslash E_1$, and each $S_d(z, \varepsilon)$ for $z \in A \backslash (E_1 \cup E_2)$ contains at most n - 2 points of $A \backslash (E_1 \cup E_2)$. Continuing this process we may write after n steps $A = \cup_{j=1}^n E_j$ where each E_j is ε-discrete, for $A \backslash \cup_{j=1}^{n-1} E_j$ will itself be ε-discrete. $\qquad\square$

14. Selected Topological Properties of the One-Point Extension

Theorem 14.1. *Let $\langle X, \tau \rangle$ be a Hausdorff space that is first countable (resp. second countable). Let $\mathscr{B}$ be a nontrivial bornology on X with a closed base. Then $o(\mathscr{B}, p)$ is first countable (resp. second countable) if and only if $\mathscr{B}$ has a countable base.*

Proof. We only deal with second countability. Let $\{V_n : n \in \mathbb{N}\}$ be a countable base for the topology τ. Suppose $\{B_n : n \in \mathbb{N}\}$ is a countable base for $\mathscr{B}$. Then it is easy to verify that

$$\{V_n : n \in \mathbb{N}\} \cup \{(X \cup \{p\}) \backslash B_n : n \in \mathbb{N}\}$$

is a base for the extension topology.

Conversely, if $o(\mathscr{B}, p)$ is second countable, then it is first countable. As a result, there is a countable local base at p. This means we can find a sequence $\langle B_n \rangle$ of closed subsets in $\mathscr{B}$ such that for each $B \in \mathscr{B} \cap \mathscr{C}(X)$, $\exists n \in \mathbb{N}$ such that

$$(X \cup \{p\}) \backslash B_n \subseteq (X \cup \{p\}) \backslash B.$$

It follows that $B \subseteq B_n$ so that $\langle B_n \rangle$ is a (closed) base for the bornology. $\square$

Our second result of the section, like the first, appears in [151]. Note that either regularity or normality of the one-point extension implies that the $o(\mathscr{B}, p)$ is Hausdorff as well since points are closed in the extension.

Theorem 14.2. *Let $\langle X, \tau \rangle$ be a regular (resp. normal) Hausdorff space, and let $\mathscr{B}$ be a nontrivial bornology with closed base on X. Then $o(\mathscr{B}, p)$ is a regular (resp. normal) space if and only if $\mathscr{B}$ has an open base.*

Proof. Note that $\mathscr{B}$ having an open base already implies that the extension topology is Hausdorff. We only deal with regularity, leaving normality to the reader.

Suppose $o(\mathscr{B}, p)$ is regular. Let $B \in \mathscr{B}$ be arbitrary. Then $\text{cl}(B)$ is closed in $o(\mathscr{B}, p)$ and so there exist $V \in \tau$ and $B_0 \in \mathscr{B}$ such that $B \subseteq V$ and $(X \cup \{p\}) \backslash B_0$ are disjoint. As $V \subseteq B_0$, we have $A \subseteq V \in \tau \cap \mathscr{B}$.

Conversely, suppose $\mathscr{B}$ has an open base. If $A \subseteq X$ is closed in the extension and $x \in X \backslash A$, then A is closed in X, so by regularity of $\langle X, \tau \rangle$, x and A can be separated by disjoint members of τ and these sets are also open in the extension. Now suppose $A \subseteq X$ is closed in $o(\mathscr{B}, p)$. This means $A \in \mathscr{B}$ so we can choose $V \in \mathscr{B} \cap \tau$ with $A \subseteq V$. Then V and $(X \cup \{p\}) \backslash \text{cl}(V)$ are disjoint neighborhoods of A and p. As a final case, suppose A is closed in the extension with $p \in A$ and $x \notin A$. Since the bornology is local, we can find $V \in \mathscr{B} \cap \tau$ with $x \in V$. Choose by regularity of $\langle X, \tau \rangle$ $W \in \tau$ with $x \in W \subseteq \text{cl}(W) \subseteq X \backslash A$. Then W and $(X \cup \{p\}) \backslash \text{cl}(W)$ are disjoint neighborhoods of x and A. $\square$

The Tychonoff separation property for the one-point extension, as established by Caterino, Panduri and Vipera [64], is less transparent. We need a definition.

Definition 14.3. We say that a bornology $\mathscr{B}$ on a Hausdorff space $\langle X, \tau \rangle$ is *functionally open*
if for each nonempty $B \in \mathscr{B}$ there exists $f_B \in C(X, [0,1])$ such that $B \subseteq f_B^{-1}(\{0\})$ and
$f_B^{-1}([0,1)) \in \mathscr{B}$.

Evidently a bornology that is functionally open must have a closed base and an open base so
that the associated one-point extension is regular and Hausdorff.

Theorem 14.4. *Let $\langle X, \tau \rangle$ be a Tychonoff space and $\mathscr{B}$ be a nontrivial bornology on X with
closed base. The following conditions are equivalent:*

  (1) *$\mathscr{B}$ is functionally open;*

  (2) *the one-point extension $o(\mathscr{B}, p)$ is Tychonoff.*

Proof. $(1) \Rightarrow (2)$. First, suppose $x \in X$ and A is closed with respect to $o(\mathscr{B}, p)$ and $x \notin A$. Since
the bornology has an open base we can choose $U \in \tau \cap \mathscr{B}$ such that $x \in U \subseteq X \backslash A$. Since X is
Tychonoff, we can find $f \in C(X, [0,1])$ with $f(x) = 0$ and $f(X \backslash U) = 1$. Extending f to $o(\mathscr{B}, p)$
by putting $f(p) = 1$ we get a continuous extension because f is identically equal to one on the
neighborhood $(X \cup \{p\}) \backslash \mathrm{cl}(U)$ of p.

Next, suppose $x = p$ and A is a closed subset of the extension with $p \notin A$. Then $A \in \mathscr{B}$, so
since the bornology is functionally open, we can choose $f \in C([X, [0,1])$ with $A \subseteq f^{-1}(\{0\})$ and
$f^{-1}([0,1)) \in \mathscr{B}$. Again, putting $f(p) = 1$ results in a continuous extension, as f is identically
equal to one on the neighborhood $(X \cup \{p\}) \backslash \mathrm{cl}(f^{-1}([0,1))$ of p.

$(2) \Rightarrow (1)$. Assume the extension is Tychonoff; we show that $\mathscr{B}$ is functionally open. Let
$B \in \mathscr{B}$; since the bornology has a closed base, $\mathrm{cl}(B)$ is closed with respect to $o(\mathscr{B}, p)$. Choose
$h \in C(X \cup \{p\}, [0,1])$ with $h(\mathrm{cl}(B)) = \{0\}$ and $h(p) = 1$ and put $g = h|_X$. Since $h^{-1}((\frac{1}{2}, 1])$ is a
neighborhood of p in the extension, we have

$$h^{-1}([0, \tfrac{1}{2}]) = g^{-1}([0, \tfrac{1}{2}]) \in \mathscr{B},$$

and so $g^{-1}([0, \tfrac{1}{2})) \in \mathscr{B}$ as well. The function f_B as stipulated in Definition 14.3 is $f_B = \min \{2g, 1\}$. $\qquad\square$

We leave the proof of the next result to the reader.

Theorem 14.5. *Let $\mathscr{B}$ be a nontrivial bornology with closed base on a Hausdorff space $\langle X, \tau \rangle$.
Then $o(\mathscr{B}, p)$ is compact if and only if $\mathscr{B} = \mathscr{K}(X)$.*

For metrizability, we use the Nagata-Smirnov metrization theorem: a Tychonoff space is metrizable if and only if the topology has a base that can be written as a countable union of locally
finite families of open subsets [155, p. 166].

Lemma 14.6. *Let $\langle X, \tau \rangle$ be a Hausdorff space and let $\mathscr{B}$ be a nontrivial bornology with a
countable base. The following conditions are equivalent:*

  (1) *$\mathscr{B}$ has both a closed base and an open base;*

  (2) *given an $B \in \mathscr{B}, \exists B_0 \in \mathscr{B}$ with $\mathrm{cl}(B) \subseteq \mathrm{int}(B_0)$;*

  (3) *there exists a base for the bornology $\{B_n : n \in \mathbb{N}\}$ such that $\forall n, \mathrm{cl}(B_n) \subseteq \mathrm{int}(B_{n+1})$.*

Proof. The proof of (1) $\Rightarrow$ (2) is immediate. For (2) $\Rightarrow$ (3), let $\{A_n : n \in \mathbb{N}\}$ be a countable base for $\mathscr{B}$. Let $B_1 = A_1$; having defined $B_1, B_2, \ldots, B_n$ choose $E_n \in \mathscr{B}$ with $\mathrm{cl}(\cup_{i=1}^n B_i) \subseteq \mathrm{int}(E_n)$ and finally put $B_{n+1} = E_n \cup A_{n+1}$. Finally for (3) $\Rightarrow$ (1) the desired closed and open bases are $\{\mathrm{cl}(B_n) : n \in \mathbb{N}\}$ and $\{\mathrm{int}(B_n) : n \in \mathbb{N}\}$ where $\{B_n : n \in \mathbb{N}\}$ is as described in condition (3). $\square$

Theorem 14.7. *Let $\mathscr{B}$ be a nontrivial bornology with closed base on a metrizable space $\langle X, \tau \rangle$. The following conditions are equivalent:*

(1) *$o(\mathscr{B}, p)$ is metrizable;*

(2) *$\mathscr{B}$ has both an open base and a countable base;*

(3) *$o(\mathscr{B}, p)$ is both first countable and regular.*

Proof. Conditions (2) and (3) are equivalent in view of what we have already shown in this section, and (1) $\Rightarrow$ (3) is obvious. To finish the proof, we verify (2) $\Rightarrow$ (1) using the Nagata-Smirnov theorem.

If our bornology is trivial, then p is an isolated point of the extension. Let d be a compatible metric for τ and define ρ for the extension by

$$
\rho(x, w) = \begin{cases} 0 & if \ x = w = p \\ \min\{d(x, w), 1\} & if \ \{x, w\} \subseteq X \\ 1 & if \ x = p \ and \ w \neq x. \end{cases}
$$

Otherwise, let $\{B_n : n \in \mathbb{N}\}$ be a base for the bornology as described in Lemma 14.6. Fix a compatible metric d for τ and consider for each integer $n \geq 2$,

$$
\mathscr{U}_n := \{S_d(x, \tfrac{1}{k}) : k \geq n \ \text{and} \ S_d(x, \tfrac{1}{k}) \subseteq \mathrm{int}(B_n)\}.
$$

Take from a locally finite open refinement of the open cover $\mathscr{U}_n \cup \{X \backslash \mathrm{cl}(B_{n-1})\}$ of X those sets contained in some ball from $\mathscr{U}_n$ and adjoin $(X \cup \{p\}) \backslash \mathrm{cl}(B_n)$ to this refinement to obtain a locally finite family of open subsets $\mathscr{V}_n$ of $o(\mathscr{B}, p)$. We claim that that

$$
\mathscr{V} := \cup_{n=2}^\infty \mathscr{V}_n
$$

is a base for the topology of $o(\mathscr{B}, p)$. By construction $\mathscr{V}$ obviously contains a local base at p. Let $x \in X$ be arbitrary and let W be a neighborhood of x in the extension. Without loss of generality, we may assume $W \in \tau$. Choose $n \geq 2$ with $x \in B_{n-1}$ and then $k > n$ with $S_d(x, 1/k) \subseteq \mathrm{int}(B_n) \cap W$. Since $2k - 1 > n$, we have $x \notin X \backslash \mathrm{cl}(B_{2k-1})$. By the definition of $\mathscr{V}_{2k}$ there exist $j \geq 2k, y \in \mathrm{int}(B_{2k})$ and $V \in \mathscr{V}_{2k}$ with $x \in V \subseteq S_d(y, \tfrac{1}{j})$ (where $S_d(x, \tfrac{1}{j}) \subseteq \mathrm{int}(B_{2k})$). As a result, $V \in \mathscr{V}$ and $x \in V \subseteq S_d(x, \tfrac{1}{k}) \subseteq W$. Apply the Smirnov-Nagata metrization theorem. $\square$

In the next section we will show that if $\mathscr{B}$ is a bornology in a metrizable space $\langle X, \tau \rangle$ with a countable base, a closed base, and an open base, then there is metric d on X for which $\tau_d = \tau$ and $\mathscr{B}_d(X) = \mathscr{B}$.

Recall that a metrizable space $\langle X, \tau \rangle$ is completely metrizable if it can be topologically embedded as a G_δ-subset of some completely metrizable space. Conversely, if $\langle X, \tau \rangle$ is completely metrizable, then it is a G_δ-subset of any metrizable space in which it is topologically embedded [155, p. 179]. From this, we obtain the last result of this section.

Theorem 14.8. *Let* $\langle X, \tau \rangle$ *be a metrizable space and let* $\mathscr{B}$ *be a nontrivial bornology on* X *with a closed base, an open base, and a countable base. Then* $\langle X, \tau \rangle$ *is completely metrizable if and only if* $o(\mathscr{B}, p)$ *is completely metrizable.*

Proof. If $o(\mathscr{B}, p)$ is completely metrizable, then since $\langle X, \tau \rangle$ is an open subspace of the one-point extension, it is completely metrizable as well. Conversely, suppose $\langle X, \tau \rangle$ is completely metrizable. Let ρ be a compatible metric for $o(\mathscr{B}, p)$ and denote the completion of the space $\langle X \cup \{p\}, \rho \rangle$ by $\langle \widehat{X \cup \{p\}}, \hat{\rho} \rangle$. Then $\langle X, \tau \rangle$ sits as a G_δ-subset of the completion $\langle \widehat{X \cup \{p\}}, \hat{\rho} \rangle$ and $\{p\}$ is a G_δ-subset as well. It follows from the general formula

$$(\cap_{n=1}^{\infty} V_n) \cup (\cap_{n=1}^{\infty} W_n) = \bigcap_{(j,k) \in \mathbb{N} \times \mathbb{N}} V_j \cup W_k$$

that $X \cup \{p\}$ is a G_δ-subset of a completely metrizable space. $\qquad\square$

15. Bornologies of Metrically Bounded Sets

Perhaps the first major result in the theory of bornological universes, discovered by Hu [102, 103], addresses this question: given a bornology $\mathscr{B}$ on a metrizable topological space $\langle X, \tau \rangle$, when does there exists a metric d compatible with τ such that $\mathscr{B} = \mathscr{B}_d(X)$? For short, we call such a bornology a *metric bornology*. As we saw in an earlier example, such a bornology has a closed base, an open base, and a countable base. In view of Lemma 14.6, Hu's theorem says that these conditions are also sufficient.

Theorem 15.1. *Let $\langle X, \tau \rangle$ be a metrizable space. Then a bornology $\mathscr{B}$ on X is a metric bornology if and only if $\mathscr{B}$ has a countable base and for each $B \in \mathscr{B}, \exists B_0 \in \mathscr{B}$ with $cl(B) \subseteq int(B_0)$.*

Proof. We need only prove sufficiency. Let d be a compatible bounded metric for the topology. If $\mathscr{B}$ is the trivial bornology, the metric d does the job. Otherwise, let $\{B_n : n \in \mathbb{N}\}$ be as described in condition (3) of Lemma 14.6. For each $n \in \mathbb{N}$, let $f_n \in C(X, [0,1])$ map $cl(B_n)$ to zero and the nonempty closed set $X \backslash int(B_{n+1})$ to one. For each $x \in X$, let $n(x)$ be the smallest integer such that $x \in B_{n(x)}$. Then only $f_1, f_2, \ldots, f_{n(x)}$ can take on values other than zero in the neighborhood $int(B_{n(x)+1})$ of x. As a result, $f := f_1 + f_2 + f_3 + \cdots$ is a well-defined and continuous function on X. Notice that f is bounded on a subset E of X if and only if E is contained in some B_n, i.e., $E \in \mathscr{B}$. As the metric ρ defined by

$$\rho(x, w) := d(x, w) + |f(x) - f(w)|$$

is equivalent to d, it is a compatible metric, and a set E is ρ-bounded if and only if $f(E)$ is bounded, i.e., $E \in \mathscr{B}$. $\square$

With Lemma 14.6 and Theorem 14.7 in mind, we get this important corollary to Theorem 15.1.

Corollary 15.2. *Let $\mathscr{B}$ be bornology with closed base on a metrizable space $\langle X, \tau \rangle$. Then $\mathscr{B}$ is a metric bornology corresponding to some compatible metric if and only if $o(\mathscr{B}, p)$ is metrizable.*

A nonnegative continuous real-valued function on a bornological universe $\langle X, \tau, \mathscr{B} \rangle$ such that $\{A \subseteq X : f(A)$ is bounded$\}$ agrees with $\mathscr{B}$ as constructed in the last proof is called a *forcing function* for the universe. A forcing function is both bornological and coercive with respect to $\mathscr{B}$ and the usual metric bornology on $\mathbb{R}$.

Corollary 15.3. *Let $\mathscr{B}$ be a bornology on a metrizable space $\langle X, \tau \rangle$. Then $\mathscr{B}$ is a metric bornology if and only if $\langle X, \tau, \mathscr{B} \rangle$ has a forcing function.*

Proof. If $\mathscr{B} = \mathscr{B}_d(X)$, then $x \mapsto d(p, x)$ is a forcing function whenever $p \in X$ is fixed. Conversely, if f is a forcing function for $\mathscr{B}$, then $\{f^{-1}([0, n]) : n \in \mathbb{N}\}$ is a countable base for the bornology. If $B \in \mathscr{B}$, we can find a closed interval $[0, \beta]$ with $f(B) \subseteq [0, \beta]$. Then by continuity, $f^{-1}([0, \beta+1))$ is an open member of $\mathscr{B}$ containing $cl(B)$. Apply Hu's theorem. $\square$

As an application of Hu's theorem, we obtain a comprehensive result that includes a classical result of Vaughan [149].

Theorem 15.4. *Let $\langle X, \tau \rangle$ be a metrizable space. The following conditions are equivalent.*

(1) *there is a compatible metric d such that $\mathscr{B}_d(X) = \mathscr{K}(X)$;*

(2) *there is a compatible metric d for which closed and d-bounded subsets are compact;*

(3) *the bornology of relatively compact subsets has a countable base;*

(4) *$\langle X, \tau \rangle$ is separable and locally compact.*

Proof. (1) $\Rightarrow$ (2) and (2) $\Rightarrow$ (3) are obvious.

(3) $\Rightarrow$ (4). By passing to closures, $\mathscr{K}(X)$ has a base $\{K_n : n \in \mathbb{N}\}$ consisting of compact sets. Compact subsets are separable, and since X is a countable union of compact sets, X is separable. Suppose $p \in X$ has no neighborhood with compact closure. Then for each positive integer n the ball $S_d(p, \frac{1}{n})$ does not lie in K_n. Pick $x_n \in S_d(p, \frac{1}{n}) \cap K_n^c$; as $\langle x_n \rangle$ converges to p, the set $\{p\} \cup \{x_n : n \in \mathbb{N}\}$ is compact but lies in no K_n, a contradiction. We have shown that $\langle X, d \rangle$ is locally compact.

(4) $\Rightarrow$ (1). The compact bornology always has a closed base, and by local compactness it has an open base as well. We intend to show that $\mathscr{K}(X)$ has a countable base, from which (1) follows from Hu's Theorem.

Let $\{p_n : n \in \mathbb{N}\}$ be a countable dense subset. Let ρ be any compatible metric and $\forall n \in \mathbb{N}$, let $A_n = \{k \in \mathbb{N} : S_\rho(p_n, 1/k) \text{ has compact closure}\}$. Whenever E is a finite set of positive integers and $k_n \in A_n$ for each $n \in E$, $\cup_{n \in E} S_\rho(p_n, \frac{1}{k_n})$ lies in the compact bornology. Clearly, there are countable many such finite unions. It remains to show that each nonempty compact set C lies in one.

For each $c \in C$ choose $\delta > 0$ such that $\mathrm{cl}(S_\rho(c, \delta))$ is compact. Then choose $k \in \mathbb{N}$ such that $2/k < \delta$ and p_n such that $d(p_n, c) < \frac{1}{k}$. Then $c \in S_\rho(p_n, 1/k)$ and $S_\rho(p_n, 1/k)$ being a subset of $S_\rho(c, \delta)$ has compact closure. By compactness, it is clear that C is contained in a finite union of such balls, completing the proof. $\qquad\square$

Metric spaces in which the compact bornology and the metric bornology coincide are of course the boundedly compact metric spaces. Hausdorff spaces for which the compact bornology has a countable base are called *hemicompact* [155]. A hemicompact space is one in which the compact subsets have a countable subfamily that is cofinal with respect to inclusion.

If $\langle X, \tau \rangle$ is a compact metrizable space, then each compatible metric d makes X a metrically bounded set, i.e., $\forall d \in \mathbf{D}(X), \mathscr{B}_d(X)$ is trivial. On the other hand, we have this result of Beer [18]:

Theorem 15.5. *Let $\langle X, \tau \rangle$ be a noncompact metrizable space. Then there is an uncountable family of metrics $\{d_i : i \in I\}$ compatible with τ such that whenever $i \neq j$ in I, then $\mathscr{B}_{d_i}(X) \neq \mathscr{B}_{d_j}(X)$.*

Proof. Let d be a bounded compatible metric. Let $\langle x_n \rangle$ be a sequence in X with distinct terms that fails to cluster. Then any subset of the set of terms is closed. For each $n \in \mathbb{N}$ put $\varepsilon_n = \frac{1}{3} d(x_n, \{x_j : j \neq n\}) > 0$. Clearly the family of balls $\{S_d(x_n, \varepsilon_n) : n \in \mathbb{N}\}$ is pairwise disjoint. Next, let Ω denote the uncountable family of infinite subsets of $\mathbb{N}$ and consider this equivalence relation on Ω:

$$E \equiv F \text{ provided the symmetric difference } E\Delta F \text{ is finite.}$$

As each equivalence class is countable, there must be uncountably many equivalence classes. Let $\{E_i : i \in I\}$ contain exactly one representative from each equivalence class. For each $i \in I$, let $f_i : X \to \mathbb{R}$ be a continuous function mapping x_n to n if $n \in E_i$ and mapping $X \backslash \cup_{n \in E_i} S_d(x_n, \varepsilon_n)$ to zero. Now define a compatible metric d_i on X by

$$d_i(x, w) := d(x, w) + |f_i(x) - f_i(w)|.$$

Clearly $A \in \mathscr{B}_{d_i}(x)$ if and only f_i is bounded on A. Our family of metrics with distinct metric bornologies is $\{d_i : i \in I\}$. To see this, note that if $i \neq j$, then either $E_i \backslash E_j$ is infinite or $E_j \backslash E_i$ is infinite. In the first case f_i is unbounded on $\{x_n : n \in E_i \backslash E_j\}$ whereas f_j maps $\{x_n : n \in E_i \backslash E_j\}$ to $\{0\}$. In the second, f_j is unbounded on $\{x_n : n \in E_j \backslash E_i\}$ whereas f_i maps $\{x_n : n \in E_j \backslash E_i\}$ to $\{0\}$. $\qquad\square$

We now look at a more refined question. Suppose $\langle X, d \rangle$ is a metric space, and $\mathscr{B}$ is bornology on X. When does there exist a uniformly equivalent metric ρ on X such that $\mathscr{B} = \mathscr{B}_\rho(X)$? If $\mathscr{B} = \mathscr{P}(X)$, we can take $\rho = \min\{d, 1\}$. Also, any bounded uniformly continuous function serves as a uniformly continuous forcing function. So we confine our attention to nontrivial bornologies, which must have by Hu's theorem a countable base. Additionally, the bornology must be *uniformly stable under small enlargments* [18, 85].

Theorem 15.6. *Let $\langle X, d \rangle$ be a metric space and $\mathscr{B}$ be a nontrivial bornology. The following conditions are equivalent:*

(1) *there is a metric ρ uniformly equivalent to d such that $\mathscr{B}_\rho(X) = \mathscr{B}$;*

(2) *$\mathscr{B}$ has a countable base and $\exists \delta > 0 \ \forall B \in \mathscr{B}, \ S_d(B, \delta) \in \mathscr{B}$;*

(3) *$\langle X, \tau_d, \mathscr{B} \rangle$ has a d-uniformly continuous forcing function.*

Proof. (1) $\Rightarrow$ (2) That the bornology has a countable base follows from the fact that it is a metric bornology. By uniform equivalence of the metrics, choose $\delta > 0$ such that $d(x, w) < \delta \Rightarrow \rho(x, w) < 1$. Suppose $B \in \mathscr{B} = \mathscr{B}_\rho(X)$. Fixing $p \in X$, for some $\alpha > 0$ we have $B \subseteq S_\rho(p, \alpha)$. Since $S_\rho(S_\rho(p, \alpha), 1) \subseteq S_\rho(p, \alpha + 1)$, it follows that

$$S_d(B, \delta) \subseteq S_d(S_\rho(p, \alpha), \delta) \subseteq S_\rho(p, \alpha + 1) \in \mathscr{B}.$$

(2) $\Rightarrow$ (3). From condition (2), we can find a base $\{B_n : n \in \mathbb{N}\}$ for the bornology such $\forall n, S_d(B_n, \delta) \subseteq B_{n+1}$. For each $n \in \mathbb{N}$, define $f_n : X \mapsto [0, \delta]$ by $f_n(x) = \min\{d(x, B_n), \delta\}$. Clearly, f_n is nonexpansive and f_n maps B_n to zero. As before, write $f = \sum_{n=1}^\infty f_n$; for each $x \in X, f_n$ vanishes eventually on $S_d(x, \delta)$, so f is finite-valued and continuous.

For each $x \in X$, let $j(x)$ be the smallest positive integer j such that $x \in B_j$. Suppose $d(x, w) < \delta$. We intend to show that $|f(x) - f(w)| < 2d(x, w)$; that is, we will show not just that f is uniformly continuous, rather, that f is Lipschitz in the small.

Set $k = \max\{j(x), j(w)\}$; without loss of generality we may assume that $j(x) = k$. Since $d(x, w) < \delta$, it follows that $k - 1 \leq j(w) \leq k$.

We see that $f_n(x)$ and $f_n(w)$ can differ only when $n = k - 2$ or $n = k - 1$, and in the case $j(x) = j(w)$, only when $n = k - 1$. Since each f_n is nonexpansive,

$$|f(x) - f(w)| \leq |f_{k-2}(x) - f_{k-2}(w)| + |f_{k-1}(x) - f_{k-1}(w)| \leq 2d(x, w)$$

as required. Finally, if $x \in B_n$, then $f(x) \leq (n-1)\delta$, whereas if $x \notin B_n$, then $f(x) \geq (n-1)\delta$. From these facts, it follows that f is a forcing function for $\mathscr{B}$.

$(3) \Rightarrow (1)$. By uniform continuity of f, the metric defined by $\rho(x,w) = \min\{d(x,w),1\} + |f(x) - f(w)|$ is uniformly equivalent to $\min\{d,1\}$ which in turn is uniformly equivalent to d. Since f is a forcing function for $\mathscr{B}$, and E is ρ-bounded if and only if $f(E)$ is bounded, condition (1) follows. $\qquad\square$

As a first consequence of Theorem 15.6 we present another result of Hejcman [94]. The proof we give follows Garrido and Meroño [36, Corollary 5.5]. Recall that a metric space $\langle X, d\rangle$ is called B-simple if $\mathscr{B}\mathscr{B}_d(X) = \mathscr{B}_\rho(X)$ for some uniformly equivalent metric ρ.

Theorem 15.7. *Let $\langle X, d\rangle$ be a metric space. Then $\langle X, d\rangle$ is $B - simple$ if and only if X is a countable union of d-Bourbaki bounded subsets and $\mathscr{B}\mathscr{B}_d(X)$ is uniformly stable under small enlargements.*

Proof. Since a countable base for a bornology on X is a countable cover of X, the stated conditions are necessary. For sufficiency, by Theorem 15.6, we need only produce a countable base for $\mathscr{B}\mathscr{B}_d(X)$.

Let $\langle A_n\rangle$ be a sequence in $\mathscr{B}\mathscr{B}_d(X)$ such that $X = \cup_{n=1}^\infty A_n$. Without loss of generality we may assume that the sequence is increasing. Choose $\delta > 0$ such that for each Bourbaki bounded set B, $S_d(B,\delta)$ is Bourbaki bounded. Put $B_1 = A_1$ and then inductively put $B_{n+1} = S_d(B_n,\delta) \cup A_{n+1}$ for $n = 1,2,3,\ldots$. Since $\mathscr{B}\mathscr{B}_d(X)$ is stable under finite unions and under enlargements of radius δ, each B_n is Bourbaki bounded. By construction, the sequence $\langle B_n\rangle$ is increasing and whenever n and k are positive integers, we have $S_d^n(B_k,\delta) \subseteq B_{n+k}$. We intend to show that $\{B_n : n \in \mathbb{N}\}$ is a base for $\mathscr{B}\mathscr{B}_d(X)$.

Let B be an arbitrary Bourbaki bounded subset. Choose $F \in \mathscr{F}(X)$ and $n \in \mathbb{N}$ such that $B \subseteq S_d^n(F,\delta)$. Now since $\langle A_k\rangle$ is increasing, we can find $k \in \mathbb{N}$ with $F \subseteq A_k$. We compute

$$B \subseteq S_d^n(A_k,\delta) \subseteq S_d^n(B_k,\delta) \subseteq B_{n+k}$$

as required. $\qquad\square$

A general bornology $\mathscr{B}$ on $\langle X, d\rangle$ such that X is a countable union of its members and that is stable under *all* enlargements need not have a countable base and so need not be a metric bornology.

Example 15.8. Let $X = \mathbb{N} \times \mathbb{R}$, equipped with the usual metric d of the plane, and let $\mathscr{B}$ be the bornology

$$\mathscr{B} := \{B \subseteq X : \forall n \in \mathbb{N},\ (\{n\} \times \mathbb{R}) \cap B \in \mathscr{B}_d(X)\}.$$

Note that $X = \cup_{k=1}^\infty \{(n,\alpha) : n \in \mathbb{N}, -k \leq \alpha \leq k\}$ and $B \in \mathscr{B} \Rightarrow S_d(B,\alpha) \in \mathscr{B}$ for all positive α. However, this bornology fails to have a countable base. Suppose $\{B_n : n \in \mathbb{N}\}$ is any countable subfamily of $\mathscr{B}$. To show that this cannot be a base we may assume without loss of generality that each B_n contains $\{(n,0) : n \in \mathbb{N}\}$. For each $n \in \mathbb{N}$, let $A_n = S_d(B_n \cap (\{n\} \times \mathbb{R}), \frac{1}{2})$. Then $\cup_{n=1}^\infty A_n$ belongs to $\mathscr{B}$ but is a subset of no B_n.

As a second consequence of Theorem 15.6, we rephrase [85, Theorem 4.2] as to when the totally bounded subsets for some initial metric can be realized as the metrically bounded subsets with respect to some uniformly equivalent metric.

Theorem 15.9. *Let $\langle X, d \rangle$ be a metric space. Then there exists a uniformly equivalent metric ρ for which $\mathscr{T}\mathscr{B}_d(X) = \mathscr{B}_\rho(X)$ if and only if the space is separable and there exists $\delta > 0$ such that $S_d(x, \delta)$ is totally bounded for each $x \in X$.*

Proof. For sufficiency, we show that conditions of statement (2) of Theorem 15.6 are met. Let P be a countable dense subset of X. We claim that

$$\{ S_d(F, \delta) : F \text{ is a finite subset of } P \}$$

is a (countable) base for $\mathscr{T}\mathscr{B}_d(X)$. To see this, let B be a nonempty totally bounded subset and choose $\{ x_1, x_2, \ldots, x_n \} \subseteq X$ such that $B \subseteq \cup_{j=1}^n S_d(x_j, \delta/2)$. Then choosing for each $j \le n$, $p_j \in P$ with $d(p_j, x_j) < \delta/2$, we have $B \subseteq \cup_{j=1}^n S_d(p_j, \delta) \in \mathscr{T}\mathscr{B}_d(X)$ as required. It also follows from this argument that

$$S_d(B, \frac{\delta}{2}) \subseteq \cup_{j=1}^n S_d(x_j, \delta) \in \mathscr{T}\mathscr{B}_d(X),$$

that is, $\mathscr{T}\mathscr{B}_d(X)$ is stable under enlargements of size $\delta/2$.

For necessity, if the space is not separable, then X is not a countable union of totally bounded sets as such a union would form a separable subspace. Thus, $\mathscr{T}\mathscr{B}_d(X)$ could not have a countable base as the union of the base would be X. The second condition follows from the uniform stability of $\mathscr{T}\mathscr{B}_d(X)$ under small enlargements and the fact that $\mathscr{F}(X) \subseteq \mathscr{T}\mathscr{B}_d(X)$. □

It can be shown that for an unbounded metric space $\langle X, d \rangle$, there is actually an uncountable family of uniformly equivalent metrics that give rise to distinct metric bornologies. We refer the interested reader to [28, Theorem 3.5].

The following result with respect to the existence of a Cauchy equivalent metric ρ for which a given bornology $\mathscr{B}$ can be realized as $\mathscr{B}_\rho(X)$ was obtained by Aggarwal and Kundu [3]. The proof is omitted.

Theorem 15.10. *Let $\langle X, d \rangle$ be a metric space and $\mathscr{B}$ be a nontrivial bornology. The following conditions are equivalent:*

(1) *there is a metric ρ Cauchy equivalent to d such that $\mathscr{B}_\rho(X) = \mathscr{B}$;*

(2) *$\mathscr{B}$ has a countable base $\{ B_n : n = -1, 0, 1, 2, \ldots \}$ where $B_{-1} = B_0 = \emptyset$ such that for each $n \in \mathbb{N}$, $cl(B_n) \subseteq B_{n+1}$ and given any d-Cauchy sequence $\langle x_n \rangle$ in X, there exists $\{ k, n_0 \} \subseteq \mathbb{N}$ such that for all $m \ge n_0$, $x_m \in B_k \backslash cl(B_{k-2})$;*

(3) *$\langle X, \tau_d, \mathscr{B} \rangle$ has a d-Cauchy continuous forcing function.*

If $\{ \langle X_j, d_j \rangle : j = 1, 2, \ldots n \}$ is a finite family of metric spaces, then it is easy to find a forcing function for the product bornology $\mathscr{B}_{\text{prod}}$ on $\prod_{j=1}^n X_j$ determined by the n metric bornologies: fix $p_j \in X_j$ for $1 \le j \le n$ and define $f : \prod_{j=1}^n X_j \to [0, \infty)$ by $f(x) := \min\{ d_j(x_j, p_j) : j = 1, 2, \ldots, n \}$. In fact, this function is nonexpansive with respect to the box metric on the product. Thus the product bornology is a metric bornology with a metric that is uniformly equivalent to d_{box}.

The standard construction of the product metric ρ for a denumerable product of metric spaces that we outlined earlier,

$$\rho(x, w) := \sum_{n=1}^\infty 2^{-n} \min\{ 1, d_n(x_n, w_n) \},$$

is not a happy one with respect to large structure, as $\mathrm{diam}_\rho(\prod_{n\in\mathbb{N}} X_n) \leq 1$, so that the metric bornologies for the factors is completely ignored with respect to the trivial metric bornology which ensues for the product.

It turns out that the product bornology $\mathscr{B}_{\mathrm{prod}}$ corresponding to a denumerable family of metric bornologies is always a metric bornology. Suppose our family of metric spaces is $\{\langle X_n, d_n\rangle : n \in \mathbb{N}\}$. If at least one of the factor bornologies is trivial, i.e., for some n, $\mathrm{diam}_{d_n}(X_n) < \infty$, then the product bornology is as well, and the standard metric ρ for the product yields the trivial bornology as its metric bornology. That the product bornology is a metric bornology when all of the metrics are unbounded, i.e., $\forall n \in \mathbb{N}$, $\mathrm{diam}_{d_n}(X_n) = \infty$, is a consequence of Hu's theorem.

For each $n \in \mathbb{N}$ fix $p_n \in X_n$, and for E a nonempty finite subset of $\mathbb{N}$ and for each $k \in \mathbb{N}$, put $C(E, k) := \cup_{n\in E} \pi_n^{-1}(S_{d_n}(p_n, k))$. Evidently, the family of all such $C(E, k)$ is a base for the product bornology, and since the product bornology has an open base and a closed base, it satisfies the conditions of Theorem 15.1 and is thus a metric bornology. Choosing a metric d for which $\mathscr{B}_d(\prod_{n\in\mathbb{N}} X_n)$ agrees with the product bornology, we see that a subset A of the product is d-unbounded if and only if for each finite subset E of $\mathbb{N}$ and $k \in \mathbb{N}$, there exists $a \in A$ such that $\forall n \in E$, $d_n(a_n, p_n) \geq k$. To summarize, we may state this attractive result [21].

Proposition 15.11. *Let $\{\langle X_n, d_n\rangle : n \in \mathbb{N}\}$ be a denumerable family of unbounded metric spaces. Then the product bornology on $\prod_{n\in\mathbb{N}} X_n$ as determined by the metric bornologies for the factors is a nontrivial metric bornology for some metric compatible with the product topology. A net $\langle x_\lambda\rangle$ is convergent to ∞ with respect to the product bornology if and only if for each n, we have $\langle \pi_n(x_\lambda)\rangle$ convergent to ∞ with respect to $\mathscr{B}_{d_n}(X_n)$.*

The box bornology for a finite product of metric spaces is induced by the box metric. As to the realization of the box bornology as a metric bornology for a denumerable product where the metric is compatible with the product topology, we have this result.

Proposition 15.12. *Let $\{\langle X_n, d_n\rangle : n \in \mathbb{N}\}$ be a denumerable family of metric spaces. Then the box bornology as determined by $\{\mathscr{B}_{d_n}(X_n) : n \in \mathbb{N}\}$ is a metric bornology for some metric compatible with the product topology if and only if $\{n : \mathrm{diam}_{d_n}(X_n) = \infty\}$ is finite.*

Proof. Suppose $\mathrm{diam}_{d_k}(X_k) = \infty$ for $k \in E$ where E is infinite. Let us write $E = \{k_j : j \in \mathbb{N}\}$ where $\forall j$, $k_j < k_{j+1}$. We claim that $\mathscr{B}_{\mathrm{box}}$ cannot have a countable base. To see this suppose $\{A_n : n \in \mathbb{N}\}$ is a countable subfamily of $\mathscr{B}_{\mathrm{box}}$. For each $k_n \in E$, pick $x_{k_n} \in X_{k_n}\backslash\pi_{k_n}(A_n)$. For each $k_n \in E$, put $B_{k_n} = \{x_{k_n}\}$ and for $n \in \mathbb{N}\backslash E$, put $B_n = \{w_n\}$ where $w_n \in X_n$ is a fixed point. Then $\prod_{n\in\mathbb{N}} B_n \in \mathscr{B}_{\mathrm{box}}$ but by construction, it fails to be a subset of any A_n. Thus, the box bornology cannot be a metric bornology for any metric, compatible with the product topology or not.

The proof of the converse is left to the reader (show $\mathscr{B}_{\mathrm{box}}$ satisfies the conditions of Hu's Theorem). $\qquad\square$

16. Bornologies of Totally Bounded Sets

Bornologies of totally bounded subsets have been studied by Beer, Costantini and Levi [28]. We first give a rather transparent but nevertheless useful answer to this question: when is a bornology on a metrizable space a bornology of totally bounded sets with respect to some compatible metric?

Theorem 16.1. *Let $\mathscr{B}$ be a bornology on a metrizable space $\langle X, \tau \rangle$. Then $\mathscr{B} = \mathscr{T}\mathscr{B}_d(X)$ for some compatible metric d if and only if there exists a topological embedding ϕ of X into a completely metrizable space Y such that*

$$\mathscr{B} = \{E \subseteq X : cl(\phi(E)) \text{ is a compact subset of } Y\}.$$

Proof. For necessity, suppose $\mathscr{B} = \mathscr{T}\mathscr{B}_d(X)$. Let ϕ be an isometric embedding of $\langle X, d \rangle$ into a complete metric space $\langle Y, \rho \rangle$ (such as the bounded continuous real functions on X equipped with the supremum norm). Then $\forall B \in \mathscr{B}, \phi(B)$ is ρ-totally bounded because totally boundedness is an intrinsic property of sets within a metric space and so $\phi(B)$ has compact closure by the completeness of ρ. For the reverse inclusion, if $\phi(E)$ has compact closure, then $\phi(E)$ is ρ-totally bounded so that E is d-totally bounded because ϕ is an isometry.

For sufficiency, suppose such a topological embedding ϕ exists, and let ρ be a compatible complete metric for Y. Define a metric d for X by $d(x, w) = \rho(\phi(x), \phi(w))$. Suppose $B \in \mathscr{B}$. Since $cl(\phi(B))$ is compact, $\phi(B)$ is ρ-totally bounded, so clearly B is d-totally bounded. Conversely, if E is d-totally bounded, then $\phi(E)$ is ρ-totally bounded and thus has compact closure which ensures $E \in \mathscr{B}$. $\square$

As an application, we answer the following question: given a metric space, when can we convert the metrically bounded sets to the metrically totally bounded sets with respect to an equivalent metric? The answer does not depend on the initial metric chosen, rather on the topology of the space.

Theorem 16.2. *Let $\langle X, d \rangle$ be a metric space. Then there exists an equivalent metric ρ such that $\mathscr{B}_d(X) = \mathscr{T}\mathscr{B}_\rho(X)$ if and only if the topology of X is separable.*

Proof. Separability is necessary because if each d-bounded set is ρ-totally bounded, then for any fixed $p \in X$, $X = \cup_{n=1}^{\infty} S_d(p, n)$, which expresses X as a countable union of separable subspaces.

For the converse, we consider two cases. First, suppose $diam_d(X) < \infty$, i.e., $\mathscr{B}_d(X)$ is trivial. Since a separable metrizable space is second countable and completely regular, by the Urysohn embedding theorem [155, p. 166], we can topologically embed X into the compact metrizable space $[0, 1]^{\mathbb{N}}$. Call this embedding ψ. If ρ is any compatible metric, then each subset of the cube is ρ-totally bounded; in particular, each subset of $\psi(X)$ is ρ-totally bounded. The restriction of ρ to $\psi(X) \times \psi(X)$ is the desired metric, under the identification $x \leftrightarrow \psi(x)$.

We are left with the case that X is not d-bounded. Let $\{x_i : i \in \mathbb{N}\}$ be a countable dense subset. Let $f_i : X \to \mathbb{R}$ be $d(\cdot, x_i)$, i.e., $f_i(x) = d(x, x_i)$. Clearly, $\{f_i : i \in \mathbb{N}\}$ separates points from closed sets, for if $x \notin C$ where C is a nonempty closed set, then choosing x_i with $d(x, x_i) < d(x_i, C)$, we have $f_i(x) \notin \mathrm{cl}(f_i(C))$. As a result, $\phi : X \to \mathbb{R}^{\mathbb{N}}$ defined by $\pi_i(\phi(x)) = f_i(x)$ is a topological embedding [155, p. 57]. Now $\mathbb{R}^{\mathbb{N}}$ equipped with the product topology is completely metrizable, so it remains to show that $E \in \mathscr{B}_d(X)$ if and only if $\mathrm{cl}(\phi(E))$ is compact.

To this end, suppose E is d-bounded. Then for each $i \in \mathbb{N}$, $f_i(E)$ lies in some closed interval $[\alpha_i, \beta_i]$. Since $\prod_{i \in \mathbb{N}}[\alpha_i, \beta_i]$ is compact with the relative topology and $\phi(E) \subseteq \prod_{i \in \mathbb{N}} f_i(E)$, $\phi(E)$ has compact closure.

Conversely, suppose E is not d-bounded, yet $\mathrm{cl}(\phi(E))$ has compact closure. Let $\langle e_n \rangle$ be a sequence in E convergent to ∞ with respect to $\mathscr{B}_d(X)$. By relative compactness of $\phi(E)$, by passing to a subsequence we may assume $\langle \phi(e_n) \rangle$ is convergent in the product topology to some point y of $\mathrm{cl}(\phi(E))$. But by the definition of the product topology, this means that for each $i \in \mathbb{N}$,

$$\lim_{n \to \infty} d(e_n, x_i) = \lim_{n \to \infty} \pi_i(\phi(e_n)) = \pi_i(y).$$

But this is impossible as $\langle e_n \rangle$ has been chosen so that for each i, $\lim_{n \to \infty} d(e_n, x_i) = \infty$. □

We next come to a more sophisticated characterization of bornologies that are bornologies of totally bounded sets as established in [28, Theorem 4.5].

Theorem 16.3. *Let $\langle X, \tau \rangle$ be a metrizable space and let $\mathscr{B}$ be a family of nonempty subsets of X. Then $\mathscr{B} = \mathscr{T}\mathscr{B}_d(X)$ for some compatible metric d if and only if there is a star-development $\langle \mathscr{U}_n \rangle$ for X such that*

$$(*) \qquad \mathscr{B} = \{E \in \mathscr{P}(X) : \forall n \in \mathbb{N}, \ \mathscr{U}_n \text{ admits a finite subcover of } E\}.$$

Proof. If $\mathscr{B} = \mathscr{T}\mathscr{B}_d(X)$ for some compatible metric d, let $\mathscr{U}_n = \{S_d(x, \frac{1}{3^n}) : x \in X\}$ for every $n \in \mathbb{N}$. Evidently, $\langle \mathscr{U}_n \rangle$ is a star-development of X. If E is d-totally bounded, then for each n, E can be covered by finitely many d-balls of radius $\frac{1}{3^n}$, i.e., by finitely many elements of $\mathscr{U}_n$, and if for each n, E can be covered by finitely many members of $\mathscr{U}_n$, for each $\varepsilon > 0$, E can be covered by finitely many balls of radius ε.

For sufficiency, as we noted earlier, the existence of a star-development $\langle \mathscr{U}_n \rangle$ allows for the construction of a compatible metric d such that $\forall n \geq 2$, both of the following conditions hold [155, pg. 167]:

(i) $\mathscr{U}_n$ refines $\{S_d(x, \frac{1}{2^{n-1}}) : x \in X\}$;

(ii) $\{S_d(x, \frac{1}{2^n}) : x \in X\}$ refines $\mathscr{U}_{n-1}$.

Suppose $\mathscr{B}$ satisfies $(*)$. Fix $E \in \mathscr{B}$, let $\varepsilon > 0$ be arbitrary and choose n with $2^{1-n} < \varepsilon$. By $(*)$ choose $\{U_1, U_2, \ldots, U_k\} \subseteq \mathscr{U}_n$ with $E \subseteq \cup_{j=1}^k U_j$ and then by (i) choose $\forall j \leq k$, x_j with $U_j \subseteq S_d(x_j, \frac{1}{2^{n-1}})$. This yields $E \subseteq S_d(\{x_1, x_2, \ldots, x_k\}, \varepsilon)$ and so $E \in \mathscr{T}\mathscr{B}_d(X)$. On the other hand, if E is d-totally bounded and $n \in \mathbb{N}$ is arbitrary, we can choose a finite subset F of X with $E \subseteq S_d(F, \frac{1}{2^{n+1}})$. By condition (ii), this gives a finite subcover of E from $\mathscr{U}_n$. □

As we have mentioned, a metrizable space X is completely metrizable if and only if it sits as a countable intersection of open subsets in each metric space in which it can be topologically embedded. The proof of our last theorem shows that if $\langle \mathscr{U}_n \rangle$ is a star-development for X, then there is a compatible metric d such that $\{E \in \mathscr{P}(X) : \forall n \in \mathbb{N}, \ \mathscr{U}_n \text{ admits a finite subcover of } E\}$

describes the d-totally bounded subsets. With statement (7) of Proposition 7.1 in mind, we obtain this additional characterization of those metrizable spaces that are completely metrizable.

Corollary 16.4. *Let $\langle X, \tau \rangle$ be a metrizable space. Then $\langle X, \tau \rangle$ is completely metrizable if and only if there exists a star-development $\langle \mathscr{U}_n \rangle$ for X such that a subset A of X is relatively compact if and only if $\forall n \in \mathbb{N}$, $\mathscr{U}_n$ admits a finite subcover of A.*

Roughly following [28], we close this section by finding necessary and sufficient conditions for the totally bounded subsets as determined by $\langle X, d \rangle$ to agree with $\mathscr{B}_\rho(X)$ for some equivalent metric without directly appealing to Hu's theorem. In the case of bornologies of totally bounded sets as with bornologies of relatively compact sets, the condition that the bornology have an open base is redundant, as we intend to show. First, we give a lemma that we will use in our proof.

Lemma 16.5. *Let $\langle X, d \rangle$ be a metric space, let B be a totally bounded subset of X, and let E be a dense subset of X. Then there exists a totally bounded (countable) subset E_0 of E with $B \subseteq cl(E_0)$.*

Proof. Let C be the closure of B in the completion $\langle \hat{X}, \hat{d} \rangle$ of $\langle X, d \rangle$. Since E is also dense in $\langle \hat{X}, \hat{d} \rangle$, we can choose for each $n \in \mathbb{N}$ an irreducible finite subset E_n of E with $C \subseteq S_{\hat{d}}(E_n, \frac{1}{n})$. This means that reciprocally, $E_n \subseteq S_{\hat{d}}(C, \frac{1}{n})$. We intend to show that $\bigcup_{n=1}^\infty E_n$ is totally bounded in the completion.

To see this, let $\varepsilon > 0$ be arbitrary and choose $k \in \mathbb{N}$ with $\frac{1}{k} < \varepsilon$. Since C is compact, $S_{\hat{d}}(E_k, \frac{1}{k})$ contains some enlargement $S_{\hat{d}}(C, \delta)$ with $\delta > 0$. By irreducibility, whenever $n > \delta^{-1}$, we have $E_n \subseteq S_{\hat{d}}(C, \delta)$, and as a result, choosing $n_0 > \delta^{-1}$, we have

$$E_0 := \bigcup_{n=1}^\infty E_n \subseteq S_{\hat{d}}(E_k, \varepsilon) \cup S_{\hat{d}}\big(\bigcup_{n=1}^{n_0} E_n, \varepsilon\big).$$

This proves that $E_0 \in \mathscr{T}\mathscr{B}_{\hat{d}}(\hat{X})$, and thus E_0 is d-totally bounded. Our construction gives $B \subseteq C \subseteq cl_{\hat{X}}(E_0)$, and so $B \subseteq cl(E_0)$. $\square$

Theorem 16.6. *Let $\langle X, d \rangle$ be a metric space. The following conditions are equivalent:*

(1) *$\mathscr{T}\mathscr{B}_d(X)$ has a countable base;*

(2) *there exists a metric ρ equivalent to d for which $\mathscr{T}\mathscr{B}_d(X) = \mathscr{B}_\rho(X) = \mathscr{T}\mathscr{B}_\rho(X)$;*

(3) *there exists a metric ρ equivalent to d for which $\mathscr{T}\mathscr{B}_d(X) = \mathscr{B}_\rho(X)$;*

(4) *the one-point extension of X associated with $\mathscr{T}\mathscr{B}_d(X)$ is metrizable;*

(5) *the one-point extension of X associated with $\mathscr{T}\mathscr{B}_d(X)$ is first countable.*

Proof. The implications (2) $\Rightarrow$ (3) and (4) $\Rightarrow$ (5) are obvious, and as the one-point extension corresponding to a metric bornology is always metrizable we get (3) $\Rightarrow$ (4). Also conditions (1) and (5) are equivalent by Theorem 14.1. It remains to establish (1) $\Rightarrow$ (2).

Let $\langle \hat{X}, \hat{d} \rangle$ be the completion of $\langle X, d \rangle$ and let $\{B_n : n \in \mathbb{N}\}$ be cofinal in $\mathscr{T}\mathscr{B}_d(X)$ with respect to inclusion. We claim that the family of compact sets $\{cl_{\hat{X}}(B_n) : n \in \mathbb{N}\}$ is cofinal in $\mathscr{K}(\hat{X})$. To see this, let K be an arbitrary compact subset of $\langle \hat{X}, \hat{d} \rangle$. By the last lemma and the density of X in $\hat{X}$, there exists $E_0 \subseteq X$ with $E_0 \in \mathscr{T}\mathscr{B}_{\hat{d}}(\hat{X})$ and $K \subseteq \mathrm{cl}_{\hat{X}}(E_0)$. Clearly $E_0 \in \mathscr{T}\mathscr{B}_d(X)$, so for some n, $E_0 \subseteq B_n$, and the claim is established.

By Theorem 15.4, there is an equivalent boundedly compact metric $\hat{\rho}$ for $\hat{X}$, and if ρ is its trace on $X \times X$, then it is clear that $\langle \hat{X}, \hat{\rho} \rangle$ is the completion of $\langle X, \rho \rangle$. Since the completions for the two metrics determine the same relatively compact subsets, Proposition 11.5 yields $\mathscr{T}\mathscr{B}_d(X) = \mathscr{T}\mathscr{B}_\rho(X)$. Also, because the completion $\langle \hat{X}, \hat{\rho} \rangle$ is boundedly compact, Proposition 7.1 gives $\mathscr{T}\mathscr{B}_\rho(X) = \mathscr{B}_\rho(X)$. □

Let $\langle X, d \rangle$ be a metric space. We note without proof that if we can find an equivalent metric ρ such that $\mathscr{B}_d(X) = \mathscr{T}\mathscr{B}_\rho(X)$, we can choose ρ for which $\mathscr{B}_\rho(X) = \mathscr{T}\mathscr{B}_\rho(X)$ also holds (see Theorem 16.2 above).

17. Strong Uniform Continuity

Let $\langle X, d \rangle$ and $\langle Y, \rho \rangle$ be metric spaces. If $f \in C(X, Y)$ and K is a nonempty compact subset of X, then $f|_K$ is uniformly continuous. But the standard proof shows something more is true: $\forall \varepsilon > 0 \; \exists \delta > 0$ such that if $d(x, w) < \delta$ and $\{x, w\} \cap K \neq \emptyset$, then $\rho(f(x), f(w)) < \varepsilon$. To see this, suppose to the contrary that $\forall n \in \mathbb{N} \; \exists x_n, w_n$ such that $d(x_n, w_n) < \frac{1}{n}$, $\{x_n, w_n\} \cap K \neq \emptyset$ but $\rho(f(x_n), f(w_n)) \geq \varepsilon$. Without loss of generality, we may assume each x_n lies in K, and by compactness $\langle x_n \rangle$ has a cluster point $p \in K$. But p is also a cluster point of $\langle w_n \rangle$ and so continuity of f at p fails. We record this stronger type of uniform continuity in a definition, as introduced by Beer and Levi [45, 46].

Definition 17.1. Let $\langle X, d \rangle$ and $\langle Y, \rho \rangle$ be metric spaces and let B be a subset of X. We say that a function $f : X \to Y$ is *strongly uniformly continuous on B* if $\forall \varepsilon > 0 \; \exists \delta > 0$ such that if $d(x, w) < \delta$ and $\{x, w\} \cap B \neq \emptyset$, then $\rho(f(x), f(w)) < \varepsilon$.

In the definition it is not required that f be continuous. In fact, pointwise continuity may be formulated in terms of this definition as described in the next obvious proposition.

Proposition 17.2. *Let $\langle X, d \rangle$ and $\langle Y, \rho \rangle$ be metric spaces and let $f : X \to Y$. Then f is continuous at $x \in X$ if and only if f is strongly uniformly continuous on $\{x\}$.*

At the other extreme, strong uniform continuity of f on X is nothing more than global uniform continuity.

If $\mathscr{B}$ is family of nonempty subsets of X, a function $f : \langle X, d \rangle \to \langle Y, \rho \rangle$ is called *uniformly continuous* (resp. *strongly uniformly continuous*) *on $\mathscr{B}$* if for each $B \in \mathscr{B}$, $f|_B$ is uniformly continuous (resp. f is strongly uniformly continuous on B). Looking at the bornology $\mathscr{F}(X)$ of finite sets, we see that strong uniform continuity on $\mathscr{F}(X)$, i.e., global continuity, is a stronger requirement than uniform continuity on $\mathscr{F}(X)$, which amounts to no requirement at all.

We now write for $f \in Y^X$,

$$\mathscr{B}^f := \{E \subseteq X : f|_E \text{ is uniformly continuous}\},$$

$$\mathscr{B}_f := \{E \subseteq X : f \text{ is strongly uniformly continuous on } E\}.$$

The following facts are obvious for a general function $f : X \to Y$:

(1) $\mathscr{B}_f \subseteq \mathscr{B}^f$;

(2) $\mathscr{B}^f$ and $\mathscr{B}_f$ are both hereditary families;

(3) $\mathscr{F}(X) \subseteq \mathscr{B}^f$;

(4) $\sum(\mathscr{B}_f) = \mathscr{B}_f$;

(5) if $f \in C(X,Y)$, then $A \in \mathscr{B}^f \Rightarrow \mathrm{cl}(A) \in \mathscr{B}^f$;

(6) if $A \in \mathscr{P}_0(X)$ and $\exists \delta > 0$ such that $S_d(A, \delta) \in \mathscr{B}^f$, then $A \in \mathscr{B}_f$.

The next example shows at once that the inclusion $\mathscr{B}_f \subseteq \mathscr{B}^f$ for a continuous function can be proper, and that $\mathscr{B}^f$ need not be stable under finite unions.

Example 17.3. In the line $\mathbb{R}$ equipped with the usual metric, let $A = \mathbb{N}$ and let $B = \{n + \frac{1}{n+1} : n \in \mathbb{N}\}$. Since A and B are disjoint closed sets, we can find $f \in C(\mathbb{R}, [0,1])$ with $f(A) = \{0\}$ and $f(B) = \{1\}$. While f is uniformly continuous on A, it is not strongly uniformly continuous on A; further while f is uniformly continuous on A and on B separately, it is not uniformly continuous on $A \cup B$.

Proposition 17.4. *Let $\langle X, d \rangle$ and $\langle Y, \rho \rangle$ be metric spaces and let $f \in Y^X$. The following conditions are equivalent:*

(1) *$f \in C(X, Y)$;*

(2) *$\mathscr{B}_f$ is a bornology;*

(3) *$\mathscr{K}(X) \subseteq \mathscr{B}_f$;*

(4) *$\mathscr{K}(X) \subseteq \mathscr{B}^f$.*

Proof. Global continuity of f means that $\forall x \in X, \{x\} \in \mathscr{B}_f$. As we have noted, even without continuity, $\mathscr{B}_f$ is hereditary and is stable under finite unions. Thus, we have (1) $\Leftrightarrow$ (2). The implication (1) $\Rightarrow$ (3) was argued at the beginning of this discussion, and (3) $\Rightarrow$ (4) is trivial.

For (4) $\Rightarrow$ (1), we show that the sequential criterion for pointwise continuity is met. Fix $p \in X$ and suppose $\lim_{n \to \infty} x_n = p$. As $\{p\} \cup \{x_n : n \in \mathbb{N}\}$ is compact, f so restricted is (uniformly) continuous, and so $\lim_{n \to \infty} f(x_n) = f(p)$. $\qquad \square$

It turns out that the compact bornology is not in general the largest bornology on which each each continuous function is strongly uniformly continuous. For example any function will be strongly uniformly continuous restricted to any uniformly isolated subset A of $\langle X, d \rangle$. We return to this point in the next section.

As to when uniform continuity of a continuous function on each member of a family of subsets $\mathscr{A}$ implies strong uniform continuity on each member of $\mathscr{A}$, we offer this sufficient (but not necessary) condition [46].

Theorem 17.5. *Let $\mathscr{A}$ be a family of subsets of $\langle X, d \rangle$ that is shielded from closed sets. Let $f \in C(X, Y)$ where $\langle Y, \rho \rangle$ is an arbitrary metric space. Then $\mathscr{A} \subseteq \mathscr{B}^f \Rightarrow \mathscr{A} \subseteq \mathscr{B}_f$.*

Proof. Suppose $\mathscr{A} \subseteq \mathscr{B}^f$, yet f fails to be strongly uniformly continuous on some $A_0 \in \mathscr{A}$. Then for some $\varepsilon > 0$ and for each $n \in \mathbb{N}$ there exist $x_n \in A_0$ and $w_n \in X$ with $d(x_n, w_n) < \frac{1}{n}$ but $\rho(f(x_n), f(w_n)) \geq \varepsilon$. Note that $\langle w_n \rangle$ cannot have a cluster point by continuity of f. Let $A_1 \in \mathscr{A}$ be a shield for A_0. Since already $\{x_n : n \in \mathbb{N}\} \subseteq A_1$, by the uniform continuity of $f|_{A_1}$, we must have eventually $w_n \notin A_1$. As a result, for some $k \in \mathbb{N}$, $\{w_n : n \geq k\}$ is a closed subset of X disjoint from A_1 satisfying $D_d(\{w_n : n \geq k\}, A_0) = 0$. This contradicts A_1 serving as a shield for A_0. $\qquad \square$

The assumption that f be continuous in our last result cannot be dropped. For example, given any function $f : \mathbb{R} \to \mathbb{R}$, then f is uniformly continuous restricted to each finite subset but is strongly uniformly continuous on each finite subset if and only if f is globally continuous.

Turning things around, given a bornology $\mathscr{B}$ on $\langle X, d \rangle$ and a second metric space $\langle Y, \rho \rangle$, we write $C^s_{\mathscr{B}}(X, Y)$ for the family of (continuous) functions that are strongly uniformly continuous on $\mathscr{B}$.

We will offer multiple alternate descriptions of strong uniform continuity on a set that the reader might prefer. We first recall the classical notion of function oscillation.

Let $f \in Y^X$. For each $n \in \mathbb{N}$, define $\omega_n(f, \cdot) : X \to [0, \infty]$ by $\omega_n(f, x) = \operatorname{diam}_\rho(f(S_d(x, \frac{1}{n})))$. Note $\forall n$, $\omega_n(f, \cdot) \geq \omega_{n+1}(f, \cdot)$. In terms of this sequence of auxiliary functions, the *oscillation function* $\omega(f, \cdot)$ for f is defined by $\omega(f, x) := \inf_n \omega_n(f, x) = \lim_{n \to \infty} \omega_n(f, x)$.

Oscillation $\omega(f, \cdot)$ as we have defined it is a nonnegative extended real-valued function on X.

Example 17.6. Define $f : \mathbb{R} \to \mathbb{R}$ by

$$f(x) = \begin{cases} \frac{1}{x} & \text{if } x > 0 \\ 1 & \text{if } x \leq 0 \text{ and } x \text{ is rational} \\ 0 & \text{if } x \leq 0 \text{ and } x \text{ is irrational.} \end{cases}$$

We easily compute

$$\omega(f, x) = \begin{cases} 0 & \text{if } x > 0 \\ \infty & \text{if } x = 0 \\ 1 & \text{if } x < 0. \end{cases}$$

The proof of the following classical result is left to the reader.

Proposition 17.7. *Let $\langle X, d \rangle$ and $\langle Y, \rho \rangle$ be metric spaces and let $f \in Y^X$. Then*

(1) *f is continuous at $p \in X$ if and only if $\omega(f, p) = 0$;*

(2) *$x \mapsto \omega(f, x)$ is upper semicontinuous.*

We now extend the definition of function oscillation from points to nonempty subsets. For $f \in Y^X$, $C \in \mathscr{P}_0(X)$, and $n \in \mathbb{N}$, put

$$\Omega_n(f, C) := \sup\{\rho(f(x), f(w)) : \{x, w\} \subseteq S_d(C, \frac{1}{n}) \text{ and } d(x, w) < \frac{1}{n}\}.$$

Note that $\forall n$, $\Omega_{n+1}(f, C) \leq \Omega_n(f, C)$. From this, we define *the oscillation of f at C* by

$$\Omega(f, C) := \inf_n \Omega_n(f, C) = \lim_{n \to \infty} \Omega_n(f, C).$$

Why does this extend the classical definition? Since $\forall n \in \mathbb{N} \ \forall x \in X$,

$$\omega_{2n}(f, x) \leq \Omega_n(f, \{x\}) \leq \omega_n(f, x),$$

we get exactly what we want: $\Omega(f, \{x\}) = \omega(f, x)$.

Beer and Cao [26, Proposition 3.1] have given the following presentation of the oscillation of a function between metric spaces at a nonempty subset of the domain.

Proposition 17.8. *Let $\langle X, d \rangle$ and $\langle Y, \rho \rangle$ be metric spaces and let $f \in Y^X$. Then for each $A \in \mathscr{P}_0(X)$, we have*

$$\Omega(f, A) = \inf_{n \in \mathbb{N}} \sup_{a \in A} \omega_n(f, a).$$

Proof. Put

$$\Omega_n^*(f, A) := \sup_{a \in A} \omega_n(f, a)$$

and then put

$$\Omega^*(f, A) := \inf_{n \in \mathbb{N}} \Omega_n^*(f, A).$$

We must show that

(a) $\Omega(f, A) \leq \Omega^*(f, A)$; and

(b) $\Omega^*(f, A) \leq \Omega(f, A)$.

In (a) we may assume that $\Omega^*(f, A)$ is finite, and in (b), we may assume that $\Omega(f, A)$ is finite. For (a), suppose $\Omega^*(f, A) < \alpha < \infty$ and choose $n \in \mathbb{N}$ such that $\Omega_n^*(f, A) < \alpha$. We claim that $\Omega_{2n}(f, A) < \alpha$. Let x and w be arbitrary members of $S_d\left(A, \frac{1}{2n}\right)$ with $d(x, w) < \frac{1}{2n}$. Choosing $a \in A$ with $d(x, a) < \frac{1}{2n}$, we have $\{x, w\} \subseteq S_d\left(a, \frac{1}{n}\right)$, and so

$$\rho(f(x), f(w)) \leq \operatorname{diam}_\rho f\left(S_d\left(a, \frac{1}{n}\right)\right) \leq \Omega_n^*(f, A),$$

so that

$$\Omega_{2n}(f, A) \leq \Omega_n^*(f, A) < \alpha,$$

which establishes the claim. This yields that $\Omega(f, A) \leq \Omega^*(f, A)$.

For (b), let α satisfy $\Omega(f, A) < \alpha < \infty$ and then choose $n \in \mathbb{N}$ such that $\Omega_n(f, A) < \alpha$. Let $a \in A$ be arbitrary and choose x, w in $S_d\left(a, \frac{1}{2n}\right)$. The triangle inequality gives

$$d(x, w) < 2 \cdot \frac{1}{2n} = \frac{1}{n},$$

and since $\{x, w\} \subseteq S_d\left(A, \frac{1}{n}\right)$, we have $\rho(f(x), f(w)) \leq \Omega_n(f, A)$. This yields $\omega_{2n}(f, a) \leq \Omega_n(f, A)$ and so

$$\Omega^*(f, A) \leq \Omega_{2n}^*(f, A) \leq \Omega_n(f, A) < \alpha,$$

from which $\Omega^*(f, A) \leq \Omega(f, A)$ follows. $\square$

We list some easily verified properties of $\Omega(f, \cdot)$ as applied to nonempty subsets of X:

- $\Omega(f, C) = \Omega(f, \operatorname{cl}(C))$;

- $A \subseteq C \Rightarrow \Omega(f, A) \leq \Omega(f, C)$;

- $\Omega(f, A \cup C) = \max\{\Omega(f, A), \Omega(f, C)\}$ because $\Omega_{2n}(f, A \cup C) \leq \max\{\Omega_n(f, A), \Omega_n(f, C)\}$;

- $\Omega(f, X) = 0$ if and only if f is globally uniformly continuous;

- $A \mapsto \Omega(f, A)$ is upper semicontinuous on $\mathscr{P}_0(X)$ equipped with either the topology of Hausdorff distance or the Vietoris topology.

Example 17.9. Oscillation $\Omega(f,\cdot)$ as determined by $f \in C(X,Y)$ need not be continuous with respect to Hausdorff distance, even if we restrict our attention to $\mathscr{C}_0(X)$. With $A = \{(0,y) : y \geq 0\}$ and $A_n = \{(1/n, k/n) : k \in \mathbb{N}\}$ for each positive integer n, let $X = A \cup (\cup_{n=1}^{\infty} A_n)$ equipped with the usual metric d of the plane. Let $f : X \to \mathbb{R}$ be given by $f(x,y) = xy$. Clearly $\lim_{n\to\infty} H_d(A, A_n) = 0$ while $\Omega(f, A) = \infty$ and $\Omega(f, A_n) = 0$ for each $n \in \mathbb{N}$.

Working in the same space with the same f, we can easily find a sequence of closed subsets $\langle B_n \rangle$ convergent to A in the Vietoris topology such that for each $n \in \mathbb{N}$, $\Omega(f, B_n) = 0$: put $B_n = \{(0, k/n) : k \leq n^2 \text{ and } k \in \mathbb{N}\}$.

In the proof of the Theorem 17.10 below [45, Theorem 3.1], we will use the important Efremovic lemma Theorem 4.8.

Theorem 17.10. *Let $\langle X, d \rangle$ and $\langle Y, \rho \rangle$ be metric spaces and let $f \in Y^X$. The following conditions are equivalent for a nonempty subset A of X:*

(1) $A \in \mathscr{B}_f$;

(2) $\forall \varepsilon > 0 \; \exists \delta > 0$ *such that whenever* $\{x_1, x_2\} \subseteq S_d(A, \delta)$ *and* $d(x_1, x_2) < \delta$ *then* $\rho(f(x_1), f(x_2)) < \varepsilon$;

(3) $\Omega(f, A) = 0$;

(4) $\langle \omega_n(f, \cdot) \rangle$ *converges uniformly to 0 on A;*

(5) *whenever* $B \in \mathscr{P}_0(A), C \in \mathscr{P}_0(X)$ *and* $D_d(B, C) = 0$, *then* $D_\rho(f(B), f(C)) = 0$.

Proof. (1) $\Rightarrow$ (2). Suppose $A \in \mathscr{B}_f$ and $\varepsilon > 0$. Choose $\lambda > 0$ such that if $a \in A$ and $d(x, a) < \lambda$, then $\rho(f(x), f(a)) < \varepsilon/3$. Now set $\delta = \frac{\lambda}{3}$, and suppose $\{x_1, x_2\} \subseteq S_d(A, \delta)$ with $d(x_1, x_2) < \delta$. We can find $\{a_1, a_2\} \subseteq A$ such that $d(a_i, x_i) < \delta$ $(i = 1, 2)$. Then $d(a_1, a_2) < \lambda$, and it follows from the triangle inequality that $\rho(f(x_1), f(x_2)) < \varepsilon$.

(2) $\Rightarrow$ (3). This is obvious from the definition of $\Omega(f, A)$.

(3) $\Rightarrow$ (4). The implication is immediate from Proposition 17.8.

(4) $\Rightarrow$ (5). Let B be a nonempty subset of A and suppose $C \neq \emptyset$ satisfies $D_d(B, C) = 0$. Let $\varepsilon > 0$ and choose by condition (4) $n \in \mathbb{N}$ such that $\sup \{\omega_n(f, a) : a \in A\} < \varepsilon$. Pick $b \in B$ and $c \in C$ such that $d(b, c) < \frac{1}{n}$. Since $\omega_n(f, b) < \varepsilon$, we obtain

$$D_\rho(f(B), f(C)) \leq \rho(f(b), f(c)) < \varepsilon$$

and it follows that $D_\rho(f(B), f(C)) = 0$ as required.

(5) $\Rightarrow$ (1). Suppose condition (1) fails. Then for some positive ε we can pick for each n, $a_n \in A$ and $x_n \in X$ with $d(a_n, x_n) < \frac{1}{n}$ yet $\rho(f(a_n), f(x_n)) > \varepsilon$. Choose by the Efremovic lemma an infinite subset $\mathbb{N}_1$ of $\mathbb{N}$ such that $\{f(a_n) : n \in \mathbb{N}_1\}$ and $\{f(x_n) : n \in \mathbb{N}_1\}$ are not ρ-near. Then condition (4) fails with $B = \{a_n : n \in \mathbb{N}_1\}$ and $C = \{x_n : n \in \mathbb{N}_1\}$. $\square$

Corollary 17.11. *Let $f \in Y^X$. Then $A \in \mathscr{B}_f \Rightarrow cl(A) \in \mathscr{B}_f$.*

Proof. This is immediate from condition (2) of the last theorem as $S_d(A, \delta) = S_d(cl(A), \delta)$. $\square$

The next corollary is perhaps the best-known consequence of the Efremovic Lemma [129, p. 35].

Corollary 17.12. *Let $\langle X, d \rangle$ and $\langle Y, \rho \rangle$ be metric spaces. Suppose $f \in Y^X$. Then f is uniformly continuous if and only if f preserves nearness for nonempty subsets of X.*

Proof. That f preserves nearness is an immediate consequence of uniform continuity of f, while the converse is a special case of $(5) \Rightarrow (1)$ of our last theorem. $\square$

We invite the reader to construct a direct proof of Theorem 4.10 using Corollary 17.12.

We next characterize strong uniform continuity of a function on a nonempty subset in terms of the direct image map induced by the function. Given $f : \langle X, d \rangle \to \langle Y, \rho \rangle$, the associated *direct image map* $\tilde{f} : \mathscr{P}_0(X) \to \mathscr{P}_0(Y)$ is defined by $\tilde{f}(A) = \{f(a) : a \in A\}$.

While Hausdorff distance between nonempty subsets does not distinguish between sets and their closures and can assume values of infinity, our definition of strong uniform continuity makes sense for functions between spaces of subsets in this setting. The next result appears as [42, Theorem 3.3].

Theorem 17.13. *Let $\langle X, d \rangle$ and $\langle Y, \rho \rangle$ be metric spaces and let $f \in Y^X$. The following conditions are equivalent for a nonempty subset A of X:*

(1) $A \in \mathscr{B}_f$;

(2) $\tilde{f} : \langle \mathscr{P}_0(X), H_d \rangle \to \langle \mathscr{P}_0(Y), H_\rho \rangle$ is strongly uniformly continuous on $\mathscr{P}_0(A)$;

(3) $\tilde{f} : \langle \mathscr{P}_0(X), H_d \rangle \to \langle \mathscr{P}_0(Y), H_\rho \rangle$ is continuous at each member of $\mathscr{P}_0(A)$.

Proof. Since $(2) \Rightarrow (3)$ is trivial, we need only prove $(1) \Rightarrow (2)$ and $(3) \Rightarrow (1)$.

$(1) \Rightarrow (2)$. Let $\varepsilon > 0$ be arbitrary and choose $\delta > 0$ such that whenever $a \in A$ and $d(x, a) < \delta$, then $\rho(f(x), f(a)) < \varepsilon$. Now suppose $A_0 \in \mathscr{P}_0(A)$ and $C \in \mathscr{P}_0(X)$ with $H_d(A_0, C) < \delta$. Let $c \in C$ be arbitrary; $\exists a \in A_0$ with $d(a, c) < \delta$ and so $\rho(f(a), f(c)) < \varepsilon$. This shows that $f(C) \subseteq S_\rho(f(A_0), \varepsilon)$ and similarly $f(A_0) \subseteq S_\rho(f(C), \varepsilon)$. Together, these give $H_\rho(\tilde{f}(C), \tilde{f}(A_0)) \leq \varepsilon$.

$(3) \Rightarrow (1)$. Suppose strong uniform continuity of f on A fails. By Theorem 17.10, there exists a nonempty subset B of A and $C \in \mathscr{P}_0(X)$ such that B and C are d-near yet $D_\rho(f(C), f(B)) > 0$. For each $n \in \mathbb{N}$ there exists $c_n \in C$ with $0 < d(c_n, B) < \frac{1}{n}$. While $H_d(B, B \cup \{c_n\}) = d(c_n, B) < \frac{1}{n}$, we have for each $n \in \mathbb{N}$,

$$H_\rho(\tilde{f}(B), \tilde{f}(B \cup \{c_n\})) = \rho(f(c_n), f(B)) \geqslant D_\rho(f(C), f(B))$$

and so continuity of the direct image map at B fails. $\square$

We now look at the preservation of strong uniform continuity under standard ways of combining functions. The proof of the first result is left to the reader as an easy exercise.

Proposition 17.14. *Let $f : \langle X, d \rangle \to \langle Y, \rho \rangle$, and let $g : \langle Y, \rho \rangle \to \langle W, \gamma \rangle$. Suppose f is strongly uniformly continuous on a nonempty subset A of X and g is strongly uniformly continuous on $f(A)$. Then $g \circ f$ is strongly uniformly continuous on A. In particular a uniformly continuous function on Y following a strongly uniformly continuous function on A is strongly uniformly continuous on A.*

Proposition 17.15. *Let f and g be strongly uniformly continuous real-valued functions on a nonempty subset A of a metric space $\langle X, d \rangle$. Then each of the following functions is strongly uniformly continuous on A: $f + g, \alpha f, |f|, max\{f, g\}$, and $min\{f, g\}$. If f and g are both bounded on A, then fg is strongly uniformly continuous on A.*

Proof. We say a few words about absolute value, maximum, and multiplication only. In the case of $|f|$, we can apply the last proposition, because $\alpha \mapsto |\alpha|$ is uniformly continuous. This in turn is used along with strong uniform continuity of $f + g$ and $f - g$ to prove strong uniform continuity of the maximum, since $max\{f, g\} = \frac{1}{2}(f + g) + \frac{1}{2}|f - g|$. Continuity of the product under the boundedness assumption of course relies on the inequality $|(fg)(w) - (fg)(x)| \leq |f(w)||g(w) - g(x)| + |g(x)||f(w) - f(x)|$. $\qquad\square$

Let $\langle X, d \rangle$ be a metric space. We close this section by addressing the question: is there a nontrivial topology on $\mathscr{P}_0(X)$ such that $A \mapsto \Omega(f, A)$ is continuous for each $f \in C(X, Y)$ for any metric target space $\langle Y, \rho \rangle$? Such a topology was identified by Beer and Cao [26, Theorem 3.6], and we actually get contininuity of oscillation with respect to any compatible metric on the domain. This topology, called the *locally finite topology*, is finer than both the topology of Hausdorff distance and the Vietoris topology. If $\mathscr{V}$ is a collection of nonempty subsets of X, put

$$\mathscr{V}^- := \{E \in \mathscr{P}_0(X) : \forall V \in \mathscr{V}, \ E \cap V \neq \emptyset\}.$$

The locally finite topology on $\mathscr{P}_0(X)$ is generated by all families of the form V^+ with $V \in \tau$ and $\mathscr{V}^-$ where $\mathscr{V}$ is a locally finite subfamily of τ.

This topology attracted some attention restricted to $\mathscr{C}_0(X)$ when it was shown to be the topology generated by the union of all Hausdorff metric topologies $\{\tau_{H_d} : d \in \mathbf{D}(X)\}$ [17, 39]. This means that a net $\langle A_\lambda \rangle$ of nonempty closed subsets is convergent to $A \in \mathscr{C}_0(X)$ if and only if $\langle d(\cdot, A_\lambda) \rangle$ is uniformly convergent to $d(\cdot, A)$ for each $d \in \mathbf{D}(X)$ [17, p. 85].

18. UC-Subsets

As we mentioned earlier, strong uniform continuity of each continuous function may occur on a bornology that properly contains $\mathcal{K}(X)$. For example, Consider $X = (-\infty, 0] \cup \mathbb{N}$ equipped with the usual metric of the line. Then each continuous function on X is strongly uniformly continuous on a subset E provided $E = A \cup B$ where A has compact closure and $B \subseteq \mathbb{N}$. To initially describe this perhaps larger bornology, which will be called the bornology of UC-subsets, we call upon our isolation functional.

Definition 18.1. A subset A of a metric space $\langle X, d \rangle$ is called a *UC-set* if whenever $\langle a_n \rangle$ is a sequence in A with $\lim_{n \to \infty} I(a_n) = 0$, then $\langle a_n \rangle$ has a cluster point in X.

Obviously each relatively compact subset is a UC-set, because each sequence in A without restriction clusters. It is a straight-forward exercise to show that the family of UC-sets forms a bornology which we will denote by $\mathscr{B}_d^{uc}(X)$ in the sequel. By nonexpansiveness of $x \mapsto I(x)$, it is clear that $A \in \mathscr{B}_d^{uc}(X) \Rightarrow \mathrm{cl}(A) \in \mathscr{B}_d^{uc}(X)$.

Being a UC-subset is not an intrinsic property of metric spaces. For example, $\mathbb{N}$ is a UC-subset of $X = \mathbb{N}$ equipped with the usual metric of the line, but it is not a UC-subset of $\mathbb{R}$.

Before giving the most important characterizations of UC-subsets, we introduce a concept due to Toader [145] that is a weakening of the notion of Cauchy sequence.

Definition 18.2. A sequence $\langle x_n \rangle$ in a metric space $\langle X, d \rangle$ is called *pseudo-Cauchy* if for each $\varepsilon > 0$ and $n \in \mathbb{N}$, there exist $k > j \geq n$ with $d(x_k, x_j) < \varepsilon$.

Evidently, pseudo-Cauchyness means that we can find beyond any index two terms that are arbitrarily close. In other words: pairs of terms are arbitrarily close frequently.

Theorem 18.3. *Let A be a nonempty subset of a metric space $\langle X, d \rangle$. The following conditions are equivalent:*

(1) *$A \in \mathscr{B}_d^{uc}(X)$;*

(2) *whenever $\mathscr{V}$ is an open cover of X, there exists $\delta > 0$ such that $\{S_d(a, \delta) : a \in A\}$ refines $\mathscr{V}$;*

(3) *whenever $\langle Y, \rho \rangle$ is a metric space and $f \in C(X, Y)$, we have $A \in \mathscr{B}_f$;*

(4) *whenever $f \in C(X, \mathbb{R})$, we have $A \in \mathscr{B}_f$;*

(5) *whenever $\langle x_n \rangle$ is a pseudo-Cauchy sequence with distinct terms with $\lim_{n \to \infty} d(x_n, A) = 0$, then $\langle x_n \rangle$ clusters;*

(6) *whenever $C \neq \emptyset$ is a closed subset of A and $E \in \mathscr{C}_0(X)$ with $C \cap E = \emptyset$, then $D_d(C, E) > 0$;*

(7) *$cl(A) \cap X'$ is compact, and $\forall \delta > 0 \; \exists \lambda > 0$ such that $a \in A \backslash S_d(cl(A) \cap X', \delta) \Rightarrow I(a) > \lambda$.*

Proof. (1) $\Rightarrow$ (2). Suppose A is a UC-subset while $\mathscr{V}$ is an open cover of X that cannot be refined by balls of a fixed radius whose centers run over A. Then for each $n \in \mathbb{N}$, there exists $a_n \in A$ such that $S_d(a_n, \frac{1}{n})$ lies in no single member of $\mathscr{V}$. As $I(a_n) < \frac{1}{n}$, the sequence $\langle a_n \rangle$ has a cluster point $p \in X$. Taking $V \in \mathscr{V}$ with $p \in V$, it is routine to show that for some large $n, S_d(a_n, \frac{1}{n}) \subseteq V$, a contradiction.

(2) $\Rightarrow$ (3). Let $f \in C(X, Y)$. For each $x \in X$ choose an open neighborhood V_x of x with $f(V_x) \subseteq S_\rho(f(x), \frac{\varepsilon}{2})$. By (2) for some $\delta > 0$, $\{S_d(a, \delta) : a \in A\}$ refines $\{V_x : x \in X\}$. Let $a \in A$ be arbitrary and choose $x \in X$ with $S_d(a, \delta) \subseteq V_x$. Then if $w \in X$ with $d(a, w) < \delta$, we have $\rho(f(a), f(w)) < \varepsilon$.

(3) $\Rightarrow$ (4). This is trivial.

(4) $\Rightarrow$ (5). Suppose $\langle x_n \rangle$ is a pseudo-Cauchy sequence with distinct terms such that $d(x_n, A) \to 0$. We can clearly find a subsequence $x_{n_1}, x_{n_2}, x_{n_3}, \ldots$ such that for each $k \in \mathbb{N}, d(x_{n_{2k-1}}, x_{n_{2k}}) < \frac{1}{k}$. Suppose $\langle x_n \rangle$ has no cluster point. Then any Urysohn function for the disjoint closed subsets $\{x_{n_{2k-1}} : k \in \mathbb{N}\}$ and $\{x_{n_{2k}} : k \in \mathbb{N}\}$ must fail to be strongly uniformly continuous on A.

(5) $\Rightarrow$ (6). Suppose $C \in \mathscr{C}_0(X) \cap \downarrow \{A\}$, and $E \in \mathscr{C}_0(X)$ is disjoint from C but with $D_d(C, E) = 0$. First, choose $c_1 \in C$ and $e_1 \in E$ with $d(c_1, e_1) < 1$. Next choose $c_2 \in C$ and $e_2 \in E$ with $d(c_2, e_2) < \min \{d(c_1, E), d(e_1, C), \frac{1}{2}\}$. Continuing choose $c_3 \in C$ and $e_3 \in E$ with $d(c_3, e_3) < \min\{d(c_2, E), d(e_2, C), \frac{1}{3}\}$. In this way, we produce a pseudo-Cauchy sequence $c_1, e_1, c_2, e_2, c_3, e_3, c_4, e_4, \ldots$ with distinct terms such that for each $n, d(c_n, A) = 0$ while $d(e_n, A) \leq d(e_n, c_n) < \frac{1}{n}$ that cannot cluster. Thus condition (5) fails if (6) fails.

(6) $\Rightarrow$ (7). We show that if A fails to have either of the structural properties of (7), then (6) must also fail.

First, suppose $\mathrm{cl}(A) \cap X'$ is not compact, and therefore, nonempty. Pick a sequence $\langle p_n \rangle$ in $\mathrm{cl}(A) \cap X'$ without a cluster point and then $\forall n \in \mathbb{N}$ choose $0 < \delta_n < \frac{1}{n}$ such that $\{S_d(p_n, \delta_n) : n \in \mathbb{N}\}$ is a pairwise disjoint family of balls. Since each p_n is a limit point of X choose for each n $x_n \neq a_n$ both in $S_d(p_n, \delta_n)$ with $a_n \in A$. Then $\{a_n : n \in \mathbb{N}\}$ and $\{x_n : n \in \mathbb{N}\}$ are disjoint closed sets but the gap between them is zero, violating (6).

We move to the second part of condition (7). Suppose now that the first structural property holds but for some $\delta > 0$ we have

$$(\clubsuit) \quad \inf \{I(a) : a \in A \backslash S_d(\mathrm{cl}(A) \cap X', \delta)\} = 0.$$

We now put $E = A \backslash S_d(\mathrm{cl}(A) \cap X', \delta)$. Note that each point of E must be an isolated point of X, and no sequence in E with distinct terms can cluster, as the cluster point would be a limit point of X lying in $\mathrm{cl}(A)$. We consider two mutually exclusive and exhaustive cases for the set E:

(a) $\inf\{d(e_1, e_2) : e_1 \in E, e_2 \in E, e_1 \neq e_2\} = 0$;

(b) $\exists \varepsilon > 0$ such that $\forall e \in E, S_d(e, \varepsilon) \cap E = \{e\}$.

In case (a), we claim that we can find a sequence $\langle e_n \rangle$ in E with distinct terms such that for each $n, d(e_{2n-1}, e_{2n}) < 1/n$. To see this, choose e_1 and e_2 in E such that $0 < d(e_1, e_2) < 1$. Having chosen distinct $e_1, e_2, \ldots, e_{2n}$ in E with $d(e_{2j-1}, e_{2j}) < 1/j$ for $j = 1, 2, \ldots, n$, let $\mu_n = \min \{I(e_j) : j = 1, 2, \ldots, 2n\}$. By $(\clubsuit)$ we can find e_{2n+1} and e_{2n+2} in E with

$$0 < d(e_{2n+1}, e_{2n+2}) < \min \{\mu_n, \frac{1}{n+1}\}.$$

By construction, both $I(e_{2n+1}) < \mu_n$ and $I(e_{2n+2}) < \mu_n$ yielding $2n + 2$ distinct terms, and the claim is verified.

Since our sequence cannot cluster, $B = \{e_{2n-1} : n \in \mathbb{N}\}$ and $C = \{e_{2n} : n \in \mathbb{N}\}$ are disjoint closed subsets of A with $D_d(B, C) = 0$.

Case (b) is easier. Choose $k \in \mathbb{N}$ such that $1/k < \frac{\varepsilon}{2}$ and then by (♣) $\forall j \geq k$, pick $a_j \in E$ and $x_j \in X$ such that for each $j \geq k$ such that $0 < d(x_j, a_j) < 1/j$ and such that the terms of $\langle a_j \rangle_{j \geq k}$ are distinct. Since $\{S_d(a_j, \frac{\varepsilon}{2}) : j \in \mathbb{N}\}$ is a pairwise disjoint family of balls, the terms of the sequence

$$a_k, x_k, a_{k+1}, x_{k+1}, a_{k+2}, x_{k+2}, \ldots$$

are distinct. Further, the sequence cannot cluster, else $\langle a_j \rangle_{j \geq k}$ would as well. Since $\{a_j : j \geq k\}$ and $\{x_j : j \geq k\}$ are disjoint closed sets, condition (6) again is violated.

$(7) \Rightarrow (1)$. Suppose condition (7) holds and $\langle a_n \rangle$ is a sequence in A with $\lim_{n \to \infty} I(a_n) = 0$. We immediately see from condition (7) that for each $\delta > 0$, eventually $a_n \in S_d(\mathrm{cl}(A) \cap X', \delta)$. In particular $\mathrm{cl}(A) \cap X'$ can't be empty, as each enlargement of the empty set is again empty. By the compactness of $\mathrm{cl}(A) \cap X'$ and $\lim_{n \to \infty} d(a_n, \mathrm{cl}(A) \cap X') = 0$, we conclude that $\langle a_n \rangle$ clusters.   □

Corollary 18.4. *In each metric space* $\langle X, d \rangle$, $\mathscr{B}_d^{uc}(X) = \cap\{\mathscr{B}_f : f \in C(X, \mathbb{R})\}$.

Corollary 18.5. *In each metric space* $\langle X, d \rangle$, *the bornology of UC-subsets is shielded from closed sets.*

Proof. Let $A \in \mathscr{B}_d^{uc}(X)$; then $\mathrm{cl}(A)$ is a UC-subset as well, so that by condition (6) of Theorem 18.3, $\mathrm{cl}(A)$ is a shield for itself and thus for A.   □

Example 18.6. The condition on a subset A of a metric space $\langle X, d \rangle$ that A belong to $\mathscr{B}^f$ for each continuous function f on $\langle X, d \rangle$ is not enough to guarantee that A be a UC-subset: consider $\mathbb{N}$ as a subset of the real line equipped with the usual metric.

Example 18.7. A pseudo-Cauchy sequence with repeated terms in a UC-subset of a metric space need not cluster. In $(-\infty, 0] \cup \mathbb{N}$ equipped with the usual metric of the line, $\mathbb{N}$ is a UC-subset, but the pseudo-Cauchy sequence $1, 1, 4, 4, 9, 9, 16, 16, \ldots$ fails to cluster. Also, that each pseudo-Cauchy sequence with distinct terms in a subset A of a metric space clusters is not enough to guarantee that A is a UC-subset. For example, $\mathbb{N}$ as a subset of $\mathbb{R}$ equipped with the usual metric is not a UC-subset, yet the criterion is satisfied vacuously. This really should come as no surprise, as the property of being a UC-subset is variational in nature, i.e., points near but outside the subset must be taken into consideration.

Exactly which bornologies on a metrizable space are bornologies of UC-subsets? We give a result in the spirit of Theorem 16.3.

Theorem 18.8. *Let* $\langle X, \tau \rangle$ *be a metrizable space and let* $\mathscr{B}$ *be a bornology on* X. *Then* $\mathscr{B} = \mathscr{B}_d^{uc}(X)$ *for some compatible metric* d *if and only if there is a star-development* $\langle \mathscr{U}_n \rangle$ *for* X *such that*

$$(*) \qquad \mathscr{B} = \{E \in \mathscr{P}(X) : whenever\ \mathscr{V}\ is\ an\ open\ cover\ of\ X, \exists n \in \mathbb{N}$$

$$such\ that\ \{U \in \mathscr{U}_n : U \cap E \neq \emptyset\}\ refines\ \mathscr{V}\}.$$

Proof. For necessity, suppose $\mathscr{B} = \mathscr{B}_d^{uc}(X)$ for some compatible metric d. Put $\mathscr{U}_n := \{S_d(x, \frac{1}{3^n}) : x \in X\}$ for every $n \in \mathbb{N}$. Then $\langle \mathscr{U}_n \rangle$ is a star-development of X. Clearly, E satisfies condition (2) of Theorem 18.3 if and only if whenever $\mathscr{V}$ is an open cover of X, then for some $n \in \mathbb{N}$, $S_d(x, \frac{1}{3^n}) \cap E = \emptyset \Rightarrow \exists V \in \mathscr{V}$ with $S_d(x, \frac{1}{3^n}) \subseteq V$.

For sufficiency, as we noted earlier, the existence of a star-development $\langle \mathscr{U}_n \rangle$ allows us to find a compatible metric d such that $\forall n \geq 2$, both

(i) $\mathscr{U}_n$ refines $\{S_d(x, \frac{1}{2^{n-1}}) : x \in X\}$;

(ii) $\{S_d(x, \frac{1}{2^n}) : x \in X\}$ refines $\mathscr{U}_{n-1}$.

Suppose $\mathscr{B}$ satisfies $(*)$, and $E \in \mathscr{B}_d^{uc}(X)$. By condition (2) of Theorem 18.3, for some $n \geq 3$, whenever $x \in E$ then $S_d(x, \frac{1}{2^{n-2}})$ lies in some member of $\mathscr{V}$. By condition (i), if $U \in \mathscr{U}_n$ hits E, then it is contained in a ball of radius $\frac{1}{2^{n-1}}$ that hits E and thus in a ball of radius $\frac{1}{2^{n-2}}$ whose center lies in E. Thus, we can find $V \in \mathscr{V}$ with $U \subseteq V$. Conversely, let $E \in \mathscr{B}$, and choose $n \in \mathbb{N}$ such that whenever $U \in \mathscr{U}_n$ and $U \cap E \neq \emptyset$, there exists $V \in \mathscr{V}$ with $U \subseteq V$. Let $x \in E$ be arbitrary; by condition (ii), $S_d(x, \frac{1}{2^{n+1}})$ is contained in some $U \in \mathscr{U}_n$ and thus in some $V \in \mathscr{V}$ because $E \cap U \neq \emptyset$. Thus, $\{S_d(x, \frac{1}{2^{n+1}}) : x \in E\}$ refines $\mathscr{V}$. $\qquad\square$

As to when the bornology of UC-subsets collapses to $\mathscr{K}(X)$ is addressed by our next result.

Proposition 18.9. *Let $\langle X, d \rangle$ be a metric space. Then $\mathscr{B}_d^{uc}(X) = \mathscr{K}(X)$ if and only if for each $\lambda > 0$, $\{x \in X : I(x) > \lambda\}$ is a finite set.*

Proof. Suppose that for some $\lambda > 0$, $E_\lambda = \{x \in X : I(x) > \lambda\}$ is infinite. Then E_λ is a closed UC-set that is not compact, showing that $\mathscr{B}_d^{uc}(X)$ contains $\mathscr{K}(X)$ properly. Conversely, suppose for each $\lambda > 0$, $\{x : I(x) > \lambda\}$ is a finite set. To show $\mathscr{B}_d^{uc}(X) = \mathscr{K}(X)$ it suffices to show that each closed UC-set A is compact. We can write A as this disjoint union:

$$A = (A \cap X') \cup A \backslash (A \cap X').$$

Let $\mathscr{V}$ be an open cover of A; by the compactness of $A \cap X'$, we can find a finite subfamily $\{V_1, V_2, \ldots, V_n\}$ that covers $A \cap X'$. But by compactness, for some $\delta > 0, \cup_{i=1}^n V_i$ contains $S_d(A \cap X', \delta)$ and $\inf \{I(a) : a \in A \backslash S_d(A \cap X', \delta)\} > 0$ now guarantees that $A \backslash S_d(A \cap X', \delta)$ is a finite set. Thus adding an appropriate finite subfamily of $\mathscr{V}$ to $\{V_1, V_2, \ldots, V_n\}$ yields a finite subcover of A itself. $\qquad\square$

19. UC-Spaces

If X is a UC-subset of itself, we call $\langle X, d \rangle$ a *UC-space*. To the best of our knowledge, this class of spaces was introduced in an article by Doss [73], followed shortly thereafter by the better-known papers by Nagata [128] and Monteiro and Peixoto [124]. Studied by numerous authors following the influential paper of Atsuji on the subject [7], UC-spaces are also frequently called *Atsuji spaces* and sometimes go by *Lebesgue spaces* [127] in view condition (2) of Theorem 19.1 below, first identified in [124]. For a valuable survey, we refer the reader to Jain and Kundu [104].

Theorem 18.3 of the previous section immediately gives

Theorem 19.1. *Let $\langle X, d \rangle$ be a metric space. The following conditions are equivalent:*

(1) *X is a UC-space;*

(2) *each open cover $\mathscr{V}$ of X has a Lebesgue number, i.e., $\exists \lambda > 0$ such that whenever $A \subseteq X$ with $\mathrm{diam}_d(A) < \lambda, \exists V \in \mathscr{V}$ with $A \subseteq V$;*

(3) *whenever $\langle Y, \rho \rangle$ is a metric space and $f \in C(X, Y)$, then f is uniformly continuous;*

(4) *whenever $f \in C(X, \mathbb{R})$, then f is uniformly continuous;*

(5) *each pseudo-Cauchy sequence with distinct terms in X clusters;*

(6) *each pair of disjoint nonempty closed subsets of X have a positive gap between them;*

(7) *X' is compact, and $\forall \delta > 0 \; \exists \lambda > 0$ such that $x \notin S_d(X', \delta) \Rightarrow I(x) > \lambda$.*

Conditions (2), (3) and (6) are well-known properties of compact spaces, none of which is characteristic of compactness. It is clear that each UC-space must be complete, because if $\langle x_n \rangle$ is a Cauchy sequence without a constant subsequence, then we can find a subsequence with distinct terms, and along that, the isolation functional goes to zero. Thus each Cauchy sequence in a UC-space clusters and that guarantees convergence to the (unique) cluster point. This fact also follows easily from condition (5) of Theorem 18.3. The constellation of Example 12.3 is an example of a UC-space that fails to be locally compact.

It is clear from the Tietze extension theorem and condition (4) of Theorem 18.3 that a closed metric subspace of a UC-space is a UC-space. This fact also follows easily from conditions (5) or (6). A product of UC-metric spaces equipped with the box metric need not be a UC-space: consider $[0, 1] \times \mathbb{N}$ whose set of limit points is noncompact. We will shortly characterize those products for which the box metric is a UC-metric.

Finally, note that by condition (7), if X is a UC-space with no limit points, then X is uniformly isolated.

There are many other characterizations of UC-spaces, some of which replicate those listed in Theorem 19.1 with minor variations. At this time, we content ourselves with three additional characterizations of substance, each of a different nature.

The next result, which has the feel of the Cantor intersection Theorem [111, p. 413], gives additional characteristic properties of UC-spaces more in the spirit of completeness.

Theorem 19.2. *Let $\langle X, d \rangle$ be a a metric space. The following conditions are equivalent.*

(1) *$\langle X, d \rangle$ is a UC-space;*

(2) *whenever $\langle A_n \rangle$ is a decreasing sequence of nonempty closed subsets with $\lim_{n \to \infty} \inf\{I(x) : x \in A_n\} = 0$, then $\cap_{n=1}^{\infty} A_n \neq \emptyset$;*

(3) *whenever $\langle A_n \rangle$ is a decreasing sequence of nonempty closed subsets with $\lim_{n \to \infty} \sup\{I(x) : x \in A_n\} = 0$, then $\cap_{n=1}^{\infty} A_n \neq \emptyset$;*

Proof. (1) $\Rightarrow$ (2). For each $n \in \mathbb{N}$, choose $a_n \in A_n$ with $I(a_n) < \inf\{I(x) : x \in A_n\} + \frac{1}{n}$. Since X is a UC-space, $\langle a_n \rangle$ clusters, and each cluster point belongs to $\cap_{n=1}^{\infty} A_n$ as the sequence belongs to each A_n eventually.

(2) $\Rightarrow$ (3). This is trivial.

(3) $\Rightarrow$ (1). Let $\langle x_n \rangle$ be a sequence in X for which $\lim_{n \to \infty} I(x_n) = 0$. Put $A_n = \mathrm{cl}(\{x_j : j \geq n\})$; by the continuity of the isolation functional, $\sup\{I(x) : x \in A_n\} = \sup\{I(x_j) : j \geq n\}$, and by assumption, the latter supremum goes to zero with increasing n. By condition (3), $\cap_{n=1}^{\infty} \mathrm{cl}(\{x_j : j \geq n\}) \neq \emptyset$. But this intersection coincides with the set of cluster points of $\langle x_n \rangle$. $\square$

When are the uniformly continuous real-valued functions on a metric space stable under reciprocation? The answer was provided by Beer, Garrido and Meroño [38, Theorem 2.2]. In the next section and complementing the next result, we will describe in various ways those metric spaces for which $UC(X, \mathbb{R})$ is stable under pointwise product.

Theorem 19.3. *Let $\langle X, d \rangle$ be a metric space. The following conditions are equivalent:*

(1) *$\langle X, d \rangle$ is a UC-space;*

(2) *whenever $\{g, f\} \subseteq UC(X, \mathbb{R})$ and f is nonvanishing, then $\frac{g}{f} \in UC(X, \mathbb{R})$ as well;*

(3) *whenever $f \in UC(X, \mathbb{R})$ is nonvanishing, then $\frac{1}{f} \in UC(X, \mathbb{R})$.*

Proof. The implications (1) $\Rightarrow$ (2) and (2) $\Rightarrow$ (3) are obvious. For (3) $\Rightarrow$ (1), suppose that $UC(X, \mathbb{R})$ is closed under reciprocation of its never-zero members. We first observe that d is a complete metric. Otherwise there is some $p \in \hat{X} \setminus X$ where $\langle \hat{X}, \hat{d} \rangle$ is the completion of the metric space. Clearly, the function f that is the restriction of $\hat{d}(\cdot, p)$ to X is uniformly continuous and nonvanishing on X, but $1/f$ is not uniformly continuous because p is a limit point of X in the completion.

Suppose now that X is not a UC-space; then there exists a sequence $\langle x_n \rangle$ in X where $\lim_{n \to \infty} I(x_n) = 0$ but the sequence does not cluster. Without loss of generality, we may assume the sequence has distinct terms. By the completeness of X, this sequence has no Cauchy subsequence. Since the set of terms cannot be totally bounded, by passing through some subsequence, we can suppose there exists $\delta > 0$ such that $d(x_n, x_m) > \delta$ whenever $n \neq m$, and such that $I(x_n)$ still goes to 0. By again passing through a subsequence, we can choose another sequence $\langle w_n \rangle$ in X such that $0 < d(x_n, w_n) < \frac{\delta}{3n}$ for each $n \in \mathbb{N}$. By construction, the terms of the sequence

$$x_1, w_1, x_2, w_2, x_3, w_3, x_4, \ldots$$

are distinct and its set of terms is a closed set. Define $h : A = \{x_n : n \in \mathbb{N}\} \cup \{w_n : n \in \mathbb{N}\} \to \mathbb{R}$ by $h(x_n) = 1/n$ and $h(w_n) = 1/n^2$, for every $n \in \mathbb{N}$. It is easy to check that h is uniformly continuous and bounded on A and therefore by the McShane extension theorem, it can be extended as a nonnegative uniformly continuous function on the whole of X. Let $\tilde{h}$ be such an extension. The uniformly continuous function $f(x) = \tilde{h}(x) + d(x, A)$ is nonvanishing because A is closed, while $1/f$ fails to be uniformly continuous. $\qquad\square$

As shown by Beer, García-Lirola and Garrido [33], Theorem 19.3 remains true if $UC(X, \mathbb{R})$ is replaced by the Lipschitz in the small real-valued functions in its statement. We take this up in a subsequent section.

The next result may be a little out of place in this section, but as it continues our train of thought, we put it here. A very different proof has been given in [33].

Theorem 19.4. *Let $\langle X, d \rangle$ be a metric space. Them $\langle X, d \rangle$ is compact if and only if whenever $f \in Lip(X, \mathbb{R})$ is nonvanishing, then $\frac{1}{f} \in Lip(X, \mathbb{R})$.*

Proof. Suppose the metric space is compact. Put $\alpha := \min_{x \in X} |f(x)| > 0$ and let $\lambda > 0$ be a Lipschitz constant for f. Then if $x_1 \neq x_2$ we have

$$\left| \frac{1}{f(x_1)} - \frac{1}{f(x_2)} \right| \leq \frac{\lambda}{\alpha^2} d(x_1, x_2).$$

For the converse, we aim to show that X is complete and totally bounded assuming that the reciprocal of each nonvanishing real-valued Lipschitz function is again Lipschitz.

If the space is not complete, let $\hat{p}$ be a point of the completion $\langle \hat{X}, \hat{d} \rangle$ that is not in X. Then $x \mapsto \hat{d}(x, \hat{p})$ is strictly positive and 1-Lipschitz on X, but its reciprocal is not Lipschitz because of the density of X in $\hat{X}$.

To show X is totally bounded, we first show that it must be bounded. If $\langle X, d \rangle$ fails to be bounded, fix $x_0 \in X$ and let $f = d(\cdot, x_0)$. Define $g : \mathbb{R} \to \mathbb{R}$ by $g(t) := 1$ if $-1 \leq t \leq 1$ and $g(t) = \frac{1}{t^2}$ if $|t| > 1$. Easily, g is 2-Lipschitz and so $g \circ f$ is strictly positive and 2-Lipschitz. But its reciprocal is not Lipschitz, for if $d(x, x_0) > 1$, we have

$$\frac{1}{(g \circ f)(x)} - \frac{1}{(g \circ f)(x_0)} = d^2(x, x_0) - 1.$$

Finally, suppose X is bounded but not totally bounded. There exists $\delta > 0$ and a sequence $\langle x_n \rangle$ in X such that $n \neq j \Rightarrow d(x_n, x_j) \geq \delta$. Let E be the set of terms of the sequence. Define $h : E \to (0, \infty)$ by

$$h(x_n) = \begin{cases} 1 & \text{if } n = 1 \\ \frac{\delta}{nd(x_1, x_n)} & \text{otherwise} \end{cases}.$$

For each $n \geq 2$, we have $0 < h(x_n) \leq \frac{1}{2}$ so that

$$\frac{|h(x_1) - h(x_n)|}{d(x_1, x_n)} < \frac{1}{\delta}$$

while for $j > n \geq 2$,

$$|h(x_n) - h(x_j)| \leq \frac{1}{n} + \frac{1}{j} < \frac{1}{2} + \frac{1}{2} \leq \frac{1}{\delta} d(x_n, x_j).$$

These estimates show that $\frac{1}{\delta}$ is a Lipschitz constant for h, and so by the remark immediately following Theorem 3.5, h has a strictly positive $\frac{1}{\delta}$-Lipschitz extension f to X. Finally, $\frac{1}{f}$ fails to be Lipschitz, as for $n > 1$ we have

$$\frac{1}{f}(x_n) - \frac{1}{f}(x_1) \geq n - 1,$$

but a Lipschitz function defined on a bounded metric space must be bounded. This contradiction concludes our proof. □

Corollary 19.5. *A metric space $\langle X, d \rangle$ is compact if and only if whenever $f, g \in Lip(X, \mathbb{R})$ where f is nonvanishing, we have $\frac{g}{f} \in Lip(X, \mathbb{R})$.*

We now show that $\langle X, d \rangle$ is a UC-space precisely when the Hausdorff metric topology is finer than the Vietoris topology on $\mathscr{C}_0(X)$.

Theorem 19.6. *A metric space $\langle X, d \rangle$ is a UC-space if and only if $\tau_V \subseteq \tau_{H_d}$ on the nonempty closed subsets of X.*

Proof. For sufficiency, suppose $\langle X, d \rangle$ is not a UC-space. By condition (6) of Theorem 19.1, let A and B be disjoint nonempty closed subsets for which $D_d(A, B) = 0$. For each $n \in \mathbb{N}$, choose $b_n \in B$ with $d(b_n, A) < 1/n$. Clearly $E := \{b_n : n \in \mathbb{N}\}$ has no limit points, else $A \cap B \neq \emptyset$. The sequence $\langle A \cup \{b_n\} \rangle$ converges to A in Hausdorff distance but not in the Vietoris topology, as $A \in (E^c)^+$ while $\forall n \in \mathbb{N}$, $A \cup \{b_n\} \notin (E^c)^+$. This shows that the Hausdorff metric topology cannot be finer than the Vietoris topology.

For necessity, suppose our metric space is a UC-space and let A be a nonempty closed subset. Suppose $A \in (\cap_{i=1}^n V_i^-) \cap W^+$ where $V_1, V_2, \ldots, V_n, W$ are nonempty open subsets of X. In view of condition (6) of Theorem 19.1 we can find $\varepsilon > 0$ such that $S_d(A, \varepsilon) \subseteq W$. We can then find $\delta \in (0, \varepsilon)$ and points $a_i \in A \cap V_i$ such that $S_d(a_i, \delta) \subseteq V_i$ for $i = 1, 2, \ldots, n$. It is routine to verify

$$A \in \{E \in \mathscr{C}_0(X) : H_d(E, A) < \delta\} \subseteq (\cap_{i=1}^n V_i^-) \cap W^+$$

showing that each τ_V-neighborhood of A contains an H_d-neighborhood as well. □

We now state a companion result regarding the inclusion $\tau_{H_d} \subseteq \tau_V$.

Theorem 19.7. *A metric space $\langle X, d \rangle$ is totally bounded if and only if $\tau_{H_d} \subseteq \tau_V$ on the nonempty closed subsets of X.*

Proof. If $\langle X, d \rangle$ is not totally bounded, then we can find $\varepsilon > 0$ and a sequence $\langle x_n \rangle$ in X such that $d(x_n, x_j) \geq \varepsilon$ whenever $n \neq j$. Put $E = \{x_n : n \in \mathbb{N}\}$ and $E_n = \{x_1, x_2, \ldots, x_n\}$ for $n \in \mathbb{N}$. Then the sequence of closed sets $\langle E_n \rangle$ converges to E in the Vietoris topology but not in Hausdorff distance. Conversely, let $A \in \mathscr{C}_0(X)$ and $\varepsilon > 0$. Choose a finite subset F of A such that $A \subseteq S_d(F, \frac{\varepsilon}{2})$. Then whenever $C \in \mathscr{C}_0(X)$,

$$C \in \cap_{x \in F} S_d(x, \frac{\varepsilon}{2})^- \cap S_d(F, \frac{\varepsilon}{2})^+ \Rightarrow H_d(A, C) \leq \varepsilon,$$

and this proves $\tau_{H_d} \subseteq \tau_V$ on $\mathscr{C}_0(X)$. □

Since a metric space is compact if and only if it is complete and totally bounded, and UC-spaces are complete, we see that the Vietoris topology coincides with Hausdorff metric topology on $\mathscr{C}_0(X)$ if and only if the metric space is compact. All of this was initially proved in the setting of uniform spaces by Michael in his fundamental paper on hyperspace topologies [121].

Theorem 19.8. *Suppose $\langle X, d_1 \rangle$ and $\langle Y, d_2 \rangle$ are metric spaces and $X \times Y$ is equipped with the box metric ρ. Then the product is a UC-space if and only if the factor spaces satisfy one of the following conditions:*

(1) *either X or Y is finite and the other is a UC-space;*

(2) *both X and Y are compact;*

(3) *both X and Y are uniformly isolated.*

Proof. For sufficiency, we only look at (1) as sufficiency of (2) and (3) separately is completely transparent. Suppose $X = \{a_1, a_2, \ldots, a_n\}$ and Y is a UC-space. If $n = 1$, then Y and $X \times Y$ are isometric so that $X \times Y$ is a UC-space. Otherwise put $\delta = \min\{I(a_j) : j \leq n\}$. Suppose $\langle (x_n, y_n) \rangle$ were a sequence in the product along which $I(\cdot)$ tends to zero. By passing to a subsequence, we may assume with no loss of generality that $x_n = a_1$ for all n. Eventually, $I((a_1, y_n)) < \delta$ so that for all such n, $I((a_1, y_n)) = I(y_n)$ from which $\langle y_n \rangle$ clusters. As a result, $\langle (x_n, y_n) \rangle$ clusters.

For necessity, we assume that $X \times Y$ is a UC-space. If a product of metric spaces equipped with the box metric a UC-space, then each factor is as well, as each can be isometrically embedded as a closed subspace of the product. We show that if none of the conditions above holds, then we get a contradiction. If this is so, then all three of the conditions below must hold:

(a) X and Y are both infinite sets;

(b) if one metric space is compact, then the other is not;

(c) if one metric space is uniformly isolated, then the other is not.

From condition (c) and the fact that both spaces are UC-spaces, by condition (7) of Theorem 18.3 we must have either $X' \neq \emptyset$ or $Y' \neq \emptyset$. Without loss of generality assume $X' \neq \emptyset$. Take $x_0 \in X'$. If X is compact, then by condition (b) Y is noncompact. Since Y is a UC-space there exists a sequence $\langle y_n \rangle$ in Y with distinct terms for which $\inf\{I(y_n) : n \in \mathbb{N}\} > 0$. Then $\langle (x_0, y_n) \rangle$ is a sequence in the product along which the isolation function as determined by ρ is zero but that fails to cluster. If X is noncompact, we can find a sequence $\langle x_n \rangle$ in X with distinct terms for which $\inf\{I(x_n) : n \in \mathbb{N}\} > 0$. Now either there exists $y_0 \in Y'$ or a sequence $\langle y_n \rangle$ in Y with distinct terms for which $\inf\{I(y_n) : n \in \mathbb{N}\} > 0$. In the first case, $\langle (x_n, y_0) \rangle$ fails to cluster, and in the second $\langle (x_0, y_n) \rangle$ fails to cluster, leading to a contradiction in each. $\square$

If we equip the nonempty closed subsets $\mathscr{C}_0(X)$ of a metric space $\langle X, d \rangle$ with Hausdorff distance H_d we may ask: when is the hyperspace a UC-space? If one is bothered by the fact that H_d may assume infinite values, just replace H_d by $\min\{H_d, 1\}$. The next result appears as [11, Theorem 3].

Theorem 19.9. *Let $\langle X, d \rangle$ be a metric space. Then $\langle \mathscr{C}_0(X), H_d \rangle$ is a UC-space if and only if either X is compact or X is uniformly isolated.*

Proof. As is well-known [17, p. 82], if X is compact, then H_d is a bona fide metric on $\mathscr{C}_0(X)$ resulting in a compact metric space. On the other hand, if for some $\varepsilon > 0$ we have $I(x) \geq \varepsilon$ for all $x \in X$, then whenever E and F are distinct nonempty (closed) subsets of X, we have $H_d(E, F) \geq \varepsilon$, i.e., the hyperspace is also uniformly isolated. Conversely, if the hyperspace is a UC-space, then so is $\langle X, d \rangle$ because $x \mapsto \{x\}$ maps X isometrically onto a closed subset of the hyperspace. Suppose X is neither compact nor uniformly isolated. Since X is a UC-space but is not uniformly isolated, the metric space has a limit point p. Further, since X is noncompact we can find a sequence $\langle x_n \rangle$ in X with no cluster point. Clearly, for each $n \in \mathbb{N}, \{x_n, p\}$ belongs to $\mathscr{C}_0(X)'$ with respect to Hausdorff distance, but the sequence $\langle \{x_n, p\} \rangle$ cannot have an H_d-cluster point, else $\langle x_n \rangle$ would cluster as well. This violates the H_d-compactness of $\mathscr{C}_0(X)'$ which would ensue if the hyperspace were a UC-space. $\square$

In the last result, we did not need to work directly with the isolation functional of the hyperspace. There is an attractive formula for its values that we feel is worthwhile to present.

Proposition 19.10. *Let $\langle X, d \rangle$ be a metric space and let I^* denote the isolation functional for $\langle \mathscr{C}_0(X), H_d \rangle$. Then for each $A \in \mathscr{C}_0(X)$, we have $I^*(A) = \inf\{I(a) : a \in A\}$.*

Proof. Put $\alpha = \inf\{I(a) : a \in A\}$. We first show that $I^*(A) \leq \alpha$. Let $\varepsilon > 0$ be arbitrary. By the definition of α, there exists $a \in A$ and $x \in X$ with $0 < d(a, x) < \alpha + \frac{\varepsilon}{2}$. If $x \notin A$, then $A \cup \{x\} \neq A$ and $H_d(A, A \cup \{x\}) \leq d(x, a) < \alpha + \varepsilon$. Otherwise, $x \in A$ and we put $\delta := \min\{d(x, a)/2, \varepsilon/2\}$. Then $A \backslash S_d(a, \delta)$ contains x and is a proper subset of A. For all $w \in A \cap S_d(a, \delta)$ we have

$$d(w, x) \leq d(w, a) + d(a, x) < \delta + \alpha + \frac{\varepsilon}{2} \leq \alpha + \varepsilon.$$

From this inequality, we obtain $H_d(A, A \backslash S_d(a, \delta)) \leq \alpha + \varepsilon$. These two arguments yield $I^*(A) \leq \alpha$.

We next suppose that $I^*(A) < \alpha$ and reach a contradiction. We can find $C \in \mathscr{C}_0(X)$ such that $C \neq A$ and $H_d(A, C) < \alpha$. Now if $C \backslash A \neq \emptyset$, taking $c \in C \backslash A$ we have $d(c, A) < \alpha$ and so for some $a \in A$ we have, $d(a, c) < \alpha$. This violates $I(a) \geq \alpha$. Otherwise, C is a proper subset of A, so for some $a \in A \backslash C$ we have $d(a, C) < \alpha$ which violates $I(a) \geq \alpha$ as well. $\square$

As we noted earlier, the class of continuous functions on a metric space $\langle X, d \rangle$ coincides with formally smaller class of Cauchy continuous functions if and only if the metric space is complete. We were not then in a position to state when the class of Cauchy continuous functions coincides with the class of uniformly continuous functions. Now we are.

Theorem 19.11. *Let $\langle X, d \rangle$ be a metric space. The following conditions are equivalent:*

(1) *the completion $\langle \hat{X}, \hat{d} \rangle$ of $\langle X, d \rangle$ is a UC-space;*

(2) *each Cauchy continuous function on X is uniformly continuous;*

(3) *each pseudo-Cauchy sequence in X with distinct terms has a Cauchy subsequence;*

(4) *each sequence $\langle x_n \rangle$ in X with $\lim_{n \to \infty} I(x_n) = 0$ has a Cauchy subsequence.*

Proof. (1) $\Rightarrow$ (2). Let $\langle Y, \rho \rangle$ be a second metric space and suppose $f : X \to Y$ is Cauchy continuous. With $\langle \hat{Y}, \hat{\rho} \rangle$ denoting the completion of the target space, we can extend f to a (Cauchy) continuous function $\hat{f} : \hat{X} \to \hat{Y}$. By condition (1), $\hat{f}$ is uniformly continuous so that its restriction to X is uniformly continuous as well.

$(2) \Rightarrow (3)$. Let $\langle x_n \rangle$ be a pseudo-Cauchy sequence in X with distinct terms that has no Cauchy subsequence in X, so that is cannot cluster in $\hat{X}$. By passing to a subsequence, we may assume that $\lim_{n \to \infty} d(x_{2n-1}, x_{2n}) = 0$. As $\{x_{2n-1} : n \in \mathbb{N}\}$ and $\{x_{2n} : n \in \mathbb{N}\}$ are disjoint closed subsets in the completion, there is a Urysohn function $\hat{f} : \hat{X} \to [0,1]$ for them. As $\langle \hat{X}, \hat{d} \rangle$ is complete, $\hat{f}$ is Cauchy continuous but not uniformly continuous. Thus, its restriction to X has the same characteristics.

$(3) \Rightarrow (1)$. Suppose condition (1) fails. Let $\langle \hat{x}_n \rangle$ be a pseudo-Cauchy sequence in the completion with distinct terms that doesn't cluster. Put $\delta_n = \min\{\frac{1}{n}, \frac{1}{3}d(\hat{x}_n, \{x_j : j \neq n\})\}$. Then $\{S_d(\hat{x}_n, \delta_n) : n \in \mathbb{N}\}$ is a pairwise disjoint family of balls, and choosing $x_n \in X \cap S_d(\hat{x}_n, \delta_n)$, we get a pseudo-Cauchy sequence in X with distinct terms that cannot have a Cauchy subsequence, else $\langle \hat{x}_n \rangle$ would as well.

$(1) \Rightarrow (4)$. It is easy to check that the index of isolation is not changed by passing to the completion. Thus, $\langle x_n \rangle$ has a convergent subsequence in $\hat{X}$ and so that subsequence must be Cauchy in X.

$(4) \Rightarrow (3)$. Clearly, each pseudo-Cauchy sequence $\langle x_n \rangle$ in X with distinct terms has a subsequence $\langle x_{n_k} \rangle$ with $\lim_{k \to \infty} I(x_{n_k}) = 0$.   □

The equivalence of conditions (1) and (2) was perhaps first proved in [12]. There are many other characterizations of metric spaces $\langle X, d \rangle$ whose completion $\langle \hat{X}, \hat{d} \rangle$ is a UC-space collected by Jain and Kundu in [105]. Later in this monograph, we will give some function space characterizations of UC-spaces and spaces with UC-completion.

To end this section, we characterize those metrizable spaces that have a compatible UC-metric in a very simple way: the set of limit points of the space must be compact. This criterion was in fact identified in the first paper on the subject by Doss [73, Theorem 3]. While the construction of Doss is elementary, we provide one of the author [13]. For more sophisticated proofs, we refer the reader to [126, 128, 132].

Theorem 19.12. *Let $\langle X, \tau \rangle$ be a metrizable space. Then there is a UC-metric compatible with τ if and only if X' is compact.*

Proof. Necessity is immediate from condition (7) of Theorem 19.1. For sufficiency, if $X' = \emptyset$, the zero-one metric does the job. Otherwise, let d be any compatible metric and use $\rho : X \times X \to [0, \infty)$ defined by $\rho(x, x) = 0$ and $\rho(x, w) = d(x, w) + \max\{d(x, X'), d(w, X')\}$ if $x \neq w$.

If α, β and γ are nonnegative reals, we have

$$\max\{\alpha, \gamma\} \leq \alpha + \gamma \leq \max\{\alpha, \beta\} + \max\{\beta, \gamma\},$$

from which the triangle inequality for ρ holds. As the other two defining properties of a metric are obviously satisfied by ρ, we see that ρ is a metric on X. Since $\rho \geq d$, the ρ-topology is finer than the d topology. To see that $\tau_\rho \subseteq \tau_d$, we show that each open ρ-ball contains a concentric open d-ball.

Fix $p \in X$ and let $\varepsilon > 0$. If $p \notin X'$, then for some $\delta > 0$ we have $S_d(p, \delta) = \{p\}$ and so $S_d(p, \delta) \subseteq S_\rho(p, \varepsilon)$. Otherwise, $p \in X'$. Let $x \in S_d(p, \varepsilon/2)$; we compute

$$\rho(p, x) = d(p, x) + \max\{0, d(x, X')\} \leq d(p, x) + d(p, x) < \varepsilon.$$

Thus, $S_d(p, \varepsilon/2) \subseteq S_\rho(p, \varepsilon)$ and so ρ is a compatible metric.

To show that the metric ρ is a UC-metric, let $\langle x_n \rangle$ be a sequence in X along which the isolation function as determined by ρ tends to zero. By passing to a subsequence, we may assume that for each $n \in \mathbb{N}$, we can find w_n with $0 < \rho(x_n, w_n) < 1/n$ so that by the definition of ρ, $d(x_n, X') < 1/n$. By the compactness and nonemptyness of X', the sequence $\langle x_n \rangle$ clusters. $\square$

20. Pointwise Products of Uniformly Continuous Real-Valued Functions

In this section we introduce a bornology larger than the bornology of Bourbaki bounded subsets such that agreement of the two bornologies is necessary and sufficient for $UC(X, \mathbb{R})$ to be stable under pointwise product, that is, for $UC(X, \mathbb{R})$ to form a ring.

Definition 20.1. We say that a subset A of a metric space $\langle X, d \rangle$ is *infinitely nonuniformly isolated* if whenever $E \subseteq A$ is infinite, then $\inf_{e \in E} I(e) = 0$.

An infinitely uniformly isolated subset A can have at most countably many isolated points, as for each $n \in \mathbb{N}$, $\{a \in A : I(a) \geq 1/n\}$ must be a finite set. An infinitely nonuniformly isolated subset need not be metrically bounded; for example, each subset of a metric space $\langle X, d \rangle$ where $X' = X$ has this property. The family of infinitely nonuniformly isolated subsets forms a bornology which we denote by $\mathscr{B}_d^{ni}(X)$ going forward.

The next result is reminiscent of the classical sequential characterizations of relatively compact subsets and of totally bounded subsets and more recently of Bourbaki bounded subsets. In qualitative terms, the result says that a subset is infinitely nonuniformly isolated if and only if each sequence with distinct terms in it has a subsequence along which the isolation functional tends to zero.

Proposition 20.2. *A subset A of a metric space $\langle X, d \rangle$ is infinitely nonuniformly isolated if and only if whenever $\langle a_n \rangle$ is a sequence in A with distinct terms, there exists a subsequence $\langle a_{n_k} \rangle$ and a sequence $\langle x_k \rangle$ in X such that for each $k \in \mathbb{N}$, $a_{n_k} \neq x_k$ satisfying $\lim_{k \to \infty} d(a_{n_k}, x_k) = 0$.*

Proof. First, suppose A is infinitely nonuniformly isolated and let $\langle a_n \rangle$ be a sequence in A with distinct terms. There exists $n_1 \in \mathbb{N}$ with $I(a_{n_1}) < 1$. Having chosen $n_1 < n_2 < \cdots < n_k$ with $I(a_{n_j}) < \frac{1}{j}$ for $j = 1, 2, \ldots, k$ we can find $n_{k+1} \in \{n : n > n_k\}$ such that $I(a_{n_{k+1}}) < \frac{1}{k+1}$ because $\{a_n : n > n_k\}$ is an infinite subset of A. Finally, for each $k \in \mathbb{N}$ choose $x_k \in X$ with

$$0 < d(a_{n_k}, x_k) < I(a_{n_k}) + \frac{1}{k} < \frac{2}{k} \quad \text{for } k = 1, 2, 3, \ldots.$$

The proof of the converse is easy and is left to the reader. $\square$

The easy proof of the next proposition, which is in the spirit of our earlier results characterizing bornologies in terms of families of functions that are bounded on them, is left to the reader.

Proposition 20.3. *Let A be a nonempty subset of a metric space $\langle X, d \rangle$. The following conditions are equivalent.*

(1) *A is an infinitely nonuniformly isolated subset;*

(2) *whenever $\langle Y, \rho \rangle$ is a metric space and $f : X \to Y$ has the property that for some $\alpha > 0, f(\{a \in A : I(a) < \alpha\})$ is bounded, then $f(A) \in \mathscr{B}_\rho(Y)$;*

(3) *whenever $f \in UC(X, \mathbb{R})$ has the property that for some $\alpha > 0$, the set $f(\{a \in A : I(a) < \alpha\})$ is bounded, then $f|_A$ is bounded.*

The next result expresses $\mathscr{K}(X)$ as the intersection of two bornologies.

Proposition 20.4. *A subset of a metric space $\langle X, d \rangle$ is relatively compact if and only if it is both a UC-subset and an infinitely nonuniformly isolated subset.*

Proof. If A is relatively compact, then A is a UC-subset. If A is finite, then A is infinitely nonuniformly isolated. If A is infinite and $\langle a_n \rangle$ is a sequence in A with distinct terms, then the conditions of Proposition 20.2 are met with respect to a constant sequence $\langle x_k \rangle$. Conversely, let A be both a UC-subset and infinitely nonuniformly isolated. If A is finite, then $A \in \mathscr{K}(X)$. Otherwise, to show relative compactness of A, we need only show that each sequence $\langle a_n \rangle$ in A with distinct terms clusters. By Proposition 20.2, $\langle a_n \rangle$ has a subsequence along which the isolation functional goes to zero, and since A is a UC-subset, the subsequence clusters. Thus, $\langle a_n \rangle$ clusters. $\square$

Proposition 20.5. *Each Bourbaki bounded subset A of a metric space $\langle X, d \rangle$ is infinitely nonuniformly isolated.*

Proof. Let $E \subseteq A$ be infinite and let $\varepsilon > 0$. Choose $F \subseteq X$ finite and $n \in \mathbb{N}$ such that each point of A can be joined to some point of F by an ε-chain of length n. Fix $e \in E \backslash F$ and take an ε-chain $x_0, x_1, x_2, \ldots, x_n$ in X with $x_0 \in F$ and $x_n = e$. Noting that $x_0 \neq e$, let $j \in \{0, 1, 2, \ldots, n-1\}$ be the largest integer such that $x_j \neq e$. Then $0 < d(x_j, e) = d(x_j, x_{j+1}) < \varepsilon$ so that $I(e) < \varepsilon$ as required. $\square$

From the last result, each totally bounded subset is infinitely nonuniformly isolated and thus so is each relatively compact subset. In fact, we have this simple characterization of UC-spaces.

Theorem 20.6. *A metric space $\langle X, d \rangle$ is a UC-space if and only if $\mathscr{K}(X)$ coincides with the bornology of infinitely nonuniformly isolated subsets.*

Proof. Suppose X is a UC-space; we need only show that each infinitely nonuniformly isolated subset A is relatively compact. Let $\langle a_n \rangle$ be a sequence in an infinitely nonuniformly isolated subset A; we will show that it clusters. Now the sequence either contains a constant subsequence or by Proposition 20.2 a subsequence on which the isolation function tends to zero, and since X is a UC-space, it clusters as well. Thus, $\langle a_n \rangle$ clusters.

Conversely, suppose X is not a UC-space. Then we can find disjoint nonempty closed subsets A and B and a sequence of distinct terms $\langle a_n \rangle$ in A with $\lim_{n \to \infty} d(a_n, B) = 0$. Then $\{a_n : n \in \mathbb{N}\}$ is an infinitely nonuniformly isolated subset while $\langle a_n \rangle$ cannot cluster, else $A \cap B \neq \emptyset$. This means $\{a_n : n \in \mathbb{N}\} \notin \mathscr{K}(X)$. $\square$

We have seen that a function between metric spaces is (i) globally continuous if and only if it is uniformly continuous when restricted to each relatively compact subset, and (ii) Cauchy continuous if and only if it is uniformly continuous when restricted to each totally bounded subset. What does uniform continuity of the restriction to each infinitely nonuniformly isolated subset mean?

Proposition 20.7. *A function f between metric spaces $\langle X, d \rangle$ and $\langle Y, \rho \rangle$ belongs to $UC(X, Y)$ if and only it is uniformly continuous when restricted to each infinitely nonuniformly isolated subset.*

Proof. We need only prove sufficiency. If f fails to be uniformly continuous on X, then for some $\varepsilon > 0$ we can find for each $n \in \mathbb{N}$ points x_n and w_n in X with $0 < d(x_n, w_n) < \frac{1}{n}$ while $\rho(f(x_n), f(w_n)) \geq \varepsilon$. Clearly, f is not uniformly continuous restricted to

$$A := \{x_n : n \in \mathbb{N}\} \cup \{w_n : n \in \mathbb{N}\}$$

whereas A is infinitely nonuniformly isolated because the infimum of the isolation functional $I(\cdot)$ restricted to any infinite subset of either $\{x_n : n \in \mathbb{N}\}$ or $\{w_n : n \in \mathbb{N}\}$ must be zero. $\qquad \square$

If the pointwise product of each pair of functions in $UC(X, \mathbb{R})$ remains in $UC(X, \mathbb{R})$, then actually $UC(X, \mathbb{R})$ forms a ring. In our opinion, the decisive result in this direction is due to Cabello-Sánchez [61]. Ultimately, we will show that $UC(X, \mathbb{R})$ is a ring if and only if each infinitely nonuniformly isolated subset is Bourbaki bounded, that is, by Proposition 20.5, if $\mathscr{B}_d^{ni}(X) = \mathscr{BB}_d(X)$. As a first step, we give two other characterizations due to Bouziad and Sukhacheva [57]. While these are not really internal characterizations, they are nevertheless very attractive.

Theorem 20.8. *Let $\langle X, d \rangle$ be a metric space. The following conditions are equivalent.*

(1) *$UC(X, \mathbb{R})$ is a ring;*

(2) *the square of each real-valued uniformly continuous function on X is uniformly continuous;*

(3) *whenever $f \in UC(X, \mathbb{R})$, there exists $k \in \mathbb{N}$ such that $\{x : |f(x)| > k\}$ is uniformly isolated;*

(4) *whenever $f \in UC(X, \mathbb{R})$ and $g : \mathbb{R} \to \mathbb{R}$ is continuous, we have $g \circ f \in UC(X, \mathbb{R})$.*

Proof. (2) $\Rightarrow$ (1). Suppose $\{f, g\} \subseteq UC(X, \mathbb{R})$. Since $UC(X, \mathbb{R})$ is a vector space, by condition (2) $(f + g)^2$ is uniformly continuous, and so $\frac{1}{2}((f + g)^2 - f^2 - g^2)$ is uniformly continuous as well.

(1) $\Rightarrow$ (3). We prove that the contrapositive holds. Suppose condition (3) fails for some uniformly continuous function f. Without loss of generality, by replacing f by $|f|$, we may assume that the values of f are nonnegative. By uniform continuity, there exists a decreasing positive sequence $\langle \delta_k \rangle$ with $\lim_{k \to \infty} \delta_k = 0$ such that $\forall k \in \mathbb{N}$,

$$d(x, w) < \delta_k \Rightarrow |f(x) - f(w)| < \frac{1}{3k} \qquad (x, w \in X)$$

Since (3) fails, $\{x : f(x) > 1\}$ is not uniformly isolated, and we can choose x_1 and w_1 such that $0 < d(x_1, w_1) < \delta_1$ and $f(x_1) > 1$. In the same vein, $\{x : f(x) > f(x_1) + 1\}$ is not uniformly isolated, and we can choose x_2 and w_2 such that $0 < d(x_2, w_2) < \delta_2$ and $f(x_2) > f(x_1) + 1$.

Continuing, choose for each $k \in \mathbb{N}$, x_k and w_k such that $0 < d(x_k, w_k) < \delta_k$ and $f(x_{k+1}) > f(x_k) + 1$. By construction, it is easy to verify that the following estimates all hold:

(i) $\forall k \in \mathbb{N}$, $f(x_k) > k$;

(ii) $\forall k \in \mathbb{N}$, $f(x_k) + \frac{1}{3k}f(w_k) > f(x_k) - \frac{1}{3k} > 0$;

(iii) whenever $j \neq k$, $d(x_j, x_k) \geq \delta_1$, $d(w_j, w_k) \geq \delta_1$, and $d(x_j, w_k) \geq \delta_1$.

Define $h : \{x_k : k \in \mathbb{N}\} \cup \{w_k : k \in \mathbb{N}\} \to \mathbb{R}$ by $h(x_k) = 0$ and $h(w_k) = \frac{1}{k}$ for all k. By estimate (iii) above, h so defined is a bounded uniformly continuous function, so that by the McShane extension theorem this function has a uniformly continuous extension to X which, for convenience, we shall denote by h as well. Of course, $f + h \in UC(X, \mathbb{R})$. We will show that $(f+h)^2$ fails to be uniformly continuous by proving that for each k, $(f+h)^2(w_k) > (f+h)^2(x_k) + \frac{4}{3}$ from which (1) fails. Fix $k \in \mathbb{N}$; we compute using (i) and (ii) and the fact that $h(x_k) = 0$ that

$$(f+h)^2(w_k) = f^2(w_k) + 2f(w_k)\frac{1}{k} + \frac{1}{k^2}$$

$$> (f(x_k) - \tfrac{1}{3k})^2 + \tfrac{2}{k}(f(x_k) - \tfrac{1}{3k}) + \tfrac{1}{k^2}$$

$$> f^2(x_k) - \tfrac{2}{3k}f(x_k) + \tfrac{2}{k}f(x_k)$$

$$= f^2(x_k) + \tfrac{4}{3k}f(x_k) > (f+h)^2(x_k) + \tfrac{4}{3}.$$

$(3) \Rightarrow (4)$. Suppose $g \in C(\mathbb{R}, \mathbb{R})$ and by condition (3), $k \in \mathbb{N}$ is chosen so that for some $\lambda > 0$, $|f(x)| > k \Rightarrow I(x) > \lambda$. Since g restricted to $[-k, k]$ is uniformly continuous, there exists $\delta_1 > 0$ such that whenever $\{\alpha, \beta\} \subseteq [-k, k]$ with $|\alpha - \beta| < \delta_1$, we have $|g(\alpha) - g(\beta)| < \varepsilon$. By uniform continuity of f, pick $\delta < \lambda$ such that $d(x, w) < \delta \Rightarrow |f(x) - f(w)| < \delta_1$. It is easy to check that

$$d(x, w) < \delta \Rightarrow |(g \circ f)(x) - (g \circ f)(w)| < \varepsilon \qquad (x, w \in X).$$

$(4) \Rightarrow (2)$. Apply condition (4) where g is the squaring function. $\qquad \square$

Theorem 20.9. *Let $\langle X, d \rangle$ be a metric space. The following conditions are equivalent.*

(1) $UC(X, \mathbb{R})$ *is a ring;*

(2) *the bornology of infinitely nonuniformly isolated subsets coincides with the Bourbaki bounded subsets;*

(3) *each subset of X is either Bourbaki bounded or contains an infinite uniformly isolated subset.*

Proof. We already know that each Bourbaki bounded subset is infinitely nonuniformly isolated, and as an infinite set is not infinitely nonuniformly isolated if and only if it contains an infinite uniformly isolated subset, conditions (2) and (3) are clearly equivalent. We next prove that condition (1) implies condition (3). If (3) fails, then we can find a nonempty subset A of X that neither contains an infinite uniformly isolated subset nor is Bourbaki bounded. Since A is not Bourbaki bounded, there exists $f \in UC(X, \mathbb{R})$ that fails to be bounded when restricted to A. This means that for each $k \in \mathbb{N}$, $\{a \in A : |f(a)| > k\}$ is an infinite set that fails to be uniformly isolated. By Theorem 20.8, condition (1) fails.

Finally, we prove condition (3) implies that the square of each uniformly continuous real-valued function is again uniformly continuous. We prove the contrapositive. Suppose $f \in UC(X, \mathbb{R})$ while f^2 fails to be uniformly continuous. Then for some $\varepsilon > 0$ we can find for each $n \in \mathbb{N}$

points x_n and w_n in X such that $0 < d(x_n, w_n) < 1/n$ but $|f^2(x_n) - f^2(w_n)| \geq \varepsilon$. Since f^2 is globally continuous, by passing to a subsequence we can assume that the terms of $\langle x_n \rangle$ are distinct. Now the restriction of f to $\{x_n : n \in \mathbb{N}\}$ cannot be bounded, else by uniform continuity its restriction to $\{w_n : n \in \mathbb{N}\}$ would be bounded. Thus f^2 would be uniformly continuous on $\{x_n : n \in \mathbb{N}\} \cup \{w_n : n \in \mathbb{N}\}$ which is impossible. We conclude that $\{x_n : n \in \mathbb{N}\}$ is not Bourbaki bounded. By construction $\{x_n : n \in \mathbb{N}\}$ has no infinite uniformly isolated subset. Thus, the internal criterion (3) fails. $\qquad\square$

The equivalence of conditions (1) and (3) is due to by Cabello-Sánchez [61] in 2017; to the best of our knowledge, his result gave the first simple answer to the question: when is $UC(X, \mathbb{R})$ a ring? The bornology of infinitely nonuniformly isolated subsets was introduced soon after by Beer, Garrido and Meroño, and they recast the Cabello-Sánchez result in bornological terms with a different proof [38, Theorem 3.9]. The paper of Bouziad and Sukhacheva [57] appeared within another year; they considered this question in the context of uniform spaces. It is remarkable how such a basic question remained open for so long.

Example 20.10. Of course, $UC(X, \mathbb{R})$ is stable under pointwise product if either each real-valued uniformly continuous function on X is bounded or if each continuous real-valued function on X is already uniformly continuous. The first situation occurs exactly when X is Bourbaki bounded, and the second occurs when X is a UC-space. A metric space that satisfies the Cabello-Sánchez criterion but is not one of these types is $X := (-1, 0) \cup \mathbb{N}$ as a metric subspace of the line. A subset E of X is Bourbaki bounded if its intersection with $\mathbb{N}$ is finite and contains an infinite uniformly isolated subset otherwise.

As proved earlier, $UC(X, \mathbb{R})$ is stable under taking reciprocals of nonvanishing members if and only if $\langle X, d \rangle$ is a UC-space, a smaller class of spaces than then Cabello-Sánchez spaces. We now state without proof another attractive characterization of Bouziad and Sukhacheva [57].

Theorem 20.11. *Let $\langle X, d \rangle$ be a metric space. Then $UC(X, \mathbb{R})$ is a ring if and only if there exists $A \in \mathscr{BB}_d(X)$ such that for every $\varepsilon > 0$ there exists $n \in \mathbb{N}$ such that $X \backslash S_d^n(A, \varepsilon)$ is uniformly isolated.*

If $UC(X, \mathbb{R})$ is a ring, we may not be able to find a Bourbaki bounded subset each of whose simple enlargements (n = 1) has uniformly isolated complement, as intricate counterexamples show [38, 57].

Recall that $\mathscr{BB}_d(X)$ and $\mathscr{K}(X)$ coincide if and only if $\langle X, d \rangle$ is Bourbaki complete. With Theorem 20.6 in mind, we get this additional characterization of UC-spaces as a consequence of Theorem 20.9. It describes precisely what must be added to stability under pointwise products for uniformly continuous real-valued functions so that all continuity is uniform [38, Theorem 3.10].

Theorem 20.12. *A metric space $\langle X, d \rangle$ is a UC-space if and only if it is Bourbaki complete and $UC(X, \mathbb{R})$ is a ring.*

The next result, which complements Theorem 20.6 and Theorem 20.9, gives necessary and sufficient conditions for the agreement of $\mathscr{TB}_d(X)$ with the bornology of infinitely nonuniformly isolated subsets [38, Theorem 3.11]. We use Theorem 20.12 in its proof.

Theorem 20.13. *Let $\langle X, d \rangle$ be a metric space. The following conditions are equivalent:*

(1) *The completion $\langle \hat{X}, \hat{d} \rangle$ of $\langle X, d \rangle$ is a UC-space;*

(2) $UC(X, \mathbb{R})$ is a ring and $\mathscr{TB}_d(X) = \mathscr{BB}_d(X)$;

(3) $\mathscr{TB}_d(X)$ coincides with the bornology of infinitely nonuniformly isolated subsets.

Proof. First, since always $\mathscr{TB}_d(X) \subseteq \mathscr{BB}_d(X)$, conditions (2) and (3) are equivalent by Theorem 20.9 and Proposition 20.5. If condition (3) holds, then the class of functions that are uniformly continuous restricted to totally bounded sets agrees with the class of functions that are uniformly continuous restricted to members of our bornology of infinitely nonuniformly isolated subsets, i.e., $CC(X, \mathbb{R}) = UC(X, \mathbb{R})$, so by Theorem 19.11 the completion of $\langle X, d \rangle$ is a UC-space. Thus condition (1) holds. It remains to prove (1) $\Rightarrow$ (2).

If condition (1) holds, by Theorem 20.12 $\langle \hat{X}, \hat{d} \rangle$ is Bourbaki complete and $UC(\hat{X}, \mathbb{R})$ is a ring. From the second property, $UC(X, \mathbb{R})$ is a ring as well because each member of $UC(X, \mathbb{R})$ uniquely extends to a member of $UC(\hat{X}, \mathbb{R})$. On the other hand, if A is a Bourbaki bounded subset of X, it is also Bourbaki bounded in $\hat{X}$. By the Bourbaki completeness of the completion, A must have compact closure in $\hat{X}$ and this means in particular that it is totally bounded in X. We conclude that $\mathscr{TB}_d(X) = \mathscr{BB}_d(X)$ as required. $\square$

The contents of Theorem 20.8 and Theorem 20.9 taken together have a parallel for Lipschitz in the small real-valued functions as shown in [33]. We state the main result without proof.

Theorem 20.14. *Let $\langle X, d \rangle$ be a metric space. The following conditions are equivalent.*

(1) *each subset of X is either Bourbaki bounded or contains an infinite uniformly isolated subset;*

(2) *for each real-valued Lipschitz in the small function f on X, $\exists k \in \mathbb{N}$ such that $\{x \in X : |f(x)| > k\}$ is uniformly isolated;*

(3) *for each real-valued Lipschitz in the small function f on X and for each real-valued locally Lipschitz function g on $\mathbb{R}$, the composition $g \circ f$ is Lipschitz in the small;*

(4) *the Lipschitz in the small real-valued functions on X are stable under pointwise product and hence form a ring;*

(5) *whenever f is real-valued and Lipschitz in the small on X and $h \in Lip(X, \mathbb{R})$ we have fh Lipschitz in the small.*

Of course, in most metric spaces, $UC(X, \mathbb{R})$ is not stable under pointwise product. Let $\langle X, d \rangle$ be a general metric space. We conclude this section by characterizing those pairs $(f, g) \in UC(X, \mathbb{R}) \times UC(X, \mathbb{R})$ such that fg is back in $UC(X, \mathbb{R})$.

Theorem 20.15. *Let f and g be uniformly continuous real-valued functions on a metric space $\langle X, d \rangle$. Then fg is uniformly continuous if and only if $\forall \varepsilon > 0$, $\exists \delta > 0$ such that whenever $d(x, w) < \delta$ we have $|\frac{1}{2}(f(x)g(w) + f(w)g(x)) - f(x)g(x)| < \varepsilon$.*

Proof. For sufficiency of the mixed average condition, let $\varepsilon > 0$ and choose $\delta > 0$ such $\forall x \in X, \forall w \in X$, $d(x, w) < \delta \Rightarrow |\frac{1}{2}(f(x)g(w) + f(w)g(x)) - f(x)g(x)| < \frac{\varepsilon}{2}$. Interchanging the arbitrary points, it is clear that also $|f(w)g(w) - \frac{1}{2}(f(x)g(w) + f(w)g(x))| < \frac{\varepsilon}{2}$. Let us put $\alpha = \frac{1}{2}(f(x)g(w) + f(w)g(x)) - f(x)g(x)$ and $\beta = f(w)g(w) - \frac{1}{2}(f(x)g(w) + f(w)g(x))$. The triangle inequality $|\alpha + \beta| \leq |\alpha| + |\beta|$ gives

$$d(x, w) < \delta \Rightarrow |f(w)g(w) - f(x)g(x)| < \varepsilon.$$

For necessity, let $\varepsilon > 0$, and choose $\delta > 0$ so that by the uniform continuity of fg and the uniform continuity of f and g separately, whenever $d(x, w) < \delta$, both

(1) $|f(w)g(w) - f(x)g(x)| < \varepsilon$, and

(2) $|(f(x) - f(w))(g(x) - g(w))| < \varepsilon$

are valid. Suppose $d(x, w) < \delta$; we compute

$$|(f(x)g(w) + f(w)g(x)) - 2f(x)g(x)| = |f(x)(g(w) - g(x)) + g(x)(f(w) - f(x))|$$
$$= |f(x)(g(w) - g(x)) + g(x)(f(w) - f(x)) + g(w)(f(w) - f(x)) - g(w)(f(w) - f(x))|$$
$$\leq |f(x)(g(w) - g(x)) + g(w)(f(w) - f(x))| + |(g(x) - g(w))(f(w) - f(x))|$$
$$< |f(w)g(w) - f(x)g(x)| + \varepsilon < 2\varepsilon.$$

It now follows that $|\frac{1}{2}(f(x)g(w) + f(w)g(x)) - f(x)g(x)| < \varepsilon$ as required.   □

The proof of the last theorem shows our mixed average condition is sufficient for uniform continuity of fg with no assumptions whatsover on the pair (f, g). The next example shows that the pointwise product of two continuous functions can be uniformly continuous without the mixed average condition holding.

Example 20.16. Let $X = \mathbb{N} \cup \{n + \frac{1}{n+1} : n \in \mathbb{N}\}$ equipped with the usual metric of the line. Let f be the characteristic function of $\mathbb{N}$ and let g be the characteristic function of $\{n + \frac{1}{n+1} : n \in \mathbb{N}\}$. Since the topology of X is discrete, both functions are continuous but neither is uniformly continuous. The product fg is the zero function, while for each $n \in \mathbb{N}$ we have $d(n, n + \frac{1}{n+1}) = \frac{1}{n+1}$ whereas

$$\left|\frac{1}{2}\left(f(n)g\left(n + \frac{1}{n+1}\right) + f\left(n + \frac{1}{n+1}\right)g(n)\right) - f(n)g(n)\right| = \frac{1}{2}.$$

Thus, our mixed average condition is not satisfied.

For more detailed information on the uniform continuity of the product of a pair of real-valued functions neither of which is assumed to be uniformly continuous on a general metric space, we refer the reader to an article of Beer and Naimpally [47], where a more elaborated version of Theorem 20.15 is presented.

21. Strong Uniform Convergence on Bornologies

We now are ready to turn to function spaces. Again let Y^X denote the set of all functions from $\langle X, d \rangle$ to $\langle Y, \rho \rangle$. Let $\langle \Lambda, \succeq \rangle$ be a directed set. A net $\langle f_\lambda \rangle_{\lambda \in \Lambda}$ in Y^X is called *uniformly convergent* to $f \in Y^X$ on $A \subseteq X$ if for each $\varepsilon > 0$ there exists $\lambda_1 \in \Lambda$ such that $\forall \lambda \succeq \lambda_1, \forall a \in A$, $\rho(f_\lambda(a), f(a)) < \varepsilon$. If the restriction of each f_λ to A is continuous, then the same is true for the limit function f. Now given a cover $\mathscr{A}$ of X, the *topology of uniform convergence for Y^X on $\mathscr{A}$* has as a local base at $f \in Y^X$ all neighborhoods of the form

$$\{g \in Y^X : \forall x \in E, \ \rho(f(x), g(x)) < \varepsilon\} \quad (E \in \textstyle\sum(\mathscr{A})).$$

We denote this topology by $\mathscr{T}_{\mathscr{A}}$. Clearly, a net of functions converges in this topology if and only if the convergence is uniform on each member of $\mathscr{A}$. We insist that $\mathscr{A}$ be a cover so that convergence of nets with respect to this topology ensures pointwise convergence. Pointwise convergence is of course equivalent to uniform convergence on each member of $\mathscr{F}(X)$. More generally, if $\langle f_\lambda \rangle_{\lambda \in \Lambda}$ is pointwise convergent to f, then $\{B \subseteq X : \langle f_\lambda \rangle$ converges uniformly to f on $B\}$ is a bornology on X. In the same vein, if we wish to consider uniform convergence on each member of a cover $\mathscr{A}$ of X, there is no loss of generality in replacing $\mathscr{A}$ by $\mathrm{born}(\mathscr{A})$.

With this in mind, suppose $\mathscr{B}$ is a bornology on X. The classical uniformity $\Delta_{\mathscr{B}}$ for the *topology $\mathscr{T}_{\mathscr{B}}$ of uniform convergence on $\mathscr{B}$* for Y^X has as a base for its entourages all sets of the form

$$[B; \varepsilon] := \{(f, g) : \forall x \in B, \ \rho(f(x), g(x)) < \varepsilon\} \quad (B \in \mathscr{B}, \varepsilon > 0).$$

As the uniformity is separating, we get a completely regular Hausdorff function space topology. When $\mathscr{B} = \mathscr{F}(X)$, we get a uniformity for the topology of pointwise convergence; when $\mathscr{B} = \mathscr{K}(X)$, we get a uniformity for the topology of uniform convergence on compacta. The same uniformity results if we replace the bornology by any base for the bornology.

In the context of functions between metric spaces, the pointwise limit of a sequence of continuous functions need not be continuous. For completeness, we give a classical result that essentially says that $C(X, Y)$ is a closed subset of Y^X equipped with $\mathscr{T}_{\mathscr{K}(X)}$.

Theorem 21.1. *Let $\langle X, d \rangle$ and $\langle Y, \rho \rangle$ be metric spaces, and suppose $\langle f_\lambda \rangle_{\lambda \in \Lambda}$ is a net in $C(X, Y)$ convergent uniformly on compact subsets to $f \in Y^X$. Then $f \in C(X, Y)$.*

Proof. We need only show that whenever $\langle x_n \rangle$ is a sequence in X convergent to p, then $\langle f(x_n) \rangle$ converges to $f(p)$. Let $C = \{p\} \cup \{x_n : n \in \mathbb{N}\} \in \mathscr{K}(X)$ and let $\varepsilon > 0$. Choose $\lambda_1 \in \Lambda$ such that $\lambda \succeq \lambda_1 \Rightarrow f_\lambda \in [C, \varepsilon/3]$. Choose by the continuity of f_{λ_1} at p a positive integer k such that $n \geq k \Rightarrow \rho(f_{\lambda_1}(x_n), f_{\lambda_1}(p)) < \varepsilon/3$. The triangle inequality now yields $\rho(f(p), f(x_n)) < \varepsilon$ for $n \geq k$. $\qquad\square$

The proof actually shows that continuity is preserved if we just have uniform convergence on the bornology of countable relatively compact subsets. The next example shows that uniform convergence on each countable compact subset is not necessary for continuity of the limit.

Example 21.2. Let $X = \{0\} \cup \{\frac{1}{n} : n \in \mathbb{N}\}$ as a metric subspace of the real line and let $f_n \in C(X, \mathbb{R})$ be the characteristic function of $\{\frac{1}{n}\}$, that is, $f_n(1/n) = 1$ and $f_n(x) = 0$ otherwise. Then $\langle f_n \rangle$ converges pointwise to the continuous zero function, but the convergence is not uniform on the countable compact set X.

Given a bornology $\mathscr{B}$, we now consider a second uniformity on Y^X introduced by Beer and Levi [45] that can produce a properly finer function space topology. As we will see, this comports well with strong uniform continuity on $\mathscr{B}$. Furthermore, in the special case that $\mathscr{B} = \mathscr{F}(X)$, we get the coarsest topology on Y^X finer than the topology of pointwise convergence that preserves continuity. In other words, we will get a definitive answer to a question that arises in a first course in analysis: what minimally must be added to pointwise convergence to preserve continuity?

Definition 21.3. Let $\langle X, d \rangle$ and $\langle Y, \rho \rangle$ be metric spaces and let $\mathscr{B}$ be a bornology on X. Then *the topology of strong uniform convergence* $\mathscr{T}_{\mathscr{B}}^s$ on $\mathscr{B}$ is determined by the uniformity $\Delta_{\mathscr{B}}^s$ on Y^X having as a base all sets of the form

$$[B; \varepsilon]^s := \{(f, g) : \exists \delta > 0 \; \forall x \in S_d(B, \delta), \; \rho(f(x), g(x)) < \varepsilon\} \quad (B \in \mathscr{B}, \varepsilon > 0).$$

Strong uniform convergence of a net $\langle f_\lambda \rangle_{\lambda \in \Lambda}$ to f means that for each B in the bornology and each $\varepsilon > 0$, eventually the uniform distance between f_λ and f is less than ε on some enlargement of B, where the size of the enlargement can vary with λ. Thus, strong uniform convergence is fundamentally variational in nature. To determine whether strong uniform convergence occurs, one can just look at a base for the bornology, or more generally, at any generating set for the bornology.

Note that $\forall B \in \mathscr{B} \; \forall \varepsilon > 0, [B; \varepsilon]^s \subseteq [B; \varepsilon]$ so that the latter uniformity is larger than the classical one resulting in a finer function space topology. The coarsest such topology is $\mathscr{T}_{\mathscr{F}(X)}^s$, which we call the *topology of strong pointwise convergence*. Since the singletons generate $\mathscr{F}(X)$, strong pointwise convergence of $\langle f_\lambda \rangle_{\lambda \in \Lambda}$ to f means that for each $x \in X$ and for each $\varepsilon > 0$, eventually $\sup_{w \in V_\lambda} \rho(f_\lambda(w), f(w)) < \varepsilon$ where V_λ is a neighborhood of x dependent on λ. This formulation makes sense for domains that are topological spaces and was first considered by Bouleau [55].

The topology $\mathscr{T}_{\mathscr{F}(X)}^s$ can be properly finer than the topology of pointwise convergence $\mathscr{T}_{\mathscr{F}(X)}$ on Y^X - but not on $C(X, Y)$.

Example 21.4. On $[0, 1]$, let f be the characteristic function of the origin, and let $f_n \in C([0,1], \mathbb{R})$ be defined by

$$f_n(x) = \begin{cases} 1 - nx & \text{if } 0 \leq x \leq \frac{1}{n} \\ 0 & \text{otherwise.} \end{cases}$$

Clearly, $\langle f_n \rangle$ converges pointwise to f, but $\mathscr{T}_{\mathscr{F}(X)}^s$ convergence fails because the neighborhood $[\{0\}; \frac{1}{2}]^s[f]$ fails to contain any f_n.

Our next result describes at once when these topologies and their defining uniformities coincide on Y^X [45, Theorem 6.2].

Theorem 21.5. *Let $\mathscr{B}$ be a bornology on a metric space $\langle X, d \rangle$. The following conditions are equivalent:*

(1) *$\mathscr{B}$ is stable under small enlargements;*

(2) *for each metric space $\langle Y, \rho \rangle$, the uniformities $\Delta_{\mathscr{B}}$ and $\Delta_{\mathscr{B}}^s$ agree on Y^X;*

(3) *for each metric space $\langle Y, \rho \rangle$, $\mathscr{T}_{\mathscr{B}}^s = \mathscr{T}_{\mathscr{B}}$ on Y^X;*

(4) *$\mathscr{T}_{\mathscr{B}}^s = \mathscr{T}_{\mathscr{B}}$ on $\mathbb{R}^X$ where $\mathbb{R}$ is equipped with the usual metric.*

Proof. (1) $\Rightarrow$ (2). Let $\langle Y, \rho \rangle$ be an arbitrary metric target space. We always have $\Delta_{\mathscr{B}} \subseteq \Delta_{\mathscr{B}}^s$. For the reverse inclusion, fix $A \in \mathscr{B}$ and $\varepsilon > 0$. Choose $\delta > 0$ such that $S_d(A, \delta) \in \mathscr{B}$. Clearly, $[S_d(A, \delta); \varepsilon] \subseteq [A; \varepsilon]^s$.

The implications (2) $\Rightarrow$ (3) and (3) $\Rightarrow$ (4) are trivial. For (4) $\Rightarrow$ (1), suppose $B_0 \in \mathscr{B}$ but the bornology contains no enlargement of B_0. For each superset B of B_0 in the bornology and each $n \in \mathbb{N}$, choose $x_{B,n}$ with $d(x_{B,n}, B_0) < 1/n$ but $x_{B,n} \notin B$, and put $E_B = \{x_{B,n} : n \in \mathbb{N}\}$. For each such B, define $f_B \in \mathbb{R}^X$ by

$$f_B(x) = \begin{cases} 0 & \text{if } x \in E_B \\ 1 & \text{otherwise} \end{cases}.$$

Directing $\{B \in \mathscr{B} : B_0 \subseteq B\}$ by inclusion, the net $B \mapsto f_B$ is $\mathscr{T}_{\mathscr{B}}$-convergent to the constant function $f \equiv 1$, because if $B \in \mathscr{B}$ and $B_0 \subseteq B$, then $\forall B_1$ containing B we have $f_{B_1}(x) = 1$ whenever $x \in B$. However, the net fails to be $\mathscr{T}_{\mathscr{B}}^s$-convergent to f, as $[B_0; \frac{1}{2}]^s[f]$ contains no term of the net. $\square$

By the last result, the topology of pointwise convergence on Y^X agrees with the topology of strong pointwise convergence if and only if $X' = \emptyset$. Another consequence: if X is locally compact, then the topology of uniform convergence on compacta contains $\mathscr{T}_{\mathscr{K}(X)}^s$ and thus the topology of strong pointwise convergence, whatever the target space Y may be. Conversely, using the proof (4) $\Rightarrow$ (1) above, one can easily show that if the inclusion holds for the topologies, then X must be locally compact.

We now write $C_{\mathscr{B}}(X, Y)$ for the functions uniformly continuous on the bornology $\mathscr{B}$ and $C_{\mathscr{B}}^s(X, Y)$ for the functions that are strongly uniformly continuous on $\mathscr{B}$. We always have $C_{\mathscr{B}}^s(X, Y) \subseteq C_{\mathscr{B}}(X, Y) \cap C(X, Y)$. Because continuity is sequentially determined, whenever $\mathscr{K}(X) \subseteq \mathscr{B}$, we have $C_{\mathscr{B}}(X, Y) \subseteq C(X, Y)$.

We state the following classical result without proof.

Proposition 21.6. *Let $\langle X, d \rangle$ and $\langle Y, \rho \rangle$ be metric spaces and let $\mathscr{B}$ be a bornology on X. Then $C_{\mathscr{B}}(X, Y)$ is closed in Y^X equipped with $\mathscr{T}_{\mathscr{B}}$.*

A parallel result (that we do prove) holds for strong uniform continuity and strong uniform convergence on bornologies.

Proposition 21.7. *Let $\langle X, d \rangle$ and $\langle Y, \rho \rangle$ be metric spaces and let $\mathscr{B}$ be a bornology on X. Then $C_{\mathscr{B}}^s(X, Y)$ is closed in Y^X equipped with $\mathscr{T}_{\mathscr{B}}^s$.*

Proof. We show the complement of $C_{\mathscr{B}}^s(X, Y)$ is open. To this end, suppose $f \in Y^X$ fails to be strongly uniformly continuous on some $A \in \mathscr{B}$. Choose $\varepsilon > 0$ such that $\forall n \, \exists \{x_n, w_n\} \subseteq S_d(A, \frac{1}{n})$ with $d(x_n, w_n) < \frac{1}{n}$ but $\rho(f(x_n), f(w_n)) \geq 3\varepsilon$. But then $(f, g) \in [A; \varepsilon]^s$ yields $\rho(g(x_n), g(w_n)) \geq \varepsilon$ for all n sufficiently large, and so g is not strongly uniformly continuous on A. Thus, $g \notin C_{\mathscr{B}}^s(X, Y)$. $\square$

From the last result, we get continuity of the strong pointwise limit of a net of continuous functions.

122 Bornologies and Lipschitz Analysis

Corollary 21.8. *Let $\langle X,d \rangle$ and $\langle Y,\rho \rangle$ be metric spaces. Then $C(X,Y)$ is closed in Y^X equipped with $\mathscr{T}^s_{\mathscr{F}(X)}$.*

Proof. This is immediate from $C(X,Y) = C^s_{\mathscr{F}(X)}(X,Y)$. □

Proposition 21.9. *Let $\langle X,d \rangle$ and $\langle Y,\rho \rangle$ be metric spaces and let $\mathscr{B}$ be a bornology on X. Then restricted to $C^s_{\mathscr{B}}(X,Y)$, $\Delta^s_{\mathscr{B}}$ collapses to $\Delta_{\mathscr{B}}$. In particular, $\mathscr{T}^s_{\mathscr{B}} = \mathscr{T}_{\mathscr{B}}$ on the strongly uniformly continuous functions with respect to the bornology $\mathscr{B}$.*

Proof. It suffices to show for each $B \in \mathscr{B}$ and for each $\varepsilon > 0$, we have

$$[B; \frac{\varepsilon}{3}] \subseteq [B; \varepsilon]^s$$

with respect to their traces on $C^s_{\mathscr{B}}(X,Y) \times C^s_{\mathscr{B}}(X,Y)$. Suppose $(f,g) \in [B; \frac{\varepsilon}{3}]$ where both functions are strongly uniformly continuous on $\mathscr{B}$. There exists $\delta > 0$ such that if $x \in B$ and $d(x,w) < \delta$, then both $\rho(f(x),f(w)) < \frac{\varepsilon}{3}$ and $\rho(g(x),g(w)) < \frac{\varepsilon}{3}$. By the definition of $[B; \frac{\varepsilon}{3}]$, $\forall x \in B$, $\rho(f(x),g(x)) < \frac{\varepsilon}{3}$. It now follows that $\forall w \in S_d(B,\delta)$, $\rho(f(w),g(w)) < \varepsilon$ and so $(f,g) \in [B; \varepsilon]^s$. □

Since each continuous function on a UC-subset is strongly uniformly continuous, we obtain

Corollary 21.10. *Let $\langle X,d \rangle$ and $\langle Y,\rho \rangle$ be metric spaces, and let $\mathscr{B}$ be a bornology consisting of UC-subsets of X. Then on $C(X,Y)$, $\mathscr{T}^s_{\mathscr{B}} = \mathscr{T}_{\mathscr{B}}$.*

By Corollary 21.10, the topologies $\mathscr{T}^s_{\mathscr{B}}$ and $\mathscr{T}_{\mathscr{B}}$ can agree on $C(X,Y)$ without the bornology being stable under small enlargements, as is often the case when the bornology is either $\mathscr{F}(X)$ or $\mathscr{K}(X)$.

Putting together Propositions 21.7 and 21.9, we get our main result regarding strong uniform continuity and strong uniform convergence [45, Theorem 6.7].

Theorem 21.11. *Let $\langle X,d \rangle$ and $\langle Y,\rho \rangle$ be metric spaces, and let $\mathscr{B}$ be a bornology on X. Suppose $\langle f_\lambda \rangle$ is a net in $C^s_{\mathscr{B}}(X,Y)$ that is $\mathscr{T}_{\mathscr{B}}$-convergent to $f \in Y^X$. The following conditions are equivalent:*

(1) $f \in C^s_{\mathscr{B}}(X,Y)$;

(2) $\langle f_\lambda \rangle$ is $\mathscr{T}^s_{\mathscr{B}}$-convergent to f.

Specializing to the bornology of finite subsets, we see that the limit of a pointwise convergent net of continuous functions is continuous if and only if the net is strongly pointwise convergent. It is remarkable that this fact has not made its way into basic analysis texts. We state this fact in more formal terms. Again, this was first observed in a more general setting by Bouleau [55].

Corollary 21.12. *Let $\langle X,d \rangle$ and $\langle Y,\rho \rangle$ be metric spaces. Suppose $\langle f_\lambda \rangle$ is a net in $C(X,Y)$ that is pointwise convergent to $f \in Y^X$. The following conditions are equivalent:*

(1) $f \in C(X,Y) = C^s_{\mathscr{F}(X)}(X,Y)$;

(2) $\langle f_\lambda \rangle$ is $\mathscr{T}^s_{\mathscr{F}(X)}$-convergent to f.

Our final result characterizes coincidence of $\mathscr{T}_{\mathscr{B}}^s$ and $\mathscr{T}_{\mathscr{B}}$ on $C(X,Y)$ [46, Theorem 4.1]. It is anticipated by Corollary 21.10 because according to condition (6) of Theorem 18.3, each closed member of $\mathscr{B}_d^{uc}$ shields itself from closed sets.

Theorem 21.13. *Let $\mathscr{B}$ be a bornology on $\langle X, d \rangle$. The following conditions are equivalent:*

(1) $\overline{\mathscr{B}}$ *is shielded from closed sets;*

(2) *for each metric space $\langle Y, \rho \rangle$, the uniformities $\Delta_{\mathscr{B}}$ and $\Delta_{\mathscr{B}}^s$ agree on $C(X,Y)$;*

(3) *for each metric space $\langle Y, \rho \rangle$, the topologies $\mathscr{T}_{\mathscr{B}}$ and $\mathscr{T}_{\mathscr{B}}^s$ agree on $C(X,Y)$;*

(4) $\mathscr{T}_{\mathscr{B}} = \mathscr{T}_{\mathscr{B}}^s$ *on $C(X, \mathbb{R})$.*

Proof. We first observe that by continuity of all functions, neither of the function space topologies nor their standard uniformities are altered by replacing $\mathscr{B}$ by $\overline{\mathscr{B}}$.

(1) $\Rightarrow$ (2). With no assumptions $\Delta_{\mathscr{B}} \subseteq \Delta_{\mathscr{B}}^s$. For the reverse inclusion, since $[B; \varepsilon]^s = [\mathrm{cl}(B); \varepsilon]^s$ for continuous functions, it suffices to show that whenever $B \in \overline{\mathscr{B}}$ and $B_1 \in \overline{\mathscr{B}}$ is a shield for B and $\varepsilon > 0$, we have
$$[B_1; \varepsilon] \subseteq [\mathrm{cl}(B); \varepsilon]^s.$$

Suppose this inclusion fails; pick $(f, g) \in [B_1; \varepsilon] \backslash [\mathrm{cl}(B); \varepsilon]^s$. Then $\forall n \in \mathbb{N}$, $\exists x_n \in S_d(B, 1/n)$ with $\rho(f(x_n), g(x_n)) \geq \varepsilon$. Clearly, each x_n lies outside B_1. If $\langle x_n \rangle$ had a cluster point p, then $p \in \mathrm{cl}(B) \subseteq B_1$, and we have a contradiction as by continuity we would have $\rho(f(p), g(p)) \geq \varepsilon$. Otherwise, $\{x_n : n \in \mathbb{N}\}$ is a closed set disjoint from B_1, yet $D_d(\{x_n : n \in \mathbb{N}\}, B) = 0$, contradicting B_1 being a shield for B. We conclude that $\Delta_{\mathscr{B}} \supseteq \Delta_{\mathscr{B}}^s$ is also valid.

(2) $\Rightarrow$ (3) and (3) $\Rightarrow$ (4) are trivial.

(4) $\Rightarrow$ (1). Suppose $B_0 \in \overline{\mathscr{B}}$ has no shield in $\overline{\mathscr{B}}$. We may assume without loss of generality that B_0 is closed. Direct $\Sigma := \{B \in \overline{\mathscr{B}} : B_0 \subseteq B \text{ and } B = \mathrm{cl}(B)\}$ by inclusion. For each $B \in \Sigma$, pick C_B closed and disjoint from B yet with $D_d(C_B, B_0) = 0$. Then pick a Urysohn function $f_B \in C(X, [0, 1])$ with $f_B(B) = \{0\}$ and $f_B(C_B) = \{1\}$. If g is the zero function on X, the net $\langle f_B \rangle_{B \in \Sigma}$ $\mathscr{T}_{\mathscr{B}}$-converges to g, whereas $\mathscr{T}_{\mathscr{B}}^s$-convergence fails: the neighborhood $[B_0; \frac{1}{2}]^s[g]$ contains no f_B for $B \in \Sigma$. $\square$

We note in closing that if $\overline{\mathscr{B}}$ is shielded from closed sets, we cannot conclude that $\mathscr{B}$ itself is shielded from closed sets. Let $\langle X, d \rangle$ be a metric space containing a dense subset E such that $X \backslash E$ is infinite. Let $\mathscr{B} = \{A \cup F : A \subseteq E \text{ and } F \in \mathscr{F}(X)\}$. As the bornology $\mathscr{B}$ fails to have a closed base, by Proposition 12.12 it cannot be shielded from closed sets. Clearly $\overline{\mathscr{B}} = \mathscr{P}(X)$ is shielded from closed sets as X is a shield for any subset.

22. Uniform Convergence on Totally Bounded Subsets

We begin by showing that a Cauchy continuous function is strongly uniformly continuous on each totally bounded subset using a direct proof unlike the one used in Proposition 7.2 to obtain uniform continuity of the restrictions to totally bounded subsets. The bornology $\mathscr{T}\mathscr{B}_d(X)$ of totally bounded subsets is an example of a bornology on which strong uniform continuity collapses to uniform continuity of restrictions which is not in general shielded from closed sets (see Theorem 17.5 above).

Proposition 22.1. *Let $\langle X, d \rangle$ and $\langle Y, \rho \rangle$ be metric spaces, and let $f : X \to Y$. The following conditions are equivalent:*

(1) *f maps Cauchy sequences of X to Cauchy sequences of Y;*

(2) *f is strongly uniformly continuous on $\mathscr{T}\mathscr{B}_d(X)$;*

(3) *f is uniformly continuous on $\mathscr{T}\mathscr{B}_d(X)$.*

Proof. $(1) \Rightarrow (2)$. Suppose $f \in CC(X, Y)$ and A is a nonempty totally bounded subset of X. Let $\hat{A}$ be the closure of A in the completion $\langle \hat{X}, \hat{d} \rangle$ of $\langle X, d \rangle$. As X is dense in $\hat{X}$, there is a Cauchy continuous extension $\hat{f}$ of f from $\langle \hat{X}, \hat{d} \rangle$ to the completion of $\langle Y, \rho \rangle$. As $\hat{f}$ is strongly uniformly continuous on the compact set $\hat{A}$, it follows that f is strongly uniformly continuous on A.

The implication $(2) \Rightarrow (3)$ is trivial and $(3) \Rightarrow (1)$ was executed in the proof of Proposition 7.2. $\qquad \square$

From the last result and Theorem 21.11, we know that the Cauchy continuous functions are closed in Y^X with respect to the topology of strong uniform convergence on $\mathscr{T}\mathscr{B}_d(X)$, and that this topology agrees with the formally weaker topology of uniform convergence on $\mathscr{T}\mathscr{B}_d(X)$ when we restrict both to the Cauchy continuous functions. In this section we show that the Cauchy continuous functions are closed with respect to $\mathscr{T}_{\mathscr{T}\mathscr{B}_d(X)}$ and describe in concrete terms when this topology agrees with $\mathscr{T}^s_{\mathscr{T}\mathscr{B}_d(X)}$ on $C(X, Y)$. We also show that the Cauchy continuous functions are closed with respect to $\mathscr{T}_{\mathscr{K}(X)}$ on $C(X, Y)$ if and only if $\langle X, d \rangle$ is a complete metric space, where the Cauchy continuous functions coincide with $C(X, Y)$.

The first claim above is easy to verify.

Theorem 22.2. *Let $\langle X, d \rangle$ and $\langle Y, \rho \rangle$ be metric spaces, and let $\langle f_\lambda \rangle_{\lambda \in \Lambda}$ be a net of Cauchy continuous functions $\mathscr{T}_{\mathscr{T}\mathscr{B}_d(X)}$-convergent to $f \in Y^X$. Then f is Cauchy continuous.*

Proof. Let $\langle x_n \rangle$ be a Cauchy sequence in X and let $\varepsilon > 0$. Since $\{x_n : n \in \mathbb{N}\}$ is totally bounded, there exist $\lambda_0 \in \Lambda$ such that $\lambda \succeq \lambda_0 \Rightarrow \sup_{n \in \mathbb{N}} \rho(f_\lambda(x_n), f(x_n)) < \frac{\varepsilon}{3}$. Choose $k \in \mathbb{N}$, such that

$\forall n > j \geq k$, $\rho(f_{\lambda_0}(x_n), f_{\lambda_0}(x_j)) < \frac{\varepsilon}{3}$. By the triangle inequality, we get $\rho(f(x_n), f(x_j)) < \varepsilon$ whenever $n > j \geq k$.

We indicate a second proof: if $\mathscr{B}$ is a bornology on X and we have a net of functions $\langle f_\lambda \rangle$ each member of which is uniformly continuous on each member of $\mathscr{B}$, then if $f = \tau_{\mathscr{B}} - \lim f_\lambda$, then f is uniformly continuous on each member of $\mathscr{B}$ as well. Apply Proposition 22.1. $\qquad\square$

By Theorem 21.13, with respect to $\mathscr{T}\mathscr{B}_d(X)$, the topology of strong uniform convergence collapses to the topology of uniform convergence on $C(X, Y)$ exactly when $\mathscr{T}\mathscr{B}_d(X)$ is shielded from closed sets. As we have seen in Section 12, this may not happen. We next give concrete necessary and sufficient conditions for this to occur.

Theorem 22.3. *For a metric space $\langle X, d \rangle$, the following conditions are equivalent:*

(1) *$\mathscr{T}\mathscr{B}_d(X)$ is shielded from closed sets;*

(2) *whenever $B \in \mathscr{C}_0(X) \cap \mathscr{T}\mathscr{B}_d(X)$, there exists a totally bounded superset B_0 such that each Cauchy sequence $\langle x_n \rangle$ in $X \backslash B_0$ with $\lim_{n\to\infty} d(x_n, B) = 0$ converges.*

Proof. (1) $\Rightarrow$ (2). Suppose B_0 is a totally bounded shield for B, yet $\langle x_n \rangle$ is a nonconvergent Cauchy sequence in $X \backslash B_0$ with $\lim_{n\to\infty} d(x_n, B) = 0$. As the sequence cannot cluster as well, $C := \{x_n : n \in \mathbb{N}\}$ is a closed set disjoint from B_0 with $D_d(C, B) = 0$, a contradiction.

(2) $\Rightarrow$ (1). Assume (2) holds. To show $\mathscr{T}\mathscr{B}_d(X)$ is shielded from closed sets, as the bornology has a closed base, it suffices to show that each closed nonempty totally bounded set has a totally bounded shield. Fix $B \in \mathscr{T}\mathscr{B}_d(X) \cap \mathscr{C}_0(X)$, and let B_0 satisfy condition (2). Suppose $C \in \mathscr{C}_0(X)$ with $C \cap B_0 = \emptyset$ yet $D_d(B, C) = 0$. For each $n \in \mathbb{N}$ choose $c_n \in C$ with $d(c_n, B) < \frac{1}{n}$. Since $\langle c_n \rangle$ is asymptotic to a sequence in B which must have a Cauchy subsequence, $\langle c_n \rangle$ has a Cauchy subsequence as well. By condition (2), the subsequence converges to a point of $B \cap C$, violating $C \cap B_0 = \emptyset$. This shows that B_0 is a shield for B. $\qquad\square$

Corollary 22.4. *Let $\langle X, d \rangle$ and $\langle Y, \rho \rangle$ be metric spaces. Then $\mathscr{T}_{\mathscr{T}\mathscr{B}_d(X)} = \mathscr{T}^s_{\mathscr{T}\mathscr{B}_d(X)}$ on $C(X, Y)$ if and only if whenever B is a nonempty closed and totally bounded subset of X, there exists a totally bounded superset B_0 such that each Cauchy sequence $\langle x_n \rangle$ in $X \backslash B_0$ with $\lim_{n\to\infty} d(x_n, B) = 0$ converges.*

Noting that the Bourbaki bounded subsets of a metric space form a bornology with closed base, the reader should be able to easily prove the next two parallel results.

Theorem 22.5. *For a metric space $\langle X, d \rangle$, the following conditions are equivalent:*

(1) *$\mathscr{B}\mathscr{B}_d(X)$ is shielded from closed sets;*

(2) *whenever $B \in \mathscr{C}_0(X) \cap \mathscr{B}\mathscr{B}_d(X)$, there exists a Bourbaki bounded superset B_0 such that each Bourbaki-Cauchy sequence $\langle x_n \rangle$ in $X \backslash B_0$ with $\lim_{n\to\infty} d(x_n, B) = 0$ has a cluster point.*

Corollary 22.6. *Let $\langle X, d \rangle$ and $\langle Y, \rho \rangle$ be metric spaces. Then $\mathscr{T}_{\mathscr{B}\mathscr{B}_d(X)} = \mathscr{T}^s_{\mathscr{B}\mathscr{B}_d(X)}$ on $C(X, Y)$ if and only if whenever B is a nonempty closed and Bourbaki bounded subset of X, there exists a Bourbaki bounded superset B_0 such that each Bourbaki-Cauchy sequence $\langle x_n \rangle$ in $X \backslash B_0$ with $\lim_{n\to\infty} d(x_n, B) = 0$ has a cluster point.*

We have observed in Theorem 22.2 that uniform convergence on totally bounded subsets preserves Cauchy continuity. Our next result shows that convergence with respect to the coarser topology of uniform convergence on relatively compact subsets preserves Cauchy continuity if and only each continuous function on $\langle X, d \rangle$ is already Cauchy continuous.

Theorem 22.7. *Let $\langle X, d \rangle$ be a metric space. The following conditions are equivalent:*

(1) *the metric d is complete;*

(2) *for each metric space $\langle Y, \rho \rangle$, the Cauchy continuous functions are closed in Y^X with respect to $\tau_{\mathscr{K}(X)}$;*

(3) *the Cauchy continuous real-valued functions are closed in $C(X, \mathbb{R})$ with respect to $\tau_{\mathscr{K}(X)}$.*

Proof. (1) $\Rightarrow$ (2). With completeness of d, $\mathscr{K}(X) = \mathscr{T}\mathscr{B}_d(X)$ and so uniform convergence on relatively compact subsets coincides with uniform convergence on totally bounded subsets. Condition (2) now follows from Theorem 22.2. The implication (2) $\Rightarrow$ (3) is trivial. For (3) $\Rightarrow$ (1), if d is not complete, we produce a sequence of Cauchy continuous real-valued functions uniformly convergent on compact subsets to a continuous function that is not Cauchy continuous.

Let $\langle \hat{X}, \hat{d} \rangle$ be the completion of $\langle X, d \rangle$ and fix $p \in \hat{X} \backslash X$. Let $f : X \to \mathbb{R}$ be defined by $f(x) = \hat{d}(x, p)$. Clearly, $\frac{1}{f}$ is continuous but not Cauchy continuous, but if $g_n = \frac{1}{\sup\{f, 1/n\}}$, then $\langle g_n \rangle$ is a sequence of Cauchy continuous functions - in fact uniformly continuous functions - that is $\tau_{\mathscr{K}(X)}$-convergent to $\frac{1}{f}$ because if C is a nonempty compact subset of X, then $\inf\{\hat{d}(c, p) : c \in C\} > 0$.   □

In the next result, we replace the Cauchy continuous functions in the last theorem by the uniformly continuous functions

Theorem 22.8. *Let $\langle X, d \rangle$ be a metric space. The following conditions are equivalent:*

(1) *$\langle X, d \rangle$ is a UC-space;*

(2) *for each metric space $\langle Y, \rho \rangle$, the uniformly continuous functions are closed in Y^X with respect to $\tau_{\mathscr{K}(X)}$;*

(3) *the uniformly continuous real-valued functions are closed in $C(X, \mathbb{R})$ with respect to $\tau_{\mathscr{K}(X)}$.*

Proof. (1) $\Rightarrow$ (2). If a net of uniformly continuous functions is $\tau_{\mathscr{K}(X)}$-convergent to $f \in Y^X$, then f is continuous. But condition (1) means that f is then uniformly continuous. The implication (2) $\Rightarrow$ (3) is trivial. For (3) $\Rightarrow$ (1), we prove the contrapositive. Suppose X is not a UC-space. Then we can find disjoint nonempty closed subsets A and B for which $\inf\{d(a, b) : a \in A, b \in B\} = 0$. Let f be a Urysohn function for the two closed sets which must fail be uniformly continuous. But since f is lower bounded, by Proposition 2.2, f is the pointwise limit of an increasing sequence of Lipschitz functions, and an easy calculation shows that the convergence must be uniform on compact subsets (in fact, one can just apply the classical theorem of Dini [90, p. 258]). Thus condition (3) fails.   □

One wonders if there is a parallel theorem for preservation of uniform continuity with respect to the topology of uniform convergence on totally bounded subsets. Indeed, there is.

Theorem 22.9. *Let $\langle X, d \rangle$ be a metric space. The following conditions are equivalent:*

(1) *the completion $\langle \hat{X}, \hat{d} \rangle$ of $\langle X, d \rangle$ is a UC-space;*

(2) *for each metric space $\langle Y, \rho \rangle$, the uniformly continuous functions are closed in Y^X with respect to $\tau_{\mathscr{T}\mathscr{B}_d(X)}$;*

(3) *the uniformly continuous real-valued functions are closed in $CC(X, \mathbb{R})$ with respect to $\tau_{\mathscr{T}\mathscr{B}_d(X)}$.*

Proof. If condition (1) holds, then by Theorem 19.11, each Cauchy continuous function on X is uniformly continuous, and by Theorem 22.2 the Cauchy continuous functions are closed in Y^X equipped with the topology $\tau_{\mathscr{T}\mathscr{B}_d(X)}$. Thus (2) holds. The assertion (2) $\Rightarrow$ (3) obviously holds. For (3) $\Rightarrow$ (1), we prove the contrapositive.

If the completion $\langle \hat{X}, \hat{d} \rangle$ fails to be a UC-space, then by Theorem 22.8, $UC(\hat{X}, \mathbb{R})$ is not closed in $C(\hat{X}, \mathbb{R})$ equipped with the topology $\tau_{\mathscr{K}(\hat{X})}$; that is, there exists a net of uniformly continuous functions on the completion converging uniformly on relatively compact subsets of the completion to a function f which is continuous but not uniformly continuous. Since the completion is a complete metric space, f is Cauchy continuous. Obviously, the restriction of f to X while Cauchy continuous cannot be uniformly continuous by the density of X in $\hat{X}$, and the convergence of the net of functions restricted to X is uniform on totally bounded subsets of X because each totally bounded subset of X has compact closure in $\langle \hat{X}, \hat{d} \rangle$. This shows that condition (3) fails. $\square$

23. Where Must Each Member of a Class of Locally Lipschitz Functions be Lipschitz?

We next present a folk theorem as our first result [118].

Theorem 23.1. *Let $\langle X, d \rangle$ be a metric space and let A be a nonempty subset. The following statements are equivalent:*

(1) *A is a finite set;*

(2) *for each function f from $\langle X, d \rangle$ to an arbitrary metric space $\langle Y, \rho \rangle$, $f|_A$ is Lipschitz;*

(3) *for each continuous real-valued function on X, $f|_A$ is Lipschitz.*

Proof. The implication string (1) $\Rightarrow$ (2) $\Rightarrow$ (3) clearly holds. For (3) $\Rightarrow$ (1) we prove the contrapositive. Suppose $A \subseteq X$ is infinite. We consider two mutually exclusive and exhaustive cases: (a) A is relatively compact; (b) there exists a sequence $\langle a_n \rangle$ in A with distinct terms that fails to cluster. In (a), let $\langle a_n \rangle$ be a sequence in A with distinct terms convergent to some $p \in X$. Then $x \mapsto \sqrt{d(x, p)}$ belongs to $C(X, \mathbb{R})$ - in fact to $UC(X, \mathbb{R})$ - but when restricted to A, it fails to be Lipschitz. In case (b), $\{a_n : n \in \mathbb{N}\}$ is a closed set without limit points, so by the Tietze extension theorem, there exists $f \in C(X, \mathbb{R})$ such that $f(a_n) = n \cdot d(a_n, a_1)$ and $f(a_1) = 0$, and again we have a globally continuous function that fails to be Lipschitz when restricted to A. $\square$

Example 23.2. In the third condition of Theorem 23.1, $C(X, \mathbb{R})$ can neither be replaced by the bounded continuous real-valued functions nor by $UC(X, \mathbb{R})$. For example, if $A = \mathbb{N}$ as a subset of the real line equipped with the usual metric, then each bounded continuous real-valued function on $\mathbb{R}$ and each member of $UC(\mathbb{R}, \mathbb{R})$ is Lipschitz restricted to A.

We include the next result for completeness; its simple proof is left to the reader.

Proposition 23.3. *Let $\langle X, d \rangle$ and $\langle Y, \rho \rangle$ be metric spaces and let A be a nonempty subset of X. If $f \in C(X, Y)$ and $f|_A$ is λ-Lipschitz, then $f|_{cl(A)}$ is λ-Lipschitz.*

Theorem 23.4. *Let $\langle X, d \rangle$ be a metric space and let A be a nonempty subset. The following statements are equivalent:*

(1) *A is relatively compact;*

(2) *for each locally Lipschitz function f from $\langle X, d \rangle$ to an arbitrary metric space $\langle Y, \rho \rangle$, $f|_A$ is Lipschitz;*

(3) *for each real-valued locally Lipschitz function f on X, $f|_A$ is Lipschitz.*

Proof. We first look at (1) $\Rightarrow$ (2). Suppose A is relatively compact and f is locally Lipschitz, yet $f|_A$ is not Lipschitz. We consider two possible exhaustive cases for A and f:

(a) $\exists\, \delta > 0$ such that $\forall n \in \mathbb{N}\ \exists\{a_n, b_n\} \subseteq A$ with $d(a_n, b_n) \geq \delta$ and $\rho(f(a_n), f(b_n)) \geq nd(a_n, b_n)$;

(b) $\forall n \in \mathbb{N}$, $\exists\{a_n, b_n\} \subseteq A$ with $0 < d(a_n, b_n) < \frac{1}{n}$ but $\rho(f(a_n), f(b_n)) \geq nd(a_n, b_n)$.

In case (a) it is clear that $f|_A$ is unbounded which is impossible since $f \in C(X, Y)$ and in case (b), f will fail to be locally Lipschitz at any cluster point of $\langle a_n \rangle$. Thus if condition (2) fails, then condition (1) fails.

The implication (2) $\Rightarrow$ (3) is trivial. Finally, for (3) $\Rightarrow$ (1), suppose A is not relatively compact, and let $\langle a_n \rangle$ be a sequence of distinct terms in A that has no cluster point in X. For each $n \geq 2$, put $\delta_n := \min\{\frac{1}{n}, \frac{1}{3}d(a_n, \{a_j : j \neq n\})\}$, and let $f_n : S_d(a_n, \delta_n) \to \mathbb{R}$ be a Lipschitz function mapping a_n to $nd(a_1, a_n)$ and which is zero on $\{x : \frac{1}{2}\delta_n \leq d(x, a_n) \leq \delta_n\}$. Finally, put

$$f(x) := \begin{cases} f_n(x) & \text{if } \exists n \geq 2 \text{ with } d(x, a_n) \leq \delta_n \\ 0 & \text{otherwise.} \end{cases}$$

Arguing using the discreteness of the family $\{S_d(a_n, \delta_n) : n \geq 2\}$, it is easy to see that f is well-defined and locally Lipschitz, while for each $n \geq 2, |f(a_n) - f(a_1)|/d(a_n, a_1) = n$. $\square$

Corollary 23.5. *Let f be a function from $\langle X, d\rangle$ to $\langle Y, \rho \rangle$. Then f is locally Lipschitz if and only if for each nonempty relatively compact subset A of X, $f|_A$ is Lipschitz.*

Proof. Necessity follows from the last result. For sufficiency, f restricted to the range of each convergent sequence is Lipschitz. Apply Proposition 8.5. $\square$

Recall from Proposition 7.2 that the Cauchy continuous functions are those that are uniformly continuous on totally bounded subsets of the metric space. However, even if a function is globally uniformly continuous, it need not be Lipschitz when restricted to totally bounded subsets (consider the square-root function on $[0, \infty)$).

Theorem 23.6. *Let $\langle X, d\rangle$ be a metric space and let A be a nonempty subset. The following statements are equivalent:*

(1) *A is totally bounded;*

(2) *for each Cauchy-Lipschitz function f from $\langle X, d\rangle$ to an arbitrary metric space $\langle Y, \rho \rangle$, $f|_A$ is Lipschitz;*

(3) *for each uniformly locally Lipschitz function f from $\langle X, d\rangle$ to an arbitrary metric space $\langle Y, \rho \rangle$, $f|_A$ is Lipschitz;*

(4) *for each real-valued uniformly locally Lipschitz function f on X, $f|_A$ is Lipschitz.*

Proof. (1) $\Rightarrow$ (2). Suppose f is Cauchy-Lipschitz yet f fails to be Lipschitz on some totally bounded subset B. Choose for each $n \in \mathbb{N}$ points $x_n \neq w_n$ in B with $\rho(f(x_n), f(w_n)) > nd(x_n, w_n)$. By passing to a subsequence, we may assume that $\langle x_n \rangle$ is a Cauchy sequence. Also, since f is Cauchy continuous, $f(B)$ is a bounded subset of Y.

We now consider two mutually exclusive and exhaustive cases for the sequences $\langle x_n \rangle$ and $\langle w_n \rangle$: (a) $\delta := \limsup_{n \to \infty} d(x_n, w_n) > 0$; (b) $\lim_{n \to \infty} d(x_n, w_n) = 0$. In the first case, we have for infinitely many n that $\rho(f(x_n), f(w_n)) > \frac{n\delta}{2}$ which violates the boundedness of $f(B)$. In the second case, $x_1, w_1, x_2, w_2, x_3, w_3, \ldots$ is a Cauchy sequence in X on which f fails to be Lipschitz, and again we have a contradiction.

The implications $(2) \Rightarrow (3)$ and $(3) \Rightarrow (4)$ are obviously true. For $(4) \Rightarrow (1)$, suppose A is not totally bounded. For some $\delta > 0$, there is a sequence of distinct terms $\langle a_n \rangle$ in A such that $d(a_k, a_n) \geq \delta$ whenever $k \neq n$. By construction, for each $x \in X, S_d(x, \frac{\delta}{3})$ can intersect at most one ball in the family $\{S_d(a_n, \frac{\delta}{3}) : n \in \mathbb{N}\}$. For each $n \geq 2$, let $f_n : S_d(a_n, \frac{\delta}{3}) \to \mathbb{R}$ be defined by

$$f_n(x) = (\frac{\delta}{3} - d(x, a_n))\frac{3n}{\delta}d(a_1, a_n).$$

Then f_n is Lipschitz on $S_d(a_n, \frac{\delta}{3})$ and $f_n(a_n) = nd(a_1, a_n)$. It follows that

$$f(x) := \begin{cases} f_n(x) & \text{if } \exists n \geq 2 \text{ with } d(x, a_n) < \frac{\delta}{3} \\ 0 & \text{otherwise} \end{cases}$$

is Lipschitz on each open ball of radius $\frac{\delta}{3}$ as f can be realized as the maximum of the zero function and at most one other Lipschitz function on the ball. However, as $f(a_1) = 0$ so that $|f(a_1) - f(a_n)| = nd(a_1, a_n)$ for each $n \geq 2$, $f \notin \text{Lip}(A, \mathbb{R})$. □

Corollary 23.7. *Let f be a function from $\langle X, d \rangle$ to $\langle Y, \rho \rangle$. Then f is Cauchy-Lipschitz if and only if for each nonempty totally bounded subset A of X, $f|_A$ is Lipschitz.*

Proof. Necessity follows from Theorem 23.6. For sufficiency, simply note that the range of each Cauchy sequence in X is a totally bounded subset of X. □

With our parallel results for common sets of boundedness of locally Lipschitz functions, Cauchy-Lipschitz functions, uniformly locally Lipschitz functions, and Lipschitz in the small functions in mind, we might anticipate that each globally defined Lipschitz in the small function will be Lipschitz on a set A precisely when A is Bourbaki bounded. But the "prime mover" had other plans: while each globally defined Lipschitz in the small function is Lipschitz on each Bourbaki bounded set, the function can be Lipschitz more generally. For example, it is clear that if X a subset of a normed linear space equipped with the metric determined by the norm, then each globally defined Lipschitz in the small function is Lipschitz restricted to any convex subset of X, metrically bounded or otherwise. Much more generally, it is easily seen that a globally defined Lipschitz in the small function is Lipschitz when restricted to a nonempty subset A of the metric space $\langle X, d \rangle$ such that $d|_{A \times A}$ is an *almost convex metric*: whenever $a_1 \neq a_2$ in A and α satisfies $d(a_1, a_2) < \alpha$, then for each $\beta \in (0, \alpha)$, there exists $a_3 \in A$ with $d(a_1, a_3) < \beta$ and $d(a_2, a_3) < \alpha - \beta$. It can be shown that a metric is almost convex if and only if the enlargement operator for subsets of the metric space is an additive function of the radius [17, p. 108].

Garrido and Jaramillo [84] called a metric space $\langle X, d \rangle$ *small determined* if each Lipschitz in the small function h on X is already Lipschitz on X. By Theorem 8.11, this is equivalent to saying that the modulus of continuity function associated with each Lipschitz in the small function has an affine majorant. That this property is determined by such behavior for real-valued functions is an immediate consequence of an earlier result.

Proposition 23.8. *Let $\langle X, d \rangle$ metric space such that each real-valued Lipschitz in the small function on X is already Lipschitz. Then $\langle X, d \rangle$ is small determined.*

Proof. Suppose $\langle Y, \rho \rangle$ is a second metric space and $h : X \to Y$ is Lipschitz in the small. Whenever $f \in \mathrm{Lip}(Y, \mathbb{R})$, clearly $f \circ h$ remains Lipschitz in the small and so by assumption the composition is Lipschitz. By Theorem 4.6, h is Lipschitz. $\qquad\square$

Garrido and Jaramillo also proved in [84] that a metric space is small determined if and only if each member of $UC(X, \mathbb{R})$ is a uniform limit of a sequence of real-valued Lipschitz functions. We shall obtain their approximation result in the next section using somewhat different methods.

It is natural to call a subset A of a metric space $\langle X, d \rangle$ *small determined* if for each globally defined Lipschitz in the small function f with values in an arbitrary metric space, $f|_A$ is Lipschitz. An obvious modification of the proof of Proposition 23.8 shows that this property of subsets is determined by real-valued functions.

Small determined subsets need not be bounded; for example, if $\langle X, d \rangle$ is a metric subspace of a normed linear space, then any convex subset of X is small determined.

Proposition 23.9. *Let A be a nonempty subset of a metric space $\langle X, d \rangle$.*

(a) *if A is Bourbaki bounded, then A small determined;*

(b) *if A is both bounded and small determined, then A is Bourbaki bounded.*

Proof. For (a), let $\langle Y, \rho \rangle$ be a second metric space and suppose $f : X \to Y$ is Lipschitz in the small. Choose $\delta > 0$ and $\alpha > 0$ such that whenever $d(x_1, x_2) < \delta$, then $\rho(f(x_1), f(x_2)) \leq \alpha d(x_1, x_2)$. By Proposition 9.4 $f|_A$ is a bounded function, so with $\mu = \mathrm{diam}_\rho \, f(A)$, we see that if $\{a_1, a_2\} \subseteq A$ and $d(a_1, a_2) \geq \delta$, then $\rho(f(a_1), f(a_2)) \leq \frac{\mu}{\delta} d(a_1, a_2)$. In view of these two estimates, the restriction of f to A is Lipschitz.

For (b) let $f : X \to \mathbb{R}$ be Lipschitz in the small. Since A is small determined, $f|_A$ is Lipschitz and since A is bounded, $f(A)$ is bounded. By Proposition 9.4, $A \in \mathscr{BB}_d(X)$. $\qquad\square$

One can prove that a function between metric spaces is Lipschitz on each Bourbaki bounded subset if and only if it is Lipschitz when restricted to the range of each Bourbaki-Cauchy sequence [37], but we will not pursue this here.

The small determined subsets of a metric space, while an hereditary family, are not in general stable under finite unions. For example, consider $A = \{(0, \beta) : \beta \in \mathbb{R}\}$ and $B = \{(1, \beta) : \beta \in \mathbb{R}\}$ and let our metric space X be $A \cup B$ equipped with the Euclidean metric of the plane. While A and B being convex subsets are small determined, $A \cup B$ fails to be. Let $f : X \to \mathbb{R}$ be the zero function on A and let f map $(1, \beta)$ to β for each $\beta \in \mathbb{R}$. As $|f(x_1) - f(x_2)| \leq d(x_1, x_2)$ whenever $d(x_1, x_2) \leq \frac{1}{2}$, f is Lipschitz in the small while $f \notin \mathrm{Lip}(X, \mathbb{R})$.

We next state necessary and sufficient conditions for a subset of a metric space to be small determined as discover by Leung and Tang [114, Theorem 1]. The statement involves ε-step territories and the O'Farrell metric used in the proof Hejcman's Theorem.

Theorem 23.10. *Let A be a nonempty subset of a metric space $\langle X, d \rangle$. The following conditions are equivalent:*

(1) *whenever $\langle Y, d \rangle$ is a metric space and $f : X \to Y$ is Lipschitz in the small, then $f|_A$ is Lipschitz;*

(2) *each real-valued Lipschitz in the small function is Lipschitz when restricted to A;*

(3) *the set A has these three structural properties:*

    (a) $\forall \varepsilon > 0$, *A intersects only finitely many ε-step territories;*

    (b) $\forall \varepsilon > 0$ *there exists $L > 0$ such that $d_\varepsilon(x, w) \leq L \cdot d(x, w)$ whenever $x \simeq_\varepsilon w$*
      and $\{x, w\} \subseteq A$;

    (c) $\forall \varepsilon > 0$, *there exists $K > 0$ such that whenever E_1 and E_2 are distinct ε-step*
      territories, whenever $\{x_1, x_2\} \subseteq E_1 \cap A$ and $\{y_1, y_2\} \subseteq E_2 \cap A$ we have
$$max\,\{d(x_1, x_2), d(y_1, y_2)\} \leq K \cdot max\,\{d(x_1, y_1), d(x_2, y_2)\}.$$

Proof. As the proof as a whole is quite lengthy, we only prove (3) $\Rightarrow$ (1) here. Let $\langle Y, \rho \rangle$ be a second metric space and let $f : X \to Y$ be Lipschitz in the small. Then for some $\varepsilon > 0$ there exists $C > 0$ such that $d(x, w) \leq \varepsilon \Rightarrow \rho(f(x), f(w)) \leq Cd(x, w)$. By condition 3(a) let $E_1, E_2, \ldots, E_n$ be the ε-step territories that intersect A. Note that for each $i \leq n$, $f|_{E_i}$ is C-Lipschitz with respect to d_ε. Pick $e_i \in A \cap E_i$ for each $i \leq n$, and put

$$M := 1 \vee \frac{1}{\varepsilon} \max\{d(e_i, e_j) : i \neq j\} \vee \max\{\rho(f(e_i), f(e_j)) : i \neq j\}.$$

To show f restricted to A is Lipschitz, let x and w be distinct points of A. If $x \simeq_\varepsilon w$, using condition 3(b), we compute $\rho(f(x), f(w)) \leq Cd_\varepsilon(x, w) \leq CLd(x, w)$. Otherwise, $x \in A \cap E_i$ and $w \in A \cap E_j$ for different i and j. Since $d(x, w) \geq \varepsilon$, we obtain

$$d(e_i, e_j) \leq \frac{d(e_i, e_j)}{\varepsilon} \cdot d(x, w) \leq Md(x, w).$$

In view of this inequality, employing 3(c) and remembering that $M \geq 1$ gives

$$max\,\{d(x, e_i), d(w, e_j)\} \leq K \cdot max\,\{d(x, w), d(e_i, e_j)\} \leq KMd(x, w).$$

Thus by 3(b) we now obtain

$$max\,\{d_\varepsilon(x, e_i), d_\varepsilon(w, e_j)\} \leq L \cdot max\,\{d(x, e_i), d(w, e_j)\} \leq LKMd(x, w).$$

Since f restricted to each ε-territory is C-Lipschitz with respect to d_ε, we finally obtain

$$\rho(f(x), f(w)) \leq \rho(f(x), f(e_i)) + \rho(f(e_i), f(e_j)) + \rho(f(e_j), f(w))$$

$$\leq C[d_\varepsilon(x, e_i) + d_\varepsilon(e_j, w)] + \varepsilon M$$

$$\leq 2CLKMd(x, w) + Md(x, w) = M(2CLK + 1)d(x, w).$$

As these two cases are exhaustive for $x \neq w$ in A, we conclude that $f|_A$ is Lipschitz. $\qquad \square$

Example 23.11. We present three non-small-determined metric subspaces X of the Euclidean plane that satisfy exactly two of the subconditions comprising condition (3) of Theorem 23.10 where $A = X$.

- $X = \{0, 1\} \times [0, \infty)$ satisfies 3(a) and 3(b);
- $X = (\{0, 1\} \times [0, \infty)) \cup \{(t, 0) : 0 \leq t \leq 1\}$ satisfies 3(a) and 3(c);
- $X = \mathbb{N} \times [0, 1]$ satisfies 3(b) and 3(c).

Unlike the locally Lipschitz functions and the Cauchy-Lipschitz functions, neither the class of uniformly locally Lipschitz functions nor the class of Lipschitz in the small functions can in general be characterized as a class of functions $\mathfrak{F}$ that are Lipschitz when restricted to each member of a prescribed family of subsets $\mathscr{A}$ of the underlying metric space $\langle X, d \rangle$. This is so even if we restrict our attention to real-valued functions. In particular, a real-valued function that is Lipschitz when restricted to each small determined subset need not be Lipschitz in the small. A single construction, as provided by Beer, García-Lirola, and Garrido [33] suffices for both cases.

Example 23.12. Consider the bounded metric space $\langle X, d \rangle$ where $X = \mathbb{N}^2$ and d is given by $d((n, k), (m, l)) = 1$ if $k \neq l$ and $d((n, k), (m, k)) = \frac{1}{k}$ if $n \neq m$. As the small determined subsets coincide with the Bourbaki bounded subsets, and each Bourbaki bounded subset is finite, the family of small determined subsets is as small as it can possibly be.

Assume $\mathscr{A}$ were a family of subsets of $\mathbb{N}^2$ such that the class of real-valued uniformly locally Lipschitz functions on X coincides with the set of real-valued functions that are Lipschitz when restricted to each member of $\mathscr{A}$. We intend to show that $\mathscr{A}$ is made up of finite sets. Let $A \in \mathscr{A}$, and consider

$$I := \{k \in \mathbb{N} : (n, k) \in A \text{ for some } n \in \mathbb{N}\}.$$

Define $f : X \to \mathbb{R}$ by $f(n, k) = k$. As f is constant on each open ball of radius 1, the function is uniformly locally Lipschitz. Now if I were infinite, there would be a strictly increasing sequence $\langle k_i \rangle$ in I and for each i a positive integer n_i such that $(n_i, k_i) \in A$. We have for $i > 1$

$$|f(n_i, k_i) - f(n_1, k_1)| = k_i - k_1 = (k_i - k_1)d((n_i, k_i), (n_1, k_1)),$$

which shows that $f|_A$ is not Lipschitz, a contradiction. We conclude that I is finite.

Now fix k in our finite set I. Define the function $g_k : X \to \mathbb{R}$ by $g_k(n, k) = n$ and $g_k(n, l) = 0$ if $l \neq k$. Then g_k is constant on each open ball of radius $\frac{1}{k}$ and so g_k is uniformly locally Lipschitz. This implies that $g_k|_A$ is Lipschitz. It follows that the set $A_k := \{n \in \mathbb{N} : (n, k) \in A\}$ cannot be infinite, else A is bounded while $g_k(A)$ is unbounded. This shows that $A = \cup_{k \in I} A_k$ is also finite.

Finally, $h : X \to \mathbb{R}$ defined by $h(n, k) = n$ is not uniformly locally Lipschitz. However h is Lipschitz when restricted to each finite subset as is any real-valued function defined on X. That is a contradiction.

We note that the functions f and g_k defined above are actually Lipschitz in the small. Thus, our construction also serves to show that the real-valued Lipschitz in the small functions on $\langle X, d \rangle$ cannot in general be described as a class of functions that are Lipschitz when restricted to certain subsets.

As we have just seen, some of our classes of Lipschitz-type functions can be characterized as those that are Lipschitz when restricted to a particular family of subsets of the domain, whereas others cannot be so characterized. For classes of continuous real-valued functions of the first type, we can state a meta-theorem with respect to the stability of our function class $\mathfrak{F}$ with respect to pointwise product and reciprocation [33].

Proposition 23.13. *Let $\mathscr{A}$ be a family of nonempty subsets of a metric space $\langle X, d \rangle$ and let $\mathfrak{F}$ be the family of continuous real-valued functions that are Lipschitz when restricted to each member of $\mathscr{A}$.*

(a) *$\mathfrak{F}$ is stable under pointwise product if and only if each member of $\mathscr{A}$ is bounded;*

(b) *$\mathfrak{F}$ is stable under reciprocation if and only if each member of $\mathscr{A}$ is relatively compact.*

The straight-forward proof is left to the interested reader.

24. Real-Valued Lipschitz Functions and Classes of Locally Lipschitz Functions

In this section, we show that membership to each of our four classes of locally Lipschitz functions is determined by following compositions with Lipschitz real-valued functions. Most of our results here appear in a recent article of Beer and Garrido [37]. We start with a tool theorem that is an immediate consequence of the Garrido-Jaramillo theorem and which is adequate for three of the four classes.

Proposition 24.1. *Let* $\langle X, d \rangle$ *and* $\langle Y, \rho \rangle$ *be metric spaces. Let* $\mathscr{B}$ *be a family of nonempty subsets of* $\langle X, d \rangle$ *and put*

$$\Delta_{\mathscr{B}} := \{ h \in C(X, Y) : \forall B \in \mathscr{B}, \ h|_B \text{ is Lipschitz} \}.$$

Then $h \in C(X, Y)$ *belongs to* $\Delta_{\mathscr{B}}$ *if and only if for each* $f \in Lip(Y, \mathbb{R})$ *and for each* $B \in \mathscr{B}$ *we have* $(f \circ h)|_B \in Lip(B, \mathbb{R})$.

Proof. For necessity, let $h \in \Delta_{\mathscr{B}}$. Then for each $B \in \mathscr{B}$ the restriction of h to B is Lipschitz so that $f \circ h$ is Lipschitz when restricted to B whenever $f : Y \to \mathbb{R}$ is Lipschitz, For sufficiency, by Theorem 4.6, for each $B \in \mathscr{B}$ we see that $h|_B$ is Lipschitz. But by definition, this means that $h \in \Delta_{\mathscr{B}}$. $\square$

We get these three corollaries of Proposition 24.1. In the first, the family of subsets $\mathscr{B}$ is the family of relatively compact sets, and in the second, $\mathscr{B}$ is the family of totally bounded sets, and in the third it is the family of all open balls of a fixed radius. In each case, necessity is obvious as each class of functions is stable under composition and each contains the Lipschitz functions.

Corollary 24.2. *Let* $\langle X, d \rangle$ *and* $\langle Y, \rho \rangle$ *be metric spaces and let* $h : X \to Y$. *Then* h *is locally Lipschitz if and only if whenever* $f \in Lip(Y, \mathbb{R})$, *the function* $f \circ h$ *is locally Lipschitz.*

Proof. This follows from Proposition 24.1 in view of Corollary 23.5. $\square$

Corollary 24.3. *Let* $\langle X, d \rangle$ *and* $\langle Y, \rho \rangle$ *be metric spaces and let* $h : X \to Y$. *Then* h *is Cauchy-Lipschitz if and only if whenever* $f \in Lip(Y, \mathbb{R})$, *the function* $f \circ h$ *is Cauchy-Lipschitz.*

Proof. This follows from Proposition 24.1 in view of Corollary 23.7. $\square$

Corollary 24.4. *Let* $\langle X, d \rangle$ *and* $\langle Y, \rho \rangle$ *be metric spaces and let* $h : X \to Y$. *Then* h *is uniformly locally Lipschitz if and only if there exists* $\delta > 0$ *such that whenever* $f \in Lip(Y, \mathbb{R})$, *the function* $f \circ h$ *is uniformly Lipschitz on each open ball of radius* δ *in* X.

Proof. Suppose h is uniformly locally Lipschitz. Then there exists $\delta_0 > 0$ such that the restriction of h to each open ball of radius δ_0 is Lipschitz. The same is then true for $f \circ h$ for each $f \in \mathrm{Lip}(Y, \mathbb{R})$. The converse follows from Proposition 24.1 where $\mathscr{B}$ is the family of all open balls in X of radius δ. $\qquad\square$

In view of Example 23.12, there is no hope of using Proposition 24.1 to provide a parallel theorem for the class of Lipschitz in the small functions. Nevertheless, we can still use the Garrido-Jaramillo theorem to achieve our goal upon replacing the metric of the target space by a bounded metric. Henceforth, we will say that a function h between $\langle X, d \rangle$ and $\langle Y, \rho \rangle$ is *Lipschitz in the small with distance control* δ if there exists some positive λ such that $\rho(h(x_1), h(x_2)) \leq \lambda \cdot d(x_1, x_2)$ whenever $d(x_1, x_2) < \delta$.

We start with two basic facts about functions that are Lipschitz in the small [84, p. 283].

Proposition 24.5. *Let $\langle X, d \rangle$ and $\langle Y, \rho \rangle$ be metric spaces and let $h : X \to Y$.*

(1) *if h is Lipschitz in the small and bounded, then $h \in \mathrm{Lip}(X, Y)$;*

(2) *h is Lipschitz in the small if and only if when viewed as a function into $\langle Y, min\{\rho, 1\}\rangle$, h is Lipschitz.*

Proof. For (1), while we encourage the reader to produce a simple direct proof, we choose to invoke Theorem 8.11. From the boundedness of h, we see that its modulus of continuity function has a constant majorant, namely $t \mapsto \mathrm{diam}_\rho h(X)$.

Turning to statement (2), for additional clarity denote the function with altered target space metric by $\tilde{h}$ and to reduce cumbersome notation, put $\tilde{\rho} = \min\{\rho, 1\}$. Suppose h is Lipschitz in the small with distance control δ. Then $\tilde{h}$ is also with the same distance control (and Lipschitz constant) because $\tilde{\rho} \leq \rho$. But $\tilde{h}$ is bounded as well, so by statement (1), $\tilde{h}$ is Lipschitz.

Conversely, if $\tilde{h}$ is λ-Lipschitz and $d(x_1, x_2) < 1/\lambda$, we obtain

$$\tilde{\rho}(h(x_1), h(x_2)) \leq \lambda \cdot d(x_1, x_2) < 1,$$

and so $\rho(h(x_1), h(x_2)) < \lambda \cdot d(x_1, x_2)$. We have shown that h is Lipschitz in the small with distance control $1/\lambda$. $\qquad\square$

Theorem 24.6. *A function h from $\langle X, d \rangle$ to $\langle Y, \rho \rangle$ is Lipschitz in the small if and only if whenever whenever $f \in \mathrm{Lip}(Y, \mathbb{R})$, the function $f \circ h$ is Lipschitz in the small.*

Proof. Necessity is obvious. For sufficiency, we rely on statement (2) of Proposition 24.5. To this end, we apply the Garrido-Jaramillo result Theorem 4.6. We retain the notation introduced in the proof of Proposition 24.5.

Let f be a real-valued function on Y that is Lipschitz with respect to $\tilde{\rho}$; since $\tilde{\rho}$ is a bounded metric, f is bounded. As the identity map from $\langle Y, \rho \rangle$ to $\langle Y, \tilde{\rho} \rangle$ is 1-Lipschitz, f is Lipschitz with respect to ρ as well. By hypothesis, $f \circ h$ is Lipschitz in the small, and as $f \circ h = f \circ \tilde{h}$ we see that $f \circ \tilde{h}$ is Lipschitz because it is a bounded function. Applying Theorem 4.6, $\tilde{h}$ is Lipschitz as required. $\qquad\square$

We intend to prove a more quantitative result, namely, that if $f \circ h$ is Lipschitz in the small for each $f \in \mathrm{Lip}(Y, \mathbb{R})$ with a common distance control δ but a variable Lipschitz constant parameter

λ, then δ serves as a distance control for h. Our reasoning requires a uniform local boundedness principle of independent interest.

Definition 24.7. We call a function h from $\langle X, d \rangle$ to $\langle Y, \rho \rangle$ *bounded in the small with distance control* δ if there exists $M > 0$ such that whenever $d(x_1, x_2) < \delta$ in X, then $\rho(h(x_1), h(x_2)) < M$.

In terms of our earlier language, h is bounded in the small if and only if $m_h(\delta)$ is finite for some positive δ. Obviously a Lipschitz in the small function with distance control δ and Lipschitz constant parameter λ is bounded in the small with the same distance control: if $d(x_1, x_2) < \delta$, then $\rho(h(x_1), h(x_2)) < \lambda \delta$.

An equivalent way of formulating Definition 24.7 is that there exist $\delta > 0$ and $M > 0$ such for each $x \in X$, $\text{diam}_\rho(h(S_d(x, \delta))) < M$. This formulation is not so convenient for our immediate purposes. It is not hard to show that each continuous function on $\langle X, d \rangle$ is bounded in the small if and only if X is compact. A weaker condition than being bounded in the small is that there exists $\delta > 0$ such that h restricted to each open ball of radius δ be bounded. We call this condition *uniform local boundedness* of the function (paralleling our terminology of uniformly locally Lipschitz function). The class of metric spaces $\langle X, d \rangle$ on which each continuous function is uniformly locally bounded strictly contains the class of uniformly locally compact metric spaces; such metric spaces are called either uniformly paracompact or cofinally complete in the literature. These will be studied more carefully later in this monograph.

Proposition 24.8. *Let $\langle X, d \rangle$ and $\langle Y, \rho \rangle$ be metric spaces and let $h : X \to Y$. Then h is bounded in the small with distance control δ if and only if whenever $f \in \text{Lip}(Y, \mathbb{R})$, the function $f \circ h$ is bounded in the small with distance control δ.*

Proof. For necessity, suppose whenever $d(x_1, x_2) < \delta$, we have $\rho(h(x_1), h(x_2)) < M$. Then if $f : Y \to \mathbb{R}$ is λ-Lipschitz, we get $\rho((f \circ h)(x_1), (f \circ h)(x_2)) < M\lambda$ whenever $d(x_1, x_2) < \delta$. Conversely, suppose h fails to be bounded in the small. This means that we can find sequences $\langle x_n \rangle$ and $\langle w_n \rangle$ in X such that for each $n \in \mathbb{N}$, we have $d(x_n, w_n) < \delta$ while $\rho(h(x_n), h(w_n)) > n$. We need only consider these two mutually exclusive and exhaustive cases for our sequences:

(a) by passing to a subsequence, one of the image sequences $\langle h(x_n) \rangle$ or $\langle h(w_n) \rangle$ is bounded - say $\langle h(w_n) \rangle$;

(b) each image sequence eventually is outside of each bounded subset of Y.

In case (a) choose $y_0 \in Y$ and $\alpha > 0$ such that $\forall n \in \mathbb{N}$, $h(w_n) \in S_\rho(y_0, \alpha)$. Since $\rho(h(x_n), h(w_n)) > n$ for each n, whenever $j \neq n$, the triangle inequality yields

$$\rho(h(x_n), h(w_j)) \geq \rho(h(x_n), h(w_n)) - 2\alpha.$$

Put $E = \{h(w_n) : n \in \mathbb{N}\}$ and let $f = \rho(\cdot, E) \in \text{Lip}(Y, \mathbb{R})$. Then

$$|(f \circ h)(x_n) - (f \circ h)(w_n)| = \rho(h(x_n), E) \geq \rho(h(x_n), h(w_n)) - 2\alpha.$$

which becomes arbitrarily large with increasing n.

Case (b) is more complex. We define subsequences $\langle x_{n_k} \rangle$ and $\langle w_{n_k} \rangle$ recursively. Put $x_{n_1} = x_1$ and $w_{n_1} = w_1$. For each $k \geq 1$, let $A_k = \{h(w_{n_1}), h(w_{n_2}), \ldots, h(w_{n_k})\}$ and let $B_k = \{h(x_{n_1}), h(x_{n_2}), \ldots, h(x_{n_k})\}$. Since enlargements of bounded sets are bounded, we can find $n_{k+1} > n_k$ such that both (i) $h(x_{n_{k+1}}) \notin S_d(A_k, k+1)$ and (ii) $h(w_{n_{k+1}}) \notin S_d(B_k, k)$. By (i) if $j < k$ we get

$$\rho(h(x_{n_k}), h(w_{n_j})) \geq (k-1) + 1 = k,$$

and by (ii) if $j > k$ we get

$$\rho(h(x_{n_k}), h(w_{n_j})) \geq j - 1 \geq k.$$

By assumption, $\rho(h(x_{n_k}), h(w_{n_k})) > n_k \geq k$. With $E = \{h(w_{n_j}) : j \in \mathbb{N}\}$, for each $k \in \mathbb{N}$, we have $\rho(h(x_{n_k}), E) \geq k$ so that $\rho(h(x_{n_k}), E) - \rho(h(w_{n_k}), E) \geq k$. Thus, with $f = \rho(\cdot, E)$, whenever $k \in \mathbb{N}$

$$|(f \circ h)(x_{n_k}) - (f \circ h)(w_{n_k})| \geq k.$$

In either case (a) or (b), we have produced a real-valued Lipschitz function f for which $f \circ h$ fails to be bounded in the small with distance control δ.   □

We now come to our more quantitative result. We will use the same general approach as we did in the proof of the Garrido-Jaramillo theorem, but this time working with the bounded real-valued Lipschitz functions equipped with the Weaver norm $\|f\|_W := \max\{\|f\|_\infty, L(f)\}$. For a proof that does not use functional analysis, but that still uses the last proposition, the reader is referred to [37].

Theorem 24.9. *Let $\langle X, d \rangle$ and $\langle Y, \rho \rangle$ be metric spaces and let $h : X \to Y$. Then h is Lipschitz in the small with distance control $\delta > 0$ if and only if for each $f \in Lip(Y, \mathbb{R})$ the composition $f \circ h$ is Lipschitz in the small with distance control δ.*

Proof. One direction is clear. For the other, Proposition 24.8 guarantees that h is bounded in the small with distance control δ, because whenever $f \in Lip(Y, \mathbb{R})$ the composition $f \circ h$ is bounded in the small with distance control δ. Put $M = \sup\{\rho(h(x_1), h(x_2)) : \{x_1, x_2\} \subseteq X$ and $d(x_1, x_2) < \delta\}$.

Now if $f \in Lip_b(Y, \mathbb{R})$, then by assumption $f \circ h$ is Lipschitz in the small as well as bounded, and so $f \circ h \in Lip_b(X, \mathbb{R})$. Applying the closed graph theorem exactly as in the proof of Theorem 4.6, $T(f) = f \circ h$ defines a continuous linear operator from $\langle Lip_b(Y, \mathbb{R}), \|\cdot\|_W \rangle$ to $\langle Lip_b(X, \mathbb{R}), \|\cdot\|_W \rangle$: both domain and target space are Banach spaces, and norm convergence in each space ensures pointwise convergence of functions - in fact, uniform convergence. Again denote the operator norm of T by K. Let x_1 and x_2 be distinct points of X with $d(x_1, x_2) < \delta$ and set $y_1 = h(x_1)$ and $y_2 = h(x_2)$. Let $f \in Lip_b(Y, \mathbb{R})$ be given by

$$f = \frac{1}{2}\rho(\cdot, y_1) - \frac{1}{2}\rho(\cdot, y_2).$$

It is routine to check that $\|f\|_\infty = \frac{1}{2}\rho(y_1, y_2)$ and $L(f) = 1$. We now compute

$$\rho(h(x_1), h(x_2)) = |f(y_1) - f(y_2)| = |(f \circ h)(x_1) - (f \circ h)(x_2)|$$

$$\leq L(f \circ h) \cdot d(x_1, x_2) \leq \|f \circ h\|_W \cdot d(x_1, x_2)$$

$$\leq K \cdot \|f\|_W \cdot d(x_1, x_2) = K \cdot \max\{\|f\|_\infty, L(f)\} \cdot d(x_1, x_2)$$

$$= K \cdot \max\{\tfrac{1}{2}d(y_1, y_2), 1\} \cdot d(x_1, x_2) \leq K \cdot \max\{\tfrac{M}{2}, 1\} \cdot d(x_1, x_2)$$

Thus, h does not expand distance by more than a factor of $K \cdot \max\{\frac{M}{2}, 1\}$ if $d(x_1, x_2) < \delta$.   □

25. Metrically Convex Spaces and Coarse Maps

Recall that we called a function between metric spaces a coarse map provided its modulus of continuity is finite-valued. In this section, we characterize this property in various ways when the domain space is equipped with a convex metric.

The standard example given in an introductory course in analysis for a real-valued function defined on a subset $\mathbb{R}$ that is uniformly continuous but not Lipschitz is $f(x) = \sqrt{x}$. Of course, f fails to be even locally Lipschitz. On the other hand, if we insist that x and w are kept apart by a prescribed minimal distance, then whatever that minimal distance may be, we can find an upper bound for $\frac{|f(x)-f(w)|}{|x-w|}$. Abstracting to a function f between metric spaces $\langle X, d \rangle$ and $\langle Y, \rho \rangle$, and adopting the terminology of [53, 84], we say that f is *Lipschitz for large distances* if for each $\delta > 0$ there exists $\lambda_\delta > 0$ such that $\rho(f(x), f(w)) \leq \lambda_\delta d(x, w)$ whenever $d(x, w) \geq \delta$.

Notice that any bounded function between metric spaces has this property, so the property does not force continuity of the function at any point of its domain. As remarked by Garrido and Jaramillo [84], the property may be combined with the notion of Lipschitz in the small to guarantee that f be fully Lipschitz. It is not enough that the function be in addition locally Lipschitz: consider the locally Lipschitz function $f : \mathbb{R} \to \mathbb{R}$ defined by $f(x) = 1$ if $x < 1$, and if $x \in [n, n+1)$ for some $n \in \mathbb{N}$,

$$f(x) = \begin{cases} (n+1)(x-n) + n & \text{if } n \leq x < n + \frac{1}{n+1} \\ n+1 & \text{if } n + \frac{1}{n+1} \leq x < n+1. \end{cases}$$

While f is not Lipschitz, the reader can check that f is Lipschitz for large distances.

For the purposes of this section, we will write $LLD(X, Y)$ for the space of functions from X to Y that are Lipschitz for large distances. When $Y = \mathbb{R}$ equipped with the usual metric, it is easy to check that the functions that are Lipschitz for large distances form a vector space, and if $f \in LLD(X, \mathbb{R})$, then $|f| \in LLD(X, \mathbb{R})$ as well. As a result, $LLD(X, \mathbb{R})$ is a vector lattice (which need not be closed under pointwise product).

For the entertainment of the reader, we include an elementary proposition.

Proposition 25.1. *Let $\langle X_j, d_j \rangle$ for $j = 1, 2, 3$ be metric spaces. Suppose $f \in LLD(X_1, X_2)$ and $g \in LLD(X_2, X_3)$ where g is also bounded in the small. Then $g \circ f \in LLD(X_1, X_3)$.*

Proof. Let $\delta > 0$ be arbitrary. Since $f \in LLD(X_1, X_2), \exists \lambda > 0$ such that $d_1(x, w) \geq \delta \Rightarrow d_2(f(x), f(w)) \leq \lambda d_1(x, w)$. Since g is bounded in the small, we can choose $\varepsilon, \mu > 0$ such that $d_3(g(v), g(z)) \leq \mu$ whenever $\{v, z\} \subseteq X_2$ and $d_2(v, z) \leq \varepsilon$. Since $g \in LLD(X_2, X_3)$, we can find $\beta > 0$ such that $d_2(v, z) \geq \varepsilon \Rightarrow d_3(g(v), g(z)) \leq \beta d_2(v, z)$.

Now let x, w be fixed in X with $d_1(x, w) \geq \delta$. If $d_2(f(x), f(w)) \geq \varepsilon$, then

$$d_3(g(f(x)), g(f(w))) \leq \beta \lambda d_1(x, w).$$

On the other hand, if $d_2(f(x), f(w)) < \varepsilon$, then

$$d_3(g(f(x)), g(f(w))) \leq \mu \leq \frac{\mu}{\delta} d_1(x, w).$$

Either way, the distance between x and w is magnified in the application of $g \circ f$ by a factor of at most $\max\{\beta\lambda, \frac{\mu}{\delta}\}$ when $d_1(x, w) \geq \delta$. $\qquad\square$

Again, a metric d on a set X is called convex provided whenever $x \neq w$ in X and $\alpha \in (0, d(x, w))$ there exists $z \in X$ with $d(x, z) = \alpha$ and $d(z, w) = d(x, w) - \alpha$. We next give an example showing that $LLD(X, \mathbb{R})$ need not coincide with the family of real-valued coarse mappings on X if d is not convex.

Example 25.2. Let $X = \{2^n : n \in \mathbb{N}\}$ equipped with the usual metric d of the line, and define $f : X \to \mathbb{R}$ by $f(2^n) = 4^n$. Now given any $t > 0$, there exists only finitely many pairs $(2^n, 2^k)$ with $n < k$ with $d(2^n, 2^k) \leq t$. As a result, $m_f(t)$ is finite for all t. On the other hand, $d(2^n, 2^{n+1}) = 2^n$ becomes arbitrarily large with increasing n, while

$$\frac{f(2^{n+1}) - f(2^n)}{2^{n+1} - 2^n} = \frac{3(4^n)}{2^n} = 3 \cdot 2^n.$$

As a result, whatever $t > 0$ may be, $\sup\{\frac{d(f(x), f(w))}{d(x, w)} : d(x, w) \geq t\} = \infty$.

Later we will give an example of a function f that is Lipschitz for large distances but for which $m_f(t) = \infty$ for all $t > 0$. This is more delicate.

Let $\langle X, d \rangle$ be metrically convex, and let $\langle Y, \rho \rangle$ be a second metric space. As a first step, we characterize membership to $LLD(X, Y)$ in terms of a strong uniform boundedness condition involving annuli of prescribed inner and outer radii.

Proposition 25.3. *Let $\langle X, d \rangle$ be metrically convex and let $\langle Y, \rho \rangle$ be a second metric space. Suppose $f : X \to Y$. The following conditions are equivalent:*

(1) *$\forall \delta > 0 \; \exists M_\delta > 0$ such that $\delta \leq d(x, w) \leq 2\delta \Rightarrow \rho(f(x), f(w)) \leq M_\delta$;*

(2) *f is Lipschitz for large distances;*

(3) *whenever $0 < \delta_1 < \delta_2$, $\exists M(\delta_1, \delta_2) > 0$ such that $\delta_1 \leq d(x, w) \leq \delta_2 \Rightarrow \rho(f(x), f(w)) \leq M(\delta_1, \delta_2)$.*

Proof. $(1) \Rightarrow (2)$. Suppose condition (1) holds and $\delta > 0$ is given. Let $x, w \in X$ satisfying $d(x, w) \geq \delta$ be arbitrary. If in addition $d(x, w) \leq 2\delta$, then by (1)

$$\rho(f(x), f(w)) \leq M_\delta \leq \frac{M_\delta}{\delta} d(x, w).$$

Otherwise $d(x, w) > 2\delta$. Let k be the smallest integer for which $d(x, w) \leq k\delta$; note that $k \geq 3$ holds. Put $z_0 = w$, and by the convexity of d for $j = 1, 2, 3, \ldots, k - 1$ choose $z_j \in X$ such that for $i < j$, $d(z_j, z_i) = (j - i)\delta$ and $d(x, z_j) = d(x, w) - j\delta$. Since $\delta < d(x, z_{k-2}) \leq 2\delta$, the triangle inequality yields

$$\rho(f(x), f(w)) \leq \sum_{j=1}^{k-2} \rho(f(z_j), f(z_{j-1})) + \rho(f(x), f(z_{k-2}))$$

$$\leq (k-2)M_\delta + M_\delta = \tfrac{M_\delta}{\delta}d(z_{k-2}, w) + M_\delta$$

$$< \tfrac{M_\delta}{\delta}d(z_{k-2}, w) + \tfrac{M_\delta}{\delta}d(x, z_{k-2}) = \tfrac{M_\delta}{\delta}d(x, w).$$

These estimates show that the choice for our Lipschitz constant $\lambda_\delta := \tfrac{M_\delta}{\delta}$ does the job.

$(2) \Rightarrow (3)$. Suppose f is Lipschitz for large distances, and let $0 < \delta_1 < \delta_2$ be given. Choose $\lambda_{\delta_1} > 0$ such that $d(x, w) \geq \delta_1 \Rightarrow \rho(f(x), f(w)) \leq \lambda_{\delta_1}d(x, w)$. Then if $\delta_1 \leq d(x, w) \leq \delta_2$, we get $\rho(f(x), f(x)) \leq \lambda_{\delta_1}\delta_2$.

$(3) \Rightarrow (1)$. This is trivial. $\qquad\square$

Since a coarse map satisfies condition (3) of Proposition 25.3, in view of Example 25.2, condition (3) need not yield Lipschitz for large distances for a function defined on a general metric space. In particular, without convexity of the metric, a function that is uniformly bounded on annuli of prescribed inner and outer radii need not be Lipschitz for large distances.

Conditions (1) or (3) are often easier to use than the definition to verify properties of $LLD(X, Y)$ when d is a convex metric. As a simple example, we use (1) to check that if $f \in LLD(X, Y)$ and the uniform distance from $g \in Y^X$ to f is finite, then $g \in LLD(X, Y)$ holds. Suppose $\sup_{x \in X}\rho(f(x), g(x)) = \alpha < \infty$. Let $\delta > 0$ be arbitrary and choose $M_\delta > 0$ such that $\rho(f(x), f(w)) \leq M_\delta$ whenever $\delta \leq d(x, w) \leq 2\delta$. Then clearly $\rho(g(x), g(w)) \leq M_\delta + 2\alpha$ for such a pair.

We now come to the main result of this section.

Theorem 25.4. *Let $\langle X, d\rangle$ be metrically convex and let $\langle Y, \rho\rangle$ be a second metric space. Suppose $f : X \to Y$. The following conditions are equivalent:*

(1) *f is bounded in the small, i.e., $\exists t_0 > 0$ with $m_f(t_0) < \infty$;*

(2) *f is a coarse map, i.e., $\forall t > 0$ we have $m_f(t) < \infty$;*

(3) *f is Lipschitz for large distances;*

(4) *f is uniformly bounded on annuli of prescribed inner and outer radii.*

Proof. $(1) \Rightarrow (2)$. It follows easily from the convexity of d that m_f is subadditive, i.e., whenever $s > 0$ and $t > 0$ we have $m_f(s + t) \leq m_f(s) + m_f(t)$. It follow that $\forall n \in \mathbb{N}$, $m_f(nt_0) \leq n \cdot m_f(t_0) < \infty$. Coarseness now follows from the fact that m_f is an increasing function.

$(2) \Rightarrow (3)$. It suffices to show given (2) that condition (1) of Proposition 25.3 holds. Let $\delta > 0$; we compute $\sup\{\rho(f(x), f(w)) : \delta \leq d(x, w) \leq 2\delta\} \leq \sup\{\rho(f(x), f(w)) : d(x, w) \leq 2\delta\} = m_f(2\delta) < \infty$.

$(3) \Rightarrow (4)$. If (3) holds, then condition (3) of Proposition 25.3 holds. As a result, if A is an annulus in X with inner radius δ_1 and outer radius δ_2, the triangle inequality yields $\sup\{\rho(f(a_1), f(a_2)) : \{a_1, a_2\} \subseteq A\} \leq 2M(\delta_1, \delta_2)$.

$(4) \Rightarrow (1)$. Suppose (1) fails, i.e., f fails to be bounded in the small. Let x_1, x_2 be distinct points of X. For each $n \in \mathbb{N}$, we can find w_n, z_n in X with $d(w_n, z_n) < \frac{d(x_1, x_2)}{4n}$ while $\rho(f(w_n), f(z_n)) \geq n$. By the triangle inequality, either (i) $d(w_n, x_1) \geq \frac{1}{2}d(x_1, x_2)$ or (ii) $d(w_n, x_2) \geq \frac{1}{2}d(x_1, x_2)$. In either case, by the convexity of d we can find $p_n \in X$ with $d(p_n, w_n) = \frac{1}{2}d(x_1, x_2)$. As

$\min\{d(p_n, w_n), d(p_n, z_n)\} \geq \frac{1}{4}d(x_1, x_2)$ and $\max\{d(p_n, w_n), d(p_n, z_n)\} \leq \frac{3}{4}d(x_1, x_2)$, we conclude that f is not uniformly bounded on annuli with inner radius $\frac{1}{4}d(x_1, x_2)$ and outer radius $\frac{3}{4}d(x_1, x_2)$. $\qquad\square$

Benyamini and Lindenstrauss [53, Proposition 1.11] proved that a uniformly continuous function on a metrically convex space is Lipschitz for large distances, thus anticipating the implication $(1) \Rightarrow (3)$ of Theorem 25.4.

Suppose $\langle X, d \rangle$ is metrically convex and $f : X \to Y$ is a coarse map. Then $m_f(1)$ is finite and by Theorem 25.4, there exists $\lambda > 0$ such that $d(x, w) \geq 1 \Rightarrow \rho(f(x), f(w)) \leq \lambda d(x, w)$. Let $t \in (0, \infty)$ be arbitrary. If $d(x, w) \leq t$ we get $\rho(f(x), f(w)) \leq m_f(1)$ provided $d(x, w) \leq 1$ and $\rho(f(x), f(w)) \leq \lambda t$ provided $1 \leq d(x, w)$. From this, we see that $m_f(t) \leq \lambda t + m_f(1)$ where λ is independent of t. This of course means that the modulus of continuity has an affine majorant if it is finite-valued, provided the metric is convex. This observation has been made in [106, Lemma 1.4] and in [58].

Obviously, there is nothing special about the number 1 used in the above argument; if we just knew that for some $t_0 > 0$, $\sup\{\frac{\rho(f(x), f(w))}{d(x,w)} : d(x, w) \geq t_0\}$ were finite, then the existence of an affine majorant for m_f would follow from coarseness. Albiac and Kalton [3] would say that such a function is *coarse Lipschitz*.

We end this section by showing that the properties of a function between metric spaces of (a) being bounded in the small and (b) being Lipschitz for large distances are independent of one another. If the domain space is metrically convex, then by Theorem 25.4, one occurs if and only if the other does. We now show in two examples that exactly one of the properties can hold.

Example 25.5. Let $X = \mathbb{N}$ equipped with the zero-one metric d, and let $Y = \mathbb{N}$ equipped with the usual Euclidean distance ρ. The identity function f is bounded in the small as $m_f(\frac{1}{2}) = 0$. Yet $\sup \{\frac{\rho(f(n), f(k))}{d(n,k)} : d(n, k) \geq 1\} = \infty$. Notice that f also fails to be even bounded on annuli, much less uniformly so.

Example 25.6. In our final example, we show that $f : X \to \mathbb{R}$ can be Lipschitz for large distances without being bounded in the small. Equip the following subset X of the line with the Euclidean metric d:

$$X := \{2^n : n \in \mathbb{N}\} \cup \{2^n + 2^{-n} : n \in \mathbb{N}\}.$$

Define $f : X \to \mathbb{R}$ by $f(2^n) = 0$ and $f(2^n + 2^{-n}) = 2^n$. Evidently, f is not bounded in the small as $\lim_{n \to \infty} f(2^n + 2^{-n}) - f(2^n) = \infty$. We claim however that given $\delta > 0$, the quotients $\frac{d(f(x), f(w))}{d(x, w)}$ for $d(x, w) \geq \delta$ are bounded above. First, if $x = 2^n - 2^{-n}$ and $w = 2^n$, then $d(x, w) \geq \delta$ for at most finitely many n, so such pairs can be essentially ignored with respect to our claim. Of course, pairs of the form $x = 2^k$ and $w = 2^n$ where $k > n$ can also be ignored. There are three more cases to be considered, and we can actually forget about the constraint $d(x, w) \geq \delta$. More precisely, 2 serves as upper bound for all such quotients, even without excluding pairs for which the constraint might fail.

First, if $x = 2^k + 2^{-k}$ and $w = 2^n$ for $k > n$, then

$$\frac{d(f(x), f(w))}{d(x, w)} = \frac{2^k}{2^k + 2^{-k} - 2^n} < \frac{2^k}{2^k - 2^n} \leq \frac{2^k}{2^k - 2^{k-1}} = 2.$$

Second, if $x = 2^k$ and $w = 2^n + 2^{-n}$ for $k > n$, then as $k - 1 \geq 1$, we get

$$\frac{d(f(x), f(w))}{d(x, w)} = \frac{2^n}{2^k - 2^n - 2^{-n}} < \frac{2^n}{2^k - 2^{k-1} - 1} \leq \frac{2^{k-1}}{2^{k-1} - 1} \leq 2.$$

Finally, if $x = 2^k + 2^{-k}$ and $w = 2^n + 2^{-n}$ for $k > n$, we obtain

$$\frac{d(f(x), f(w))}{d(x, w)} = \frac{2^k - 2^n}{2^k - 2^n - (2^{-n} - 2^{-k})} < \frac{2^k - 2^n}{2^k - 2^n - 1} \leq 2.$$

We have demonstrated that there is an upper bound for the quotients $\frac{d(f(x), f(w))}{d(x, w)}$ for $d(x, w) \geq \delta$, consisting of the maximum of some finite set of numbers that includes 2.

What is even more remarkable about this example is that given $0 < \delta_1 < \delta_2$, there are only finitely many distinct nonempty annuli in $\langle X, d \rangle$ with these numbers as inner and outer radii, and each such annulus contains at most finitely many points. From this, we see that on a general metric space, uniform boundedness of a function f on annuli of prescribed inner and outer radii need not force f to be bounded in the small, much less a coarse map.

26. Some Density Results

To start this section, we prove a general result about the density of the real-valued Lipschitz functions in a class of functions equipped with a topology of uniform convergence on a bornology. We will apply this to $C(X, \mathbb{R})$, to the Cauchy continuous real-valued functions, and to the uniformly continuous real-valued functions. To execute this project, we will use our earlier uniform approximation result for bounded uniformly continuous functions by its Lipschitz regularizations.

We immediately state our omnibus density theorem. With respect to its proof, all the work has been already done.

Theorem 26.1. *Let $\mathscr{B}$ be a bornology on a metric space $\langle X, d \rangle$ and let $\mathscr{A}$ be a subfamily of $C(X, \mathbb{R})$ containing $Lip(X, \mathbb{R})$ such that whenever $B \in \mathscr{B}$ and $f \in \mathscr{A}$, then $f(B)$ is a bounded set of reals and f is uniformly continuous restricted to B. Then $Lip(X, \mathbb{R})$ is dense in $\mathscr{A}$ equipped with $\mathscr{T}_{\mathscr{B}}$.*

Proof. Fix $B \in \mathscr{B}$ and $\varepsilon > 0$. By the hypotheses of this theorem and Proposition 2.2 we can find $g \in Lip(B)$ such that $\sup_{b \in B} |g(b) - f|_B(b)| < \varepsilon$. Applying Theorem 3.5, extend g to a globally defined Lipschitz function h. $\square$

Evidently, each continuous real-valued function is uniformly continuous on each nonempty relatively compact subset of a metric space $\langle X, d \rangle$ because it is so on the closure. Of course, the converse is false, as uniform continuity of continuous real-valued functions occurs on each UC-subset. By Theorem 9.2, each real-valued continuous function is bounded on a nonempty subset of $\langle X, d \rangle$ if and only if the subset is relatively compact. Each real-valued Cauchy continuous function is uniformly continuous on each nonempty totally bounded subset; in fact this characterizes Cauchy continuity by Proposition 22.1. According to Theorem 9.3, each real-valued Cauchy continuous function is bounded on a subset B if and only if B is totally bounded. Similarly, by Theorem 9.4, each real-valued uniformly continuous function is bounded on a subset B if and only if B is Bourbaki bounded. Thus, we get three special cases of our omnibus theorem that we collect in a single theorem.

Theorem 26.2. *Let $\langle X, d \rangle$ be a metric space. Then*

(1) *$Lip(X, \mathbb{R})$ is dense in $C(X, \mathbb{R})$ equipped with the topology of uniform convergence on $\mathscr{K}(X)$;*

(2) *$Lip(X, \mathbb{R})$ is dense in the Cauchy continuous real-valued functions on X, equipped with the topology of uniform convergence on $\mathscr{T}\mathscr{B}_d(X)$;*

(3) *$Lip(X, \mathbb{R})$ is dense in the uniformly continuous real-valued functions on X, equipped with the topology of uniform convergence on $\mathscr{B}\mathscr{B}_d(X)$.*

Given a family of continuous functions $\mathscr{A}$ from $\langle X, d \rangle$ to $\langle Y, \rho \rangle$, we say that a subfamily $\mathscr{A}_0$ is *uniformly dense* in $\mathscr{A}$ provided for each $f \in \mathscr{A}$ there exists $g \in \mathscr{A}_0$ such that $\sup_{x \in X} \rho(f(x), g(x)) \leq$

ε. We first establish the uniform density of the real-valued locally Lipschitz functions within $C(X, \mathbb{R})$ for an arbitrary metric space $\langle X, d \rangle$. While this is most easily established using Lipschitz partitions of unity [36, 65], it also falls out of more general uniform density results [83]. The first proof that we know of this density result is due to Miculescu [123] who used locally Lipschitz partitions of unity.

Theorem 26.3. *Let $\langle X, d \rangle$ be a metric space and let $f \in C(X, \mathbb{R})$. Given $\varepsilon > 0$ there is a locally Lipschitz function $g : X \to \mathbb{R}$ such that $\sup_{x \in X} |f(x) - g(x)| \leq \varepsilon$.*

Proof. For each $x \in X$ choose $\delta_x > 0$ such that $f(S_d(x, \delta_x)) \subseteq S_\rho(f(x), \varepsilon)$. For each x, put $V_x := S_d(x, \delta_x)$ and let $\{p_i : i \in I\}$ be a Lipschitz partition of unity subordinated to $\{V_x : x \in X\}$ whose existence is guaranteed by Theorem 2.22 of Frolik. For each $i \in I$, set $U_i = \{x : p_i(x) > 0\}$ and choose $x(i)$ with $U_i \subseteq V_{x(i)}$. Of course, $\{U_i : i \in I\}$ is a locally finite open refinement of $\{V_x : x \in X\}$. Since the real-valued Lipschitz functions defined on a given set form a vector space, by local finiteness $g := \sum_{i \in I} f(x(i)) p_i(\cdot)$ is a well-defined locally Lipschitz function. It remains to show that $\sup_{x \in X} |f(x) - g(x)| \leq \varepsilon$.

To this end, fix $x \in X$ and choose $\alpha > 0$ such that $S_d(x, \alpha)$ hits at most finitely many U_i, say for $i_1, i_2, \ldots, i_n$ in I. By relabeling if necessary, let $U_{i_1}, U_{i_2}, \ldots, U_{i_m}$ with $1 \leq m \leq n$ be those U_i in which x lies. For each $j \leq m$, $d(x, x(i_j)) < \delta_{x(i_j)}$, so that for each $j \leq m$, $|f(x) - f(x(i_j))| < \varepsilon$. Since $p_{i_j}(x) = 0$ for $m < j \leq n$,

$$|g(x) - f(x))| = |\sum_{j=1}^{m} f(x(i_j)) p_{i_j}(x) - f(x) p_{i_j}(x)| \leq \sum_{j=1}^{m} |f(x(i_j)) - f(x)| p_{i_j}(x) < \sum_{j=1}^{m} \varepsilon \cdot p_{i_j}(x) = \varepsilon,$$

concluding the proof. $\qquad\square$

We next prove a parallel result for the Lipschitz in the small real-valued functions with respect to the uniformly continuous functions. This density result is due to Garrido and Jaramillo [84, Theorem 1], who provided a much different proof than the one we provide here. Following [36, Theorem 6.1], our proof involves regularizations of our initial function that are Lipschitz in the small rather than Lipschitz.

Theorem 26.4. *Let $\langle X, d \rangle$ be a metric space and let $f : X \to \mathbb{R}$ be a uniformly continuous function. Given $\varepsilon > 0$, there exists a Lipschitz in the small function $g : X \to \mathbb{R}$ with $\sup_{x \in X} |f(x) - g(x)| \leq \varepsilon$.*

Proof. Choose $\delta > 0$ such that for each $x, w \in X$, $d(x, w) < \delta \Rightarrow |f(x) - f(w)| < \varepsilon$. Choose $k \in \mathbb{N}$ such that ($\spadesuit$) $\frac{k\delta}{2} > 1 + \varepsilon$. Define $g : X \to \mathbb{R}$ by

$$g(x) := \inf_{w \in S_d(x, \delta)} f(w) + k d(x, w).$$

Note that for each $x \in X$,

$$f(x) \geq g(x) \geq \inf_{w \in S_d(x, \delta)} f(w) \geq f(x) - \varepsilon.$$

It remains to show that g is Lipschitz in the small. We will show that if $d(x_1, x_2) < \frac{\delta}{2}$, then $|g(x_2) - g(x_1)| \leq k d(x_2, x_1)$. By symmetry, it suffices to show that $g(x_2) - g(x_1) \leq k d(x_2, x_1)$. To this end, let $\lambda \in (0, 1)$ be arbitrary, and choose $w_1 \in S_d(x_1, \delta)$ such that $f(w_1) + k d(x_1, w_1) < g(x_1) + \lambda$. Now either $d(x_1, w_1) \geq \frac{\delta}{2}$ or $d(x_1, w_1) < \frac{\delta}{2}$. In the first case, we compute using ($\spadesuit$)

$$f(w_1) + k d(x_1, w_1) > f(x_1) - \varepsilon + 1 + \varepsilon = f(x_1) + 1 > g(x_1) + \lambda,$$

which is impossible. We are left with $d(x_1, w_1) < \frac{\delta}{2}$ so that $d(x_2, w_1) < \delta$ as well. As w_1 can be used be estimate $g(x_2)$, we have

$$g(x_2) \leq f(w_1) + kd(x_2, w_1) = f(w_1) + kd(x_1, w_1) + kd(x_2, w_1) - kd(x_1, w_1)$$
$$< g(x_1) + \lambda + kd(x_2, x_1).$$

Since $\lambda \in (0, 1)$ was arbitrary, the proof is complete.   □

As each Lipschitz in the small real-valued function is uniformly continuous, for any metric space $\langle X, d \rangle$, the class of uniformly continuous locally Lipschitz real-valued functions is uniformly dense in $UC(X, \mathbb{R})$. This may be a properly larger class. In fact, the uniformly continuous uniformly locally Lipschitz functions may properly contain the Lipschitz in the small functions.

Example 26.5. Let X be $\mathbb{N} \cup \{n + \frac{1}{n+1} : n \in \mathbb{N}\}$ as metric subspace of the real line. Since each ball in the space contains at most finitely many points, each real-valued function on the space is Lipschitz when restricted to balls of any radius. Now define $f : X \to \mathbb{R}$ by

$$f(x) = \begin{cases} \frac{1}{\sqrt{n}} & \text{if } x = n \text{ for some } n \in \mathbb{N} \\ 0 & \text{otherwise} \end{cases}.$$

Clearly, f is uniformly continuous. For each positive integer n, we have

$$\frac{|f(n) - f(n + \frac{1}{n+1})|}{|n - (n + \frac{1}{n+1})|} = \frac{n+1}{\sqrt{n}}$$

and this shows that f is not Lipschitz in the small.

The following corollary is a folk-theorem that greatly improves Corollary 2.3. The proof we supply is adapted from [33].

Corollary 26.6. *Let $\langle X, d \rangle$ be a metric space and let $f \in UC(X, \mathbb{R})$. The following conditions are equivalent.*

(1) *f can be uniformly approximated by real-valued Lipschitz functions on X;*

(2) *$\exists g \in Lip(X, \mathbb{R})$ with $\sup_{x \in X} |f(x) - g(x)| < \infty$;*

(3) *the modulus of continuity function m_f for f has an affine majorant.*

Proof. (1) $\Rightarrow$ (2) is trivial. For (2) $\Rightarrow$ (3), let λ be a Lipschitz constant for g; this means that for the modulus of continuity function m_g for g, we have $m_g(t) \leq \lambda t$ for all $t \in [0, \infty)$. Denoting $\sup_{x \in X} |f(x) - g(x)|$ by μ, we have $|m_g(t) - m_f(t)| \leq 2\mu$ for all $t \geq 0$, and so $t \mapsto \lambda t + 2\mu$ is a majorant for m_f. For (3) $\Rightarrow$ (1), let $\varepsilon > 0$ be arbitrary and by Theorem 26.4, choose a Lipschitz in the small function g with $\sup_{x \in X} |f(x) - g(x)| < \varepsilon$. Then $t \mapsto \alpha t + \beta + 2\varepsilon$ is an affine majorant for m_g. Apply Theorem 8.11.   □

The next result of Garrido and Jaramillo [84, Theorem 7] came years before [33].

Corollary 26.7. *Let $\langle X, d \rangle$ be a metric space. The following conditions are equivalent.*

(1) *$Lip(X, \mathbb{R})$ is uniformly dense in $UC(X, \mathbb{R})$;*

(2) *whenever $f : X \to \mathbb{R}$ is Lipschitz in the small, $\exists g \in Lip(X, \mathbb{R})$ with $\sup_{x \in X} |f(x) - g(x)| < \infty$;*

(3) *X is small determined.*

Proof. As each Lipschitz in the small function is uniformly continuous, (1) ⇒ (2) is obvious. For (2) ⇒ (3), by Proposition 23.8, it suffices to work with real-valued functions. Let $f : X \to \mathbb{R}$ be Lipschitz in the small and choose $g \in \mathrm{Lip}(X, \mathbb{R})$ with $\sup_{x \in X} |f(x) - g(x)| < \infty$. As a bounded Lipschitz in the small function is already Lipschitz and $f - g$ is Lipschitz in the small, $f = g + (f - g)$ is Lipschitz. Finally, (3) ⇒ (1) follows from Theorem 26.4. ☐

The next density theorem was obtained by Beer and Garrido [36, Theorem 4.5] who introduced the class of Cauchy-Lipschitz functions to this end.

Theorem 26.8. *Let $\langle X, d \rangle$ be a metric space and let $f : X \to \mathbb{R}$ be a Cauchy continuous function. Given $\varepsilon > 0$, there exists a Cauchy-Lipschitz function $g : X \to \mathbb{R}$ with $\sup_{x \in X} |f(x) - g(x)| \leq \varepsilon$.*

Proof. As usual, denote the completion of $\langle X, d \rangle$ by $\langle \hat{X}, \hat{d} \rangle$. Now f can be extended to a Cauchy continuous function $\hat{f}$ on the completion. By the continuity of $\hat{f}$ and Theorem 26.3, there exists a locally Lipschitz function $\hat{g} : \hat{X} \to \mathbb{R}$ such that $\sup_{\hat{x} \in \hat{X}} |\hat{f}(\hat{x}) - \hat{g}(\hat{x})| \leq \varepsilon$. By Theorem 8.13, $\hat{g}$ is Cauchy-Lipschitz so that its restriction g to X does the job. ☐

Since the Cauchy-Lipschitz real-valued functions are uniformly dense in the Cauchy continuous real-valued functions, the formally larger class of functions consisting of the Cauchy continuous locally Lipschitz functions will be uniformly dense as well. The next example shows that the Cauchy-Lipschitz real-valued functions may not exhaust the formally larger class.

Example 26.9. Let $X = \{2^{-n} : n \in \mathbb{N}\}$ as a metric subspace of the real line. As the topology of X is discrete, each (real-valued) function on X is locally Lipschitz. Define $f : X \to \mathbb{R}$ by $f(2^{-n}) = \frac{1}{n}$. Since f maps $\langle 2^{-n} \rangle$ to a Cauchy sequence, we have $f \in CC(X, \mathbb{R})$. However f is not Lipschitz on the range of $\langle 2^{-n} \rangle$, as

$$\left| \frac{f(2^{-n}) - f(2^{-(n+1)})}{2^{-n} - 2^{-(n+1)}} \right| = \frac{\frac{1}{n} - \frac{1}{n+1}}{2^{-n} - 2^{-(n+1)}} = \frac{2^{n+1}}{n^2 + n}$$

which produces an unbounded set of numbers as n ranges over $\mathbb{N}$.

One might hope that the smaller class of uniformly locally Lipschitz real-valued functions is also uniformly dense in $CC(X, \mathbb{R})$, but this is not to be in general.

Example 26.10. We show that for any infinite dimensional Banach space $\langle X, || \cdot || \rangle$, the uniformly locally Lipschitz real-valued functions are not uniformly dense in the Cauchy continuous functions. Let $x_1 \in X$ be a norm-one vector, and for each $n \in \mathbb{N}$, let $x_n = n^2 x_1$. Let B_n be the closed ball with radius $\frac{1}{n}$ with center x_n. By infinite dimensionality, the ball is noncompact and so we can find an unbounded $f_n \in C(B_n, \mathbb{R})$. Now $\cup_{n=1}^{\infty} B_n$ is a closed set, so by the Tietze extension theorem [155, p. 103], there exists $f \in C(X, \mathbb{R})$ such that for each $n \in \mathbb{N}$, $f|_{B_n} = f_n$. Since the metric of the norm is complete, f is Cauchy continuous.

Now let $g : X \to \mathbb{R}$ be an arbitrary uniformly locally Lipschitz function. By definition, there exists $\delta > 0$ such that $\forall x \in X$, g restricted to $S_d(x, \delta)$ is Lipschitz and is therefore bounded. Choose $n > \frac{1}{\delta}$; we compute

$$\sup_{x \in X} |f(x) - g(x)| \geq \sup_{x \in B_n} |f(x) - g(x)| = \infty,$$

and this shows that f cannot be uniformly approximated by uniformly locally Lipschitz functions.

We will display necessary and sufficient conditions on the domain space to get this kind of uniform density in a subsequent section.

We have saved the best for last in this section: a uniform closure result for restricted Lipschitz in the small functions that yields as immediate corollaries a number of uniform density results, two of which we have already encountered in this section. Without question, our analysis establishes the primacy of Lipschitz in the small functions in the study of spaces of continuous real-valued functions defined on a metric spaces.

Let $\mathscr{B}$ be a family of subsets of a metric space $\langle X, d \rangle$. We will show that

$$\nabla_{\mathscr{B}} := \{f \in C(X, \mathbb{R}) : \forall B \in \mathscr{B}, f|_B \text{ Lipschitz in the small}\}$$

is uniformly dense in

$$\Omega_{\mathscr{B}} := \{f \in C(X, \mathbb{R}) : \forall B \in \mathscr{B}, f|_B \text{ is uniformly continuous}\}$$

under the assumption that $\mathscr{B}$ be shielded from closed sets. In general, uniform continuity of a subfamily of $C(X, \mathbb{R})$ on each member of $\mathscr{B}$ implies uniform continuity on each member of $\downarrow \mathscr{B} \cup \mathscr{F}(X)$, and by Theorem 17.5, the shielded from closed sets assumption also gives uniform continuity (in fact, strong uniform continuity) on each member of $\text{born}(\mathscr{B})$. Thus, there is no loss in generality in assuming that $\mathscr{B}$ is a bornology throughout.

Our uniform closure theorem was established by Beer and Garrido [37, Theorem 6.5].

Theorem 26.11. *Let $\mathscr{B}$ be a bornology on $\langle X, d \rangle$ that is shielded from closed sets. Then the uniform closure of the real-valued continuous functions that are Lipschitz in the small on every member of $\mathscr{B}$ is the family of real-valued continuous functions that are uniformly continuous on every member of $\mathscr{B}$.*

Proof. It is clear that the family of continuous functions that are uniformly continuous on members of $\mathscr{B}$ is uniformly closed and contains all the continuous functions that are Lipschitz in the small on these subsets. Now let $f \in C(X, \mathbb{R})$ be uniformly continuous on each member of $\mathscr{B}$. Given $\varepsilon > 0$ we will produce a continuous $v : X \to \mathbb{R}$ that is Lipschitz in the small on members of $\mathscr{B}$ such that $\forall x \in X, |f(x) - v(x)| < \varepsilon$.

For each $m \in \mathbb{Z}$, put

$$C_m := \{x \in X : (m-1)\varepsilon < f(x) < (m+1)\varepsilon\}.$$

The open cover $\{C_m : m \in \mathbb{Z}\}$ of X has these properties:

- if $f(x) = \varepsilon m$ for some $m \in \mathbb{Z}$, then x belongs to exactly one C_m, otherwise x belongs to exactly two sets C_m;
- by continuity, each x has a neighborhood that meets at most three C_m;
- $C_m \cap C_n = \emptyset$ provided $|m - n| > 1$.

If some C_m is the entire space, the constant function $v(x) \equiv m\varepsilon$ does the job. Otherwise, each C_m is the cozero set of the 1-Lipschitz function $g_m(x) = \min\{1, d(x, X \backslash C_m)\}$. Put $s(x) = \sum_{m \in \mathbb{Z}} g_m(x)$; clearly if $x \in C_m$, then $s(x) = g_{m-1}(x) + g_m(x) + g_{m+1}(x)$ where in fact at most two of the summands are nonzero so that $s(x) \leq 2$.

Now put $p_m(x) = g_m(x)/s(x)$ for $x \in X$ and $m \in \mathbb{Z}$. The family of cozero sets for $\{p_m : m \in \mathbb{Z}\}$ is locally finite, as each point has a neighborhood that hits at most 3 of the sets C_m. Thus, $\{p_m :$

$m \in \mathbb{Z}\}$ is a continuous partition of unity on X subordinated to the open cover $\{C_m : m \in \mathbb{Z}\}$. As a result

$$v(x) := \sum_{m \in \mathbb{Z}} m\varepsilon \cdot p_m(x)$$

is a globally defined continuous function.

To see that v ε-approximates f, let $x \in X$ be arbitrary. If $f(x) = m\varepsilon$ for some integer m, then $f(x) = v(x)$. Otherwise, choosing the unique m with $m\varepsilon < f(x) < (m+1)\varepsilon$, we see that $v(x)$ is a convex combination of $m\varepsilon$ and $(m+1)\varepsilon$ and thus $|f(x) - v(x)| < \varepsilon$.

It remains to show that v is Lipschitz in the small on each $B \in \mathscr{B}$ that contains at least two points. To this end, it suffices to work with $g = \frac{1}{\varepsilon}v$, that is, $g(x) = \sum_{m \in \mathbb{Z}} m \cdot p_m(x)$. Note that $\forall x \in C_m$, $(\lozenge)$ $\quad g(x) = (m-1)p_{m-1}(x) + m \cdot p_m(x) + (m+1)p_{m+1}(x)$

$$= m \sum_{j=-1}^{1} p_{m+j}(x) - p_{m-1}(x) + p_{m+1}(x) = m - \frac{g_{m-1}(x)}{s(x)} + \frac{g_{m+1}(x)}{s(x)}.$$

Fix $B \in \mathscr{B}$; by Theorem 17.5, choose $\delta \in (0,1)$ such that if $b \in B$ and $x \in X$, $d(b,x) < \delta \Rightarrow$ $|f(x) - f(b)| < \varepsilon/2$. We claim that for each $b \in B$ there exists $m \in \mathbb{Z}$ such that $S_d(b,\delta) \subseteq C_m$. To see this, choose $m \in \mathbb{Z}$ with

$$m\varepsilon - \frac{\varepsilon}{2} \le f(b) \le m\varepsilon + \frac{\varepsilon}{2}.$$

Then if $x \in S_d(b,\delta)$, by strong uniform continuity on B and our choice of δ, we get

$$(m-1)\varepsilon < f(x) < (m+1)\varepsilon,$$

that is, $S_d(b,\delta) \subseteq C_m$. In particular, $g_m(b) = \min\{1, d(b, X \backslash C_m)\} \ge \delta$ so that $s(b) \ge \delta$ whatever $b \in B$ may be.

We intend to show that whenever b_1 and b_2 are distinct points of B with $d(b_1, b_2) < \delta$, we have $|g(b_1) - g(b_2)| \le \frac{10}{\delta^2}d(x,y)$. As a first step to realizing this inequality, taking m with $\{b_1, b_2\} \subseteq C_m$, we compute

$(\clubsuit)$ $\quad |s(b_1) - s(b_2)| = |\sum_{j=-1}^{1} g_{m+j}(b_1) - \sum_{j=-1}^{1} g_{m+j}(b_2)| \le 3d(b_1, b_2).$

Now using $(\lozenge)$, $(\clubsuit)$, the fact that each g_k is 1-Lipschitz, and $\forall b \in B$, $\delta \le s(b) \le 2$, we finally get

$$|g(b_1) - g(b_2)| = \left| -\frac{g_{m-1}(b_1)}{s(b_1)} + \frac{g_{m+1}(b_1)}{s(b_1)} + \frac{g_{m-1}(b_2)}{s(b_2)} - \frac{g_{m+1}(b_2)}{s(b_2)} \right|$$

$$\le \frac{1}{\delta^2}|g_{m-1}(b_1)s(b_2) - g_{m-1}(b_2)s(b_1)| + \frac{1}{\delta^2}|g_{m+1}(b_1)s(b_2) - g_{m+1}(b_2)s(b_1)|$$

$$\le \frac{1}{\delta^2}|s(b_2)(g_{m-1}(b_1) - g_{m-1}(b_2))| + \frac{1}{\delta^2}|g_{m-1}(b_2)(s(b_1) - s(b_2))|$$

$$+ \frac{1}{\delta^2}|s(b_2)(g_{m+1}(b_1) - g_{m+1}(b_2))| + \frac{1}{\delta^2}|g_{m+1}(b_2)(s(b_1) - s(b_2))|$$

$$\le \frac{1}{\delta^2}(2d(b_1, b_2) + 3d(b_1, b_2) + 2d(b_1, b_2) + 3d(b_1, b_2)) = \frac{10}{\delta^2}d(b_1, b_2)$$

as asserted. $\qquad\qquad\qquad\qquad\qquad\qquad\qquad\qquad\qquad\qquad\qquad\qquad\qquad\qquad\qquad$ $\square$

We now collect in a single corollary four consequences of Theorem 26.11; the first and last statements have been obtained earlier in this section by very different methods. The second statement is classical and was established in Section 2.

Corollary 26.12. *Let $\langle X, d \rangle$ be a metric space. Then*

    (1) *the family of Lipschitz in the small real-valued functions is uniformly dense in $UC(X, \mathbb{R})$;*

    (2) *the family of bounded real-valued Lipschitz functions is uniformly dense in the bounded uniformly continuous real-valued functions on X;*

    (3) *the family of real-valued functions that are Lipschitz in the small when restricted to bounded subsets of X is uniformly dense in the family of real-valued functions on X that are uniformly continuous on bounded subsets of X;*

    (4) *the family of locally Lipschitz real-valued functions is uniformly dense in $C(X, \mathbb{R})$.*

Proof. For statement (1), let $\mathscr{B} = \mathscr{P}(X)$. Statement (2) follows from statement (1), noting that a that a function at a finite uniform distance from a bounded function must also be bounded, and by Proposition 24.5 a bounded Lipschitz in the small function is already Lipschitz. For statement (3), let $\mathscr{B} = \mathscr{B}_d(X)$, noting that a function that is Lipschitz in the small on just the open bounded subsets is globally locally Lipschitz and thus belongs to $C(X, \mathbb{R})$. For statement (4), let $\mathscr{B} = \mathscr{K}(X)$, noting that a function that is Lipschitz in the small on each relatively compact subset is bounded and thus Lipschitz so restricted, and by Corollary 23.5 the function is thus globally locally Lipschitz. $\qquad\square$

27. More on our Four Classes of Locally Lipschitz Functions

While Cauchy continuous functions and uniformly continuous functions defined on a subset A of a metric space $\langle X, d \rangle$ with values in a complete metric space can be extended to $\mathrm{cl}(A)$ in a way such that these properties are preserved, the same cannot be said for continuous functions. A parallel situation occurs when we look at our four classes of locally Lipschitz functions. A locally Lipschitz function defined on a subset A with values in a complete metric space could have a continuous extension to its closure that fails to be locally Lipschitz: consider $f(x) = \sqrt{x}$ on $(0, \infty)$ viewed as a subset of $[0, \infty)$. However, if a function is either Cauchy-Lipschitz or uniformly locally Lipschitz or Lipschitz in the small on a subset, then it has an extension to its closure with the same property. We choose to state our results in terms of extensions from a dense subset of a metric space to the entire space, rather from subsets to their closures.

We obtain our extension theorem for Cauchy-Lipschitz functions from a fact of independent interest. For a more direct proof, the reader can consult [35].

Proposition 27.1. *Let $\langle X, d \rangle$ be a metric space and let B be a nonempty totally bounded subset of the completion $\langle \hat{X}, \hat{d} \rangle$. Then there exists $A \in \mathscr{TB}_d(X)$ such that $B \subseteq cl_{\hat{X}}(A)$.*

Proof. Let $C = \mathrm{cl}_{\hat{X}}(B)$, a compact subset of the completion. It suffices to produce a totally bounded subset A of X with $C \subseteq \mathrm{cl}_{\hat{X}}(A)$. Since X is dense in $\hat{X}$ and C is totally bounded, $\forall n \in \mathbb{N}$ there exists $F_n \in \mathscr{F}(X)$ with $C \subseteq S_{\hat{d}}(F_n, \frac{1}{n})$ and such that F_n is irreducible in this regard. Put $A = \cup_{n=1}^{\infty} F_n$. By construction, $C \subseteq \mathrm{cl}_{\hat{X}}(A)$. It remains to show that $A \in \mathscr{TB}_{\hat{d}}(\hat{X})$ because total boundedness is an intrinsic property of metric spaces. For this we need only show that $C \cup A$ is totally bounded.

Let $\varepsilon > 0$ be arbitrary, and choose a finite subset F_0 of C such that $C \subseteq S_{\hat{d}}(F_0, \varepsilon)$. By the compactness of C this enlargement of F_0 contains an enlargement of C itself, say $S_{\hat{d}}(C, \delta)$ for some $\delta \in (0, \varepsilon)$. Choose $k \in \mathbb{N}$ with $1/k < \delta$. By the irreducibility of each F_n, we have

$$\cup_{n=k}^{\infty} F_n \subseteq S_{\hat{d}}(C, \delta),$$

and so A is contained in the ε-enlargement of the finite set $F_0 \cup F_1 \cup F_2 \cup \cdots F_n$ in the completion. $\square$

Proposition 27.2. *Let $\langle X, d \rangle$ be a metric space and let $\langle Y, \rho \rangle$ be a complete metric space. Suppose f is a Cauchy-Lipschitz function from X to Y. Then f has a Cauchy-Lipschitz extension from the completion $\langle \hat{X}, \hat{d} \rangle$ to Y.*

Proof. First, since f is Cauchy continuous, f has a (Cauchy) continuous extension $\hat{f}$ to $\hat{X}$. To show this is Cauchy-Lipschitz, by Corollary 23.7, we only need to show that it is Lipschitz when restricted to each nonempty totally bounded subset B of $\hat{X}$. By the last result, let A be a totally bounded subset of X for which $B \subseteq \mathrm{cl}_{\hat{X}}(A)$. Since $\hat{f}|_A$ is Lipschitz, by Proposition 23.3, $\hat{f}|_{\mathrm{cl}_{\hat{X}}}(A)$ is Lipschitz and so $\hat{f}|_B$ is Lipschitz. $\square$

Theorem 27.3. *Let A be a dense subset of the metric space $\langle X, d \rangle$ and let $\langle Y, d \rangle$ be a complete metric space. Suppose $f : A \to Y$ is Cauchy-Lipschitz. Then f has a Cauchy-Lipschitz extension to X.*

Proof. By Proposition 27.2 f has a Cauchy-Lipschitz extension to $\langle \hat{A}, \hat{d} \rangle$, the completion of the metric subspace $\langle A, d \rangle$. Since A is dense in X, we have $\langle \hat{A}, \hat{d} \rangle = \langle \hat{X}, \hat{d} \rangle$. Thus, the restriction of $\hat{f}$ to X is the desired extension.   □

Since elements of $CC(X, \mathbb{R})$ can be extended from an arbitrary nonempty subset of X, one anticipates that an analogous result holds for Cauchy-Lipschitz real-valued functions, and this is so [33].

Theorem 27.4. *Let $\langle X, d \rangle$ be a metric space and let $A \subseteq X$ be nonempty. Suppose $f : A \to \mathbb{R}$ is Cauchy-Lipschitz. Then f has a Cauchy-Lipschitz extension to X.*

Proof. By the last theorem, f has a Cauchy-Lipschitz extension $\tilde{f}$ to $\mathrm{cl}_{\hat{X}}(A)$. By Theorem 8.3 $\tilde{f}$ being locally Lipschitz has a locally Lipschitz extension $\hat{f}$ to $\langle \hat{X}, \hat{d} \rangle$. Since $\hat{d}$ is a complete metric, this second extension is Cauchy-Lipschitz and so $\hat{f}|_X$ is the desired function.   □

We next consider extensions to the closure for Lipschitz in the small functions. This is the easiest case.

Theorem 27.5. *Let A be a dense subset of the metric space $\langle X, d \rangle$ and let $\langle Y, d \rangle$ be a complete metric space. Suppose $f : A \to Y$ is Lipschitz in the small. Then f has a Lipschitz in the small extension to X.*

Proof. Since f is uniformly continuous, f has a (uniformly) continuous extension $\tilde{f}$ to X. By definition, there exists $\delta > 0$ and $\lambda > 0$ such that whenever $\{a_1, a_2\} \subseteq A$ with $d(a_1, a_2) < \delta$, we have $\rho(f(a_1), (a_2)) \leq \lambda \cdot d(a_1, a_2)$. Let $x_1 \neq x_2$ be points of X with $d(x_1, x_2) < \delta$. We can find sequences $\langle a_{n1} \rangle$ and $\langle a_{n2} \rangle$ in A convergent to x_1 and x_2 respectively such that for each $n \in \mathbb{N}$, $d(a_{n1}, a_{n2}) < \delta$. By the continuity of $\tilde{f}$ and the fact that $\tilde{f}$ extends f, we have

$$\rho(\tilde{f}(x_1), \tilde{f}(x_2)) = \lim_{n \to \infty} \rho(f(a_{n1}), f(a_{n2})) \leq \lambda \cdot \lim_{n \to \infty} d(a_{n1}, a_{n2}) = \lambda \cdot d(x_1, x_2),$$

showing that $\tilde{f}$ is Lipschitz in the small with the same parameters that apply to f.   □

The remaining case is only slightly harder than the Lipschitz in the small case, but it is subtle.

Theorem 27.6. *Let A be a dense subset of the metric space $\langle X, d \rangle$ and let $\langle Y, d \rangle$ be a complete metric space. Suppose $f : A \to Y$ is uniformly locally Lipschitz. Then f has a uniformly locally Lipschitz extension to X.*

Proof. Since each uniformly locally Lipschitz function is Cauchy-Lipschitz, we already know that f has a Cauchy-Lipschitz extension $\tilde{f}$ to X. Suppose the restriction of f to each open ball of radius δ in A is Lipschitz. We claim that $\tilde{f}$ is Lipschitz on each open ball of radius $\frac{\delta}{2}$ in X. Fix $p \in X$ and choose $a \in A$ with $d(p, a) < \frac{\delta}{2}$. Choose $\alpha > 0$ such that if $\{a_1, a_2\} \subseteq A \cap S_d(a, \delta)$ then $\rho(f(a_1), f(a_2)) \leq \alpha \cdot d(a_1, a_2)$. Take $\{x, w\} \subseteq S_d(p, \frac{\delta}{2})$ and sequences $\langle x_n \rangle$ and $\langle w_n \rangle$ in A convergent to x and w, respectively. By the triangle inequality, eventually both sequences lie in $S_d(a, \delta)$ so that eventually $\rho(f(x_n), f(w_n)) \leq \alpha \cdot d(x_n, w_n)$. Continuity of $\tilde{f}$ now yields $\rho(\tilde{f}(x), \tilde{f}(w)) \leq \alpha \cdot d(x, w)$ as required.   □

Any positive number strictly less than δ could have been used for the common radius of the balls in X in the proof of Theorem 27.6. But it may not be possible to use δ itself as our next example shows.

Example 27.7. Let ℓ_1 be the space of absolutely summable real sequences equipped with its usual norm $||\cdot||_1$. We will construct a metric subspace X of this classical Banach space containing the origin 0_{ℓ_1}, a dense subset A of X, and a real-valued function f on A such that f restricted to each open ball of radius one is Lipschitz but the same cannot be said for its (unique) continuous extension to X.

To start things off, we define for each positive integer n elements x_n and w_n in ℓ_1. Our metric subspace will be $X = \{0_{\ell_1}\} \cup \{x_n : n \in \mathbb{N}\} \cup \{w_n : n \in \mathbb{N}\}$. For the values of x_n we prescribe $x_n(n) = \frac{1}{n+1}$ and $x_n(j) = 0$ whenever $j \neq n$. For the values of w_n we prescribe $w_n(n) = 1 - \frac{1}{n+2}$ and $x_n(j) = 0$ whenever $j \neq n$. Note that for all n,

$$||x_n - w_n||_1 = 1 - \frac{1}{n+2} - \frac{1}{n+1} < 1,$$

while whenever $n \neq m$, $||w_n - w_m||_1 = 1 - \frac{1}{n+2} + 1 - \frac{1}{m+2} > 2 - \frac{2}{3} > 1$. For $n \neq m$,

$$||x_n - w_m||_1 = \frac{1}{n+1} + 1 - \frac{1}{m+2}$$

which is less than one if and only if $m + 2 \leq n$.

For $A = \{x_n : n \in \mathbb{N}\} \cup \{w_n : n \in \mathbb{N}\}$, define $f : A \to \mathbb{R}$ by

$$f(a) = \begin{cases} n & \text{if } a = w_n \text{ for some } n \in \mathbb{N} \\ 0 & \text{otherwise} \end{cases}.$$

Noticing that $S_d(x_n, 1) \cap A = \{x_k : k \in \mathbb{N}\} \cup \{w_m : m \leq n - 2\} \cup \{w_n\}$ while $S_d(w_n, 1) \cap A = \{x_n, w_n\} \cup \{x_k : k \geq n + 2\}$, it is routine to check that f restricted to each open ball in A of radius one is Lipschitz. Now the unique continuous extension $\bar{f}$ to X must assign the value 0 to 0_{ℓ_1}. It follows that $\bar{f}$ restricted to $S_d(0_{\ell_1}, 1)$ is not Lipschitz because for each $n \in \mathbb{N}$, $||w_n||_1 < 1$ and

$$\frac{|f(w_n) - f(0_{\ell_1})|}{||w_n||_1} = \frac{n}{1 - \frac{1}{n+2}} = \frac{n^2 + 2n}{n+1}.$$

We next revisit our project to determine when our four classes of locally Lipschitz functions pairwise coincide. To this point, we have only characterized coincidence of the locally Lipschitz functions with the Cauchy-Lipschitz functions in Theorem 8.13. We next settle this for the largest and smallest classes.

As to when the locally Lipschitz functions agree with the Lipschitz in the small functions, we obtain a very satisfying answer.

Theorem 27.8. *Let $\langle X, d \rangle$ be a metric space. Then each locally Lipschitz function on X with values in an arbitrary metric space is Lipschitz in the small if and only if $\langle X, d \rangle$ is a UC-space.*

Proof. First suppose the space is UC, yet some locally Lipschitz function $f : \langle X, d \rangle \to \langle Y, \rho \rangle$ fails to be Lipschitz in the small. This means that for each $n \in \mathbb{N}$, we can find x_n and w_n in X with $0 < d(x_n, w_n) < \frac{1}{n}$ yet $\rho(f(x_n), f(w_n)) > n d(x_n, w_n)$. With $I(\cdot)$ as usual denoting the isolation functional for the metric space X, we have $\lim_{n \to \infty} I(x_n) = 0$, and since $\langle X, d \rangle$ is a UC-space,

$\langle x_n \rangle$ has a cluster point p. By construction the restriction of f to each neighborhood of p is not Lipschitz, and we obtain a contradiction to f being locally Lipschitz.

Conversely, suppose our metric space is not UC. In view of Toader's characterization of the UC property for metric spaces, we can find a pseudo-Cauchy sequence $\langle x_n \rangle$ with distinct terms that does not cluster. By passing to a subsequence we can assume that for each $n \in \mathbb{N}$, $d(x_{2n-1}, x_{2n}) < \frac{1}{n}$. By Urysohn's lemma, we can find a continuous function f with values in $[0,1]$ mapping the closed set $\{x_1, x_3, x_5, \ldots\}$ to zero and the closed set $\{x_2, x_4, x_6, \ldots\}$ to one. By Theorem 26.3, we can find a locally Lipschitz function $g : X \to \mathbb{R}$ such that $\sup\{|f(x) - g(x)| : x \in X\} < \frac{1}{4}$. It follows that for each $n \subset \mathbb{N}$, $|g(x_{2n-1}) - g(x_{2n})| \geq \frac{1}{2}$, so that

$$|g(x_{2n-1}) - g(x_{2n})| \geq \frac{n}{2} d(x_{2n-1}, x_{2n}).$$

Clearly, g cannot be Lipschitz in the small. $\qquad \square$

Proposition 27.9. *Let $\langle X, d \rangle$ be a metric space. The following conditions are equivalent.*

(1) *the metric space is a UC-space;*

(2) *the reciprocal of each nonvanishing real-valued Lipschitz in the small function is Lipschitz in the small;*

(3) *the reciprocal of each nonvanishing member of $Lip(X, \mathbb{R})$ is Lipschitz in the small.*

Proof. The implication $(1) \Rightarrow (2)$ follows from Theorem 27.8 and the fact that the reciprocal of each nonvanishing real-valued locally Lipschitz function is again locally Lipschitz. The implication $(2) \Rightarrow (3)$ is trivial. For $(3) \Rightarrow (1)$, we prove the contrapositive.

If $\langle X, d \rangle$ fails to be a UC-space, we can find disjoint nonempty closed subsets A and B such that

$$\inf\{d(a, b) : a \in A, b \in B\} = 0.$$

For each $n \in \mathbb{N}$ choose $a_n \in A$ and $b_n \in B$ with $d(a_n, b_n) < \frac{1}{n}$. As a result, for each n we have $d(a_n, B) < \frac{1}{n}$ and $d(b_n, A) < \frac{1}{n}$. By passing to a subsequence we may assume without loss of generality that $\forall n \in \mathbb{N}$, we have $d(a_n, B) \leq d(b_n, A)$. Define $f \in \mathrm{Lip}(X, \mathbb{R})$ by $f(x) = \frac{1}{2} d(x, B) + d(x, A)$. Since $A \cap B = \emptyset$, for all $x \in X$ we have $f(x) > 0$. We compute

$$\left| \frac{1}{f}(a_n) - \frac{1}{f}(b_n) \right| = \frac{2}{d(a_n, B)} - \frac{1}{d(b_n, A)} \geq \frac{2}{d(a_n, B)} - \frac{1}{d(a_n, B)} > n,$$

which shows that $\frac{1}{f}$ is not Lipschitz in the small because $\lim_{n \to \infty} d(a_n, b_n) = 0$. $\qquad \square$

We refer the reader to [33] for different proofs of the preceding and following propositions.

Proposition 27.10. *Let $\langle X, d \rangle$ be a metric space. The following conditions are equivalent.*

(1) *the metric space is compact;*

(2) *$Lip(X, \mathbb{R})$ is uniformly dense in $C(X, \mathbb{R})$;*

(3) *the metric space is both a UC-space and small determined.*

Proof. (1) $\Rightarrow$ (2). If X is compact, then by Theorem 23.4, each locally Lipschitz function on X is Lipschitz. It follows from Theorem 26.3 that $\mathrm{Lip}(X, \mathbb{R})$ is uniformly dense in $C(X, \mathbb{R})$.

(2) $\Rightarrow$ (3). By condition (2), each member of $C(X, \mathbb{R})$ is the uniform limit of a sequence in $UC(X, \mathbb{R})$, from which $C(X, \mathbb{R}) \subseteq UC(X, \mathbb{R})$. It follows that X is a UC-space. Also, since each Lipschitz in the small function on X is continuous, by Corollary 26.7, X is small determined.

(3) $\Rightarrow$ (1). Since X is a UC-space, by Theorem 27.8, the class of locally Lipschitz functions on X coincides with the class of Lipschitz in the small functions, and since X is small determined, the class of locally Lipschitz functions on X coincides with the smaller class of Lipschitz functions on X. By Theorem 23.4, the metric space is compact. $\qquad\square$

The class of metric spaces on which the uniformly locally Lipschitz functions agree with the Lipschitz in the small functions coincides with the class of metric spaces on which the Cauchy-Lipschitz functions agree with the Lipschitz in the small functions.

Theorem 27.11. *Let $\langle X, d \rangle$ be a metric space. The following conditions are equivalent.*

(1) *the metric space $\langle X, d \rangle$ has UC-completion;*

(2) *each Cauchy-Lipschitz function on X with values in an arbitrary metric space is Lipschitz in the small;*

(3) *each uniformly locally Lipschitz function on X with values in an arbitrary metric space is Lipschitz in the small.*

Proof. (1) $\Rightarrow$ (2). We rely on this characteristic property of spaces that have UC-completion established by Theorem 19.11: each sequence $\langle x_n \rangle$ in X with $\lim_{n \to \infty} I(x_n) = 0$ has a Cauchy subsequence.

Let $\langle Y, \rho \rangle$ be a second metric space and let $f : X \to Y$ fail to be Lipschitz in the small. For each $n \in \mathbb{N}$, we can find x_n, w_n in X with $0 < d(x_n, w_n) < \frac{1}{n}$ and $\rho(f(x_n), f(w_n)) > nd(x_n, w_n)$. As $I(x_n) < \frac{1}{n}$, by passing to a subsequence, we may assume $\langle x_n \rangle$ is a Cauchy sequence, so that by construction, the spliced sequence $x_1, w_1, x_2, w_2, \ldots$ is Cauchy as well. Evidently, f is not Lipschitz on the range of this spliced sequence. We conclude that f is not Cauchy-Lipschitz.

Since each uniformly locally Lipschitz function is Cauchy-Lipschitz, (2) $\Rightarrow$ (3) is immediate. For (3) $\Rightarrow$ (1), suppose we can find a sequence $\langle x_n \rangle$ in X with $\lim_{n \to \infty} I(x_n) = 0$ that has no Cauchy subsequence. Then $\{x_n : n \in \mathbb{N}\}$ is not totally bounded, and by passing to subsequence, we can assume the terms are distinct. By passing again to a subsequence, we can find $\mu > 0$ such that whenever $n \neq j$, $d(x_n, x_j) \geq \mu$. As a result, for each $x \in X, S_d(x, \frac{\mu}{3})$ intersects at most one ball from the family $\{S_d(x_n, \frac{\mu}{3}) : n \in \mathbb{N}\}$. Using the fact that $I(x_n) \to 0$, we intend to show that the uniformly locally Lipschitz function $f : X \to \mathbb{R}$ defined by

$$f(x) = \begin{cases} n - \frac{3n}{\mu} d(x, x_n) & \text{if } \exists n \text{ with } d(x, x_n) < \frac{\mu}{3} \\ 0 & \text{otherwise} \end{cases}$$

fails to be Lipschitz in the small. To see this, let $\delta > 0$ and $\alpha > 0$ be arbitrary. Choose n so large that both $I(x_n) + \frac{1}{n} < \min\{\delta, \frac{\mu}{3}\}$ and $\frac{3n}{\mu} > \alpha$. We can find $w_n \neq x_n$ with $d(x_n, w_n) < I(x_n) + \frac{1}{n} < \frac{\mu}{3}$, and as a result, $|f(x_n) - f(w_n)| = \frac{3n}{\mu} d(x_n, w_n)$. Thus, $d(x_n, w_n) < \delta$ while $|f(x_n) - f(w_n)| > \alpha d(x_n, w_n)$. $\qquad\square$

It remains to determine those metric spaces for which (a) the locally Lipschitz functions coincide with the uniformly locally Lipschitz functions; (b) the Cauchy-Lipschitz functions coincide with the uniformly locally Lipschitz function. As their resolution involves the class of uniformly paracompact metric spaces not yet introduced, we must address these questions later.

28. Real-Valued Functionals and Bornologies

The point of departure for this section is our earlier consideration of UC-subsets of a metric space, viewed as sets in which each sequence along which the isolation functional goes to zero clusters. Following Beer and Segura [49], in this section we look at a general continuous extended real-valued functional ϕ defined on $\langle X, d \rangle$ with values in $[0, \infty]$ and study the structure of the family of its subsets $\mathscr{B}(\phi)$ in which each sequence along which the functional goes to zero clusters. Members of this bornology will be called ϕ-*subsets*; if X itself is a ϕ-subset, we call X a ϕ-*space*. Thus, a subset A is a UC-subset if and only A is an $I(\cdot)$-subset where $I(\cdot)$ is the isolation functional. Notice that, by continuity, if $\lim_{n \to \infty} \phi(x_n) = 0$, then any cluster point of $\langle x_n \rangle$ is a global minimizer of ϕ.

Theorem 28.1. *Let $\langle X, d \rangle$ be a metric space and let $\phi : X \to [0, \infty]$ be continuous. Then $\mathscr{B}(\phi)$ is a bornology with closed base that contains $\mathscr{K}(X)$.*

Proof. Clearly $\mathscr{B}(\phi)$ contains $\mathscr{K}(X)$ (and thus $\mathscr{F}(X)$) and is hereditary. If each sequence along which ϕ goes to zero in B_i clusters for $i = 1, 2, \ldots, n$, then the same is true in $\cup_{i=1}^{n} B_i$. Now suppose $B \in \mathscr{B}(\phi)$ and $\langle x_n \rangle$ is a sequence in $\mathrm{cl}(B)$ with $\lim_{n \to \infty} \phi(x_n) = 0$. Without loss of generality, we may assume ϕ is real-valued along the sequence. By continuity, we can find for each $n \in \mathbb{N}$, $b_n \in B$ with $d(x_n, b_n) < \frac{1}{n}$ and $|\phi(x_n) - \phi(b_n)| < \frac{1}{n}$. By assumption $\langle b_n \rangle$ clusters and so $\langle x_n \rangle$ clusters as well. $\square$

Many interesting bornologies arise in this way from geometrically defined functionals. A major goal for us is to characterize the bornologies that can be realized as $\mathscr{B}(\phi)$ for some $\phi \in C(X, [0, \infty])$. In this effort, we could restrict our search to functionals that lie in $C(X, [0, 1])$, as replacing ϕ by $\min\{\phi, 1\}$ does not alter $\mathscr{B}(\phi)$. Actually, we will confine our search (without loss of generality) to nonnegative real-valued continuous functionals.

Of importance in our study is $\mathrm{Ker}(\phi) := \{x \in X : \phi(x) = 0\}$, which we call the *kernel* of ϕ. Obviously, if $\mathrm{Ker}(\phi) = \emptyset$, then $\mathscr{B}(\phi) = \{B : \inf_{x \in B} f(x) > 0\}$. Observe also that if $\inf_{x \in X} \phi(x) > 0$, then $\mathscr{B}(\phi) = \mathscr{P}(X)$. Finally, $\mathrm{Ker}(\phi)$ belongs to $\mathscr{B}(\phi)$ if and only if the set is compact.

Example 28.2. Suppose $\langle X, d \rangle$ is a metric space and $x_0 \in X$ is fixed. Consider $\phi \in C(X, [0, \infty))$ defined by

$$\phi(x) := \frac{1}{1 + d(x, x_0)}$$

Notice that $\inf_{x \in X} \phi(x) = 0$ if and only if X is unbounded. Whether or not this is true, we have $\mathrm{Ker}(\phi) = \emptyset$, and a subset B satisfies $\inf_{x \in B} \phi(x) > 0$ if and only if $B \in \mathscr{B}_d(X)$. Thus, $\mathscr{B}(\phi)$ consists of the metrically bounded subsets of X.

We now look at a framework in which many important nonnegative continuous functionals arise. Let P be an hereditary property of open subsets of $\langle X, d \rangle$, that is, if V, W are open sets where

$W \subseteq V$ then $P(V) \Rightarrow P(W)$. Define $\phi_P : X \to [0, \infty]$ by

$$\phi_P(x) = \begin{cases} \sup\{\alpha > 0 \ : \ P(S_d(x, \alpha))\} & \text{if } \exists \alpha > 0 \text{ with } P(S_d(x, \alpha)); \\ 0 & \text{otherwise.} \end{cases}$$

Example 28.3. First let us look at $P(V) := V$ is empty. Then ϕ_P is the zero functional, and the kernel is the entire space X. Evidently, $\mathscr{B}(\phi_P) = \mathscr{K}(X)$.

Example 28.4. Consider $P(V) := V$ contains at most one point. Then the resulting ϕ_P is the isolation functional, that is, $\phi_P(x) = I(x) = d(x, X \setminus \{x\})$, and $\mathscr{B}(\phi_P) = \mathscr{B}_d^{uc}$.

Example 28.5. Consider $P(V) := V$ contains at most finitely many points, i.e., $V \in \mathscr{F}(X)$. The resulting functional ϕ_P has been called the *local finiteness functional* [24]. The kernel of the functional agrees with the kernel of the isolation functional: it is the set of limit points X' of X. Since, $I(\cdot) \leq \phi_P(\cdot)$, if $\langle x_n \rangle$ is a sequence for $\phi_P(x_n) \to 0$, then it is so for the isolation functional as well, so any UC-subset is a ϕ_P-subset. To see that the converse fails, consider the following metric subspace of the Euclidean plane:

$$X = \{(0, \beta) : \beta \in \mathbb{R}\} \cup (\mathbb{N} \times \{0\}) \cup \{(n + \frac{1}{n+1}, 0) : n \in \mathbb{N}\}.$$

Notice that $\text{Ker}(\phi_P) = \{(0, \beta) : \beta \in \mathbb{R}\}$. While $\mathbb{N} \times \{0\}$ fails to be a UC-subset, it is ϕ_P-subset, as it contains no ϕ_P-minimizing sequence.

Example 28.6. The property $P(V) := V$ is countable is weaker still. Here, $x \in \text{Ker}(\phi_P)$ if and only if each neighborhood of X is uncountable. Such points are called *condensation points* [155, p. 90] of the space.

Example 28.7. Now consider the property $P(V) := \text{cl}(V)$ is compact, i.e., $V \in \mathscr{K}(X)$. The resulting ϕ_P is a *local compactness functional* denoted elsewhere by ν [20, 32], giving at each x having a compact neighborhood the supremum of the radii of the closed balls with center x that are compact. For this functional, $\text{Ker}(\phi_P)$ equals the points of non-local compactness of X, i.e., those points that fail to have a compact neighborhood. The bornology $\mathscr{B}(\nu)$ has a number of elegant characterizations that we take up in the next separate section. We will call a member of the bornology a *uniformly paracompact subset*, as A belongs to the bornology if and only if for each open cover of A there exists $\delta > 0$ and an open refinement of the cover such that $\forall a \in A, S_d(a, \delta)$ hits at most finitely many members of the refinement. When X itself belongs to the bornology, the space goes by either *uniformly paracompact* or *cofinally complete* in the literature [20, 31, 32, 60, 98, 100, 101, 133].

Example 28.8. Finally, let $E \in \mathscr{P}_0(X)$ be fixed and consider $P(V) := V \cap E = \emptyset$. Then the induced ϕ_P gives the distance from a variable point of X to the set E. Here, the kernel is $\text{cl}(E)$. A subset B belongs to $\mathscr{B}(\phi_P) = \mathscr{B}(d(\cdot, E))$ provided each sequence $\langle b_n \rangle$ in B where $\lim_{n \to \infty} d(b_n, E) = 0$ clusters.

Some of the properties P enumerated above determine functionals that can take on the value infinity. For example, the local finiteness functional takes on infinite values if and only if $\mathscr{B}_d(x) = \mathscr{F}(X)$. The functional determined by $P(V) := V$ is countable takes on infinite values if and only if each bounded subset of the metric space is countable. The local compactness functional takes on values of infinity if and only if $\mathscr{B}_d(x) = \mathscr{K}(X)$. The next result clarifies the situation; most importantly, in each case we get a nonnegative extended real-valued, continuous functional.

Proposition 28.9. *Let P be an hereditary property of open sets in $\langle X, d\rangle$. If $\phi_P(x_0) = \infty$ for some $x_0 \in X$, then $\phi_P(x) = \infty$ for all $x \in X$. Otherwise, if ϕ_P is finite-valued, then ϕ_P is nonexpansive.*

Proof. Suppose $\phi_P(x_0) = \infty$ for some $x_0 \in X$. Let $x \in X$ where $x \neq x_0$, and let $\alpha > 0$ be arbitrary. Since $\sup\{\mu > 0 \ : \ P(S_d(x_0, \mu))\} = \infty$, $\exists \alpha_0 > 0$ such that $P(S_d(x_0, \alpha_0))$ and $S_d(x, \alpha) \subseteq S_d(x_0, \alpha_0)$; since P is hereditary, $P(S_d(x, \alpha))$. This shows that $\phi_P(x) = \infty$ for all $x \in X$.

Otherwise, ϕ_P is finite-valued. If ϕ_P fails to be 1-Lipschitz, there exist $x, w \in X$ with $\phi_P(x) > \phi_P(w) + d(x, w)$. Take an $\alpha > 0$ where $\phi_P(x) > \alpha > \phi_P(w) + d(x, w)$, so that $P(S_d(x, \alpha))$. Since $S_d(w, \alpha - d(x, w)) \subseteq S_d(x, \alpha)$, we have $P(S_d(w, \alpha - d(x, w)))$. However, $\alpha - d(x, w) > \phi_P(w)$, which is a contradiction. Hence, ϕ_P is 1-Lipschitz. $\qquad\square$

The next theorem is anticipated by the equivalence of the first and last statements in Theorem 18.3.

Theorem 28.10. *Let $\langle X, d\rangle$ be a metric space and suppose $\phi \in C(X, [0, \infty])$. Then $B \in \mathscr{P}_0(X)$ is a ϕ-subset if and only if $cl(B) \cap Ker(\phi)$ is compact and $\forall \delta > 0$, $\exists \varepsilon > 0$ such that $b \in B \setminus S_d(cl(B) \cap Ker(\phi), \delta) \Rightarrow \phi(b) > \varepsilon$.*

Proof. We start with necessity. Let $B \in \mathscr{B}(\phi)$ and suppose $cl(B) \cap Ker(\phi)$ is not compact. Choose a sequence $\langle x_n\rangle$ in $cl(B) \cap Ker(\phi)$ with no cluster point. For each $n \in \mathbb{N}$, $\phi(x_n) = 0$, and as $\langle x_n\rangle$ has no cluster point, $cl(B)$ is not a ϕ-subset. Since the bornology of ϕ-subsets has a closed base, B canot be a ϕ-subset, which is impossible. Suppose now that for some $\delta > 0$ that $\inf\{\phi(b) : b \in B \setminus S_d(cl(B) \cap Ker(\phi), \delta)\} = 0$. Select $b_n \in B \setminus S_d(cl(B) \cap Ker(\phi), \delta)$ with $\phi(b_n) < \frac{1}{n}$. The sequence $\langle b_n\rangle$ cannot have a cluster point p, as by continuity $\phi(p) = 0$ must hold as well as $p \in cl(B)$, while $d(p, cl(B) \cap Ker(\phi)) \geq \delta$. Again, B is not a ϕ-subset, and we have a contradiction.

Conversely, suppose $cl(B) \cap Ker(\phi)$ is compact, and $\forall \delta > 0$, $\exists \varepsilon_\delta > 0$ such that
$$b \in B \setminus S_d(cl(B) \cap Ker(\phi), \delta) \Rightarrow \phi(b) > \varepsilon_\delta.$$

Let $\langle b_n\rangle$ be a sequence in B along which ϕ tends to zero. If $cl(B) \cap Ker(\phi) = \emptyset$, then $B = B \setminus S_d(cl(B) \cap Ker(\phi), 1)$ so that $\inf_n \phi(b_n) \geq \varepsilon_1 > 0$, a contradiction, and we conclude that $cl(B) \cap Ker(\phi) \neq \emptyset$. For each $\delta > 0$, $\phi(b_n) \leq \varepsilon_\delta$ eventually $\Rightarrow b_n \in S_d(cl(B) \cap Ker(\phi), \delta)$ eventually. By the compactness of $cl(B) \cap Ker(\phi)$, the sequence $\langle b_n\rangle$ has a cluster point. $\qquad\square$

We will use the following corollary in our characterization of bornologies that are bornologies of ϕ-subsets.

Corollary 28.11. *Let $\langle X, d\rangle$ be a metric space and let $\phi : X \to [0, \infty]$ be continuous. Suppose B is a nonempty closed subset of X. Then B is a ϕ-subset if and only if $B \cap Ker(\phi)$ is compact, and whenever A is a nonempty closed subset of B with $A \cap Ker(\phi) = \emptyset$, then $\inf_{a \in A} \phi(a) > 0$.*

Proof. Suppose first that B is a ϕ-subset. Since $B = cl(B)$, by the last theorem, $B \cap Ker(\phi)$ is compact. If $A \in \mathscr{C}_0(X) \cap \downarrow \{B\}$, then since
$$A \cap Ker(\phi) = A \cap (B \cap Ker(\phi)),$$
by compactness, $A \cap Ker(\phi) = \emptyset$ ensures that A is disjoint from some enlargement of $B \cap Ker(\phi)$. By the last theorem, $\inf_{a \in A} \phi(a) > 0$.

Conversely, suppose $B \in \mathscr{C}_0(X)$ satisfies the two conditions of Corollary 28.11. As $B = \mathrm{cl}(B)$, the set $\mathrm{cl}(B) \cap \mathrm{Ker}(\phi)$ is compact. On the other hand, $B \setminus S_d((\mathrm{cl}(B) \cap \mathrm{Ker}(\phi), \delta)$ is a closed subset of B disjoint from $\mathrm{Ker}(\phi)$, so the infimum of ϕ over the set is positive. Thus $\exists \varepsilon > 0 \; \forall b \in B \setminus S_d(\mathrm{cl}(B) \cap \mathrm{Ker}(\phi), \delta)$, $\phi(b) > \varepsilon$. $\qquad \square$

Proposition 28.12. *Let $\phi : X \to [0, \infty]$ be continuous. Then a ϕ-set B is relatively compact if and only if $\forall \varepsilon > 0$, $B_\varepsilon := \{b \in B : \phi(b) \geq \varepsilon\}$ is relatively compact.*

Proof. We may assume that B is nonempty. Necessity is obvious as each subset of a relatively compact set is relatively compact. For sufficiency, suppose $\langle b_n \rangle$ is an arbitrary sequence in B. If $\phi(b_n) \to 0$, then the sequence clusters because B is a ϕ-subset. Otherwise, $\exists \varepsilon > 0$ and an infinite subset $\mathbb{N}_1$ of $\mathbb{N}$ such that $\forall n \in \mathbb{N}_1$, $\phi(b_n) \geq \varepsilon$. Hence $\langle b_n \rangle_{n \in \mathbb{N}_1}$ is a sequence in the relatively compact set B_ε, and so the subsequence clusters. $\qquad \square$

Our next proposition involves ϕ-subsets and strong uniform continuity.

Proposition 28.13. *Let $\phi : \langle X, d \rangle \to [0, \infty)$ be continuous.*

(1) *If $B \neq \emptyset$ is a ϕ-subset and ϕ is strongly uniformly continuous on B, and $\langle x_n \rangle$ is a sequence in X with both $\lim_{n \to \infty} d(x_n, B) = 0$ and $\lim_{n \to \infty} \phi(x_n) = 0$, then $\langle x_n \rangle$ clusters.*

(2) *Strong uniform continuity of ϕ on each nonempty member of $\mathscr{B}(\phi)$ implies $\phi \in UC(X, \mathbb{R})$.*

(3) *If ϕ is strongly uniformly continuous on some nonempty $B \in \mathscr{B}(\phi)$, then B has a shield in $\mathscr{B}(\phi)$.*

Proof. We prove statements (2) and (3), leaving (1) to the reader. For (2), suppose ϕ fails to be uniformly continuous. Then for some $\varepsilon > 0$, there exist sequences $\langle x_n \rangle$ and $\langle w_n \rangle$ in X such that for each n, $d(x_n, w_n) < \frac{1}{n}$ yet $\phi(x_n) + \varepsilon \leq \phi(w_n)$. Since $\inf_n \phi(w_n) \geq \varepsilon$, $B := \{w_n : n \in \mathbb{N}\}$ is in $\mathscr{B}(\phi)$, and clearly ϕ is not strongly uniformly continuous on B.

Moving on to (3), without loss of generality, we may assume that (i) X is not a ϕ-space, else X would be a shield for B, and that (ii) $B \in \mathscr{C}_0(X)$. By strong uniform continuity of ϕ on B, $\forall n \in \mathbb{N}, \exists \delta_n \in (0, \frac{1}{n})$ such that $\forall b \in B, \forall x \in X$, $d(x, b) < \delta_n \Rightarrow |\phi(b) - \phi(x)| < \frac{1}{n}$. We may also assume that $\langle \delta_n \rangle$ is decreasing. Let $b \in B \setminus \mathrm{Ker}(\phi)$. There exists a smallest $n_b \in \mathbb{N}$ such that $\frac{1}{n_b} < \phi(b)$. If $x \in X$ satisfies $d(x, b) < \delta_{2n_b}$, then

$$\frac{1}{2n_b} < \phi(x) < \phi(b) + \frac{1}{2n_b}.$$

Also note that $\phi(b) \leq \frac{1}{n_b - 1}$, whenever $n_b \neq 1$. Set $\delta(b) = \delta_{2n_b}$. We claim

$$B_1 := (\mathrm{Ker}(\phi) \cap B) \cup \bigcup_{b \in B \setminus \mathrm{Ker}(\phi)} S_d(b, \delta(b))$$

is a shield for B which lies in $\mathscr{B}(\phi)$.

We first show B_1 is ϕ-subset. Let $\langle x_k \rangle$ be a sequence in B_1 with $\phi(x_k) \to 0$. If infinitely many terms of $\langle x_k \rangle$ are contained in $\mathrm{Ker}(\phi) \cap B$, then $\langle x_k \rangle$ must cluster by the compactness of $\mathrm{Ker}(\phi) \cap B$. Otherwise, by passing to a subsequence we can assume $\forall k \in \mathbb{N}$, $x_k \in \bigcup_{b \in B \setminus \mathrm{Ker}(\phi)} S_d(b, \delta(b))$

and $\phi(x_k) < \frac{1}{2}$. Pick $b_k \in B \setminus \text{Ker}(\phi)$ with $x_k \in S_d(b_k, \delta(b_k))$. Fix k and let's for the moment write $n := n_{b_k}$. We know that $\frac{1}{2n} < \phi(x_k)$, so $n \geq 2$ and $\phi(b_k) \leq \frac{1}{n-1}$. Note also $\frac{1}{n-1} \leq \frac{2}{n}$, so

$$\phi(b_k) \leq \frac{2}{n} = 4 \cdot \frac{1}{2n} < 4\phi(x_k).$$

Hence $\phi(b_k) \to 0$, so $\langle b_k \rangle$ has a cluster point p. Note that p is a cluster point of $\langle x_k \rangle$ as $\delta(b_k) \to 0$.

Now we must show whenever $C \in \mathscr{C}_0(X)$ with $C \cap B_1 = \emptyset$, then $D_d(C, B) > 0$. By the compactness of $\text{Ker}(\phi) \cap B$, we find $\mu > 0$ such that $D_d(C, \text{Ker}(\phi) \cap B) > 2\mu$. Put $T_1 := B \cap S_d(\text{Ker}(\phi) \cap B, \mu)$ and $T_2 := B \setminus S_d(\text{Ker}(\phi) \cap B, \mu)$, so that $T_1 \cup T_2 = B$. Then $D_d(C, T_1) \geq \mu > 0$. By Corollary 28.11, there exists $\varepsilon > 0$ such that $\forall b \in T_2$, $\phi(b) > \varepsilon$. Let $k \in \mathbb{N}$ satisfy $\frac{1}{k} < \varepsilon$. If $b \in T_2$, then $\phi(b) > \frac{1}{k}$ so $\delta_{2k} \leq \delta(b)$. Hence,

$$\bigcup_{b \in T_2} S_d(b, \delta_{2k}) \subseteq \bigcup_{b \in T_2} S_d(b, \delta(b)) \subseteq B_1.$$

As a result, $C \cap \bigcup_{b \in T_2} S_d(b, \delta_{2k}) = \emptyset$. Then $D_d(C, T_2) \geq \delta_{2k} > 0$. Thus, $D_d(C, B) = D_d(C, T_1 \cup T_2) = \min\{D_d(C, T_1), \ D_d(C, T_2)\} > 0$. $\square$

Corollary 28.14. *Suppose $\phi : X \to [0, \infty)$ is strongly uniformly continuous on each nonempty member of $\mathscr{B}(\phi)$. Then $\mathscr{B}(\phi)$ is shielded from closed sets.*

Corollary 28.15. *Let P be an hereditary property of open subsets of $\langle X, d \rangle$. Then $\mathscr{B}(\phi_P)$ is shielded from closed sets.*

Proof. By Proposition 28.9, either ϕ_P is identically equal to infinity or it is finite-valued and nonexpansive. In this first case, the bornology is trivial, and $X \in \mathscr{B}(\phi_P)$ is a shield for each subset. In the second, apply the immediately preceding corollary. $\square$

Example 28.16. Statements (1) and (3) of Proposition 28.13 can fail without the strong uniform continuity assumption. Let $X = [0, \infty) \times [0, \infty)$ on which we define the continuous function $\phi(\alpha, \beta) = \alpha\beta$. Then $B := \{(\alpha, \beta) : \alpha\beta = 1\}$ is a ϕ-subset on which ϕ is not strongly uniformly continuous. If $x_n = (0, n)$, then $\lim_{n \to \infty} d(x_n, A) = 0$ and $\lim_{n \to \infty} \phi(x_n) = 0$, but the sequence $\langle x_n \rangle$ does not cluster. Thus statement (1) of Proposition 28.13 is not in general valid without the strong uniform continuity assumption

This context also provides a counterexample to statement (3). Suppose $B_1 \in \mathscr{B}(\phi)$ were a shield for B. Since the functional is zero along the horizontal axis, $\exists n \in \mathbb{N}$ such that $B_1 \cap \{(\alpha, 0) : \alpha \geq 0\} \subseteq [0, n] \times \{0\}$. Then $C := [2n, \infty) \times \{0\}$ is closed and disjoint from B_1, while $D_d(C, B) = 0$. This is not compatible with B_1 being a shield for B.

To obtain our characterization of bornologies that can be realized as $\mathscr{B}(\phi)$ for some nonnegative continuous ϕ, we break our ϕ-functionals into two classes: those for which $\text{Ker}(\phi) = \emptyset$ [49, Theorem 4.17], and those for which and those for which $\text{Ker}(\phi) \neq \emptyset$ [49, Theorem 4.18]. As we have noted, there is no loss in generality in restricting our attention to real-valued continuous functionals.

Theorem 28.17. *Let $\mathscr{B}$ be a bornology on $\langle X, d \rangle$. The following conditions are equivalent:*

(1) $\mathscr{B} = \mathscr{B}(\phi)$ *for some* $\phi \in C(X, [0, \infty))$ *with* $\text{Ker}(\phi) = \emptyset$;

(2) $\mathscr{B} = \mathscr{B}_\rho(X)$ *for some metric ρ equivalent to d.*

Proof. $(1) \Rightarrow (2)$. Since $\mathrm{Ker}(\phi) = \emptyset$, $B \neq \emptyset$ is a ϕ-subset if only if ϕ has a positive infimum on B, that is,

$$\mathscr{B}(\phi) = \downarrow \{\phi^{-1}((\frac{1}{n}, \infty)) : n \in \mathbb{N}\} = \downarrow \{\phi^{-1}([\frac{1}{n}, \infty)) : n \in \mathbb{N}\}.$$

With these representations, by Hu's theorem, the bornology is a metric bornology determined by a metric equivalent to d, as it has a countable open base and a countable closed base.

$(2) \Rightarrow (1)$. Suppose $\mathscr{B} = \mathscr{B}_\rho(X)$ where ρ is equivalent to d. Then fixing $x_0 \in X$, $x \mapsto (1 + \rho(x, x_0))^{-1}$ determines a strictly positive continuous function ϕ on X for which $\mathscr{B}(\phi) = \mathscr{B}_\rho(X)$. $\qquad\square$

The characterization of bornologies of the form $\mathscr{B}(\phi)$ when ϕ has nonempty kernel is not so transparent and relies on Corollary 28.11.

Theorem 28.18. *Let $\mathscr{B}$ be a bornology on $\langle X, d \rangle$. The following conditions are equivalent:*

(1) $\mathscr{B} = \mathscr{B}(\phi)$ *for some* $\phi \in \mathcal{C}(X, [0, \infty))$ *with* $\mathrm{Ker}(\phi) \neq \emptyset$;

(2) $\mathscr{B}$ *has a closed base, and* $\exists C \in \mathscr{C}_0(X)$ *with open neighborhoods* $\{V_n : n \in \mathbb{N}\}$ *satisfying* $\cap_{n=1}^\infty V_n = C$ *and* $\forall n \in \mathbb{N}$, $\mathrm{cl}(V_{n+1}) \subseteq V_n$ *such that* $\forall B \in \mathscr{C}_0(X)$, $B \in \mathscr{B} \Leftrightarrow B \cap C$ *is compact, and whenever* A *is a nonempty closed subset of* B *disjoint from* C, *then for some* n, $A \cap V_n = \emptyset$.

Proof. $(1) \Rightarrow (2)$. By Theorem 28.1, $\mathscr{B}$ has a closed base, and by Corollary 28.11, we can take $C = \mathrm{Ker}(\phi)$ and $V_n = \phi^{-1}([0, \frac{1}{n}))$.

$(2) \Rightarrow (1)$. We consider several cases for the set C. First if $C = X$, then a nonempty closed subset B is in $\mathscr{B}$ if and only if B is compact, and since the bornology has a closed base, it is the bornology $\mathscr{K}(X)$. With $\phi(x) \equiv 0$, we get $\mathscr{B} = \mathscr{B}(\phi)$. A second possibility is that $C = V_n \subset X$ for some $n \in \mathbb{N}$. Since $\{C, X \backslash C\}$ forms a nontrivial separation of X, the function ϕ assigning 0 to each point of C and 1 to each point of $X \backslash C$ is continuous. We intend to show that $\mathscr{B} = \mathscr{B}(\phi)$.

Since both bornologies have closed bases, it suffices to show closed nonempty members of one belong to the other. If $B \in \mathscr{B} \cap \mathscr{C}_0(X)$, then any minimizing sequence in B lies eventually in C, and since $B \cap C$ is compact, it clusters. This shows $B \in \mathscr{B}(\phi)$. For the reverse inclusion, if $B \in \mathscr{B}(\phi)$ is closed and nonempty, then $B \cap \mathrm{Ker}(\phi)$ is compact, that is, $B \cap C$ is compact. Also if A is a closed nonempty subset of B disjoint from C, then $A \cap V_n = \emptyset$ because $C = V_n$ without any consideration of ϕ.

In the remaining case we may assume without loss of generality that $\forall n \in \mathbb{N}$, $C \subset V_n \subset X$. We now apply Hu's construction to the metric subspace $X \backslash C$ with respect to the metric bornology having base $\{X \backslash V_n : n \in \mathbb{N}\}$. We produce an unbounded continuous $f : X \backslash C \to [0, \infty)$ such that $\forall A \in \mathscr{P}_0(X \backslash C)$, $f(A)$ is bounded if and only if for some n, $A \subseteq X \backslash V_n$. We next define our function ϕ by

$$\phi(x) = \begin{cases} 0 & \text{if } x \in C \\ \frac{1}{1+f(x)} & \text{otherwise} \end{cases}.$$

Evidently ϕ is continuous restricted to the open set $X \backslash C$. Given $\varepsilon \in (0, 1)$, choose $n \in \mathbb{N}$ with $\{x \in X \backslash C : f(x) \leq \frac{1-\varepsilon}{\varepsilon}\} \subseteq X \backslash V_n$. It follows that $\forall x \in V_n$, we have $\phi(x) < \varepsilon$. Thus, whenever $c \in C$ and $\langle x_n \rangle$ is a sequence in X with $\lim_{n \to \infty} x_n = c$, we have $\phi(x_n) < \varepsilon$ eventually as V_n is a neighborhood of c. This proves $\phi \in C(X, [0, \infty))$.

Again, it suffices to show that $\mathscr{B} \cap \mathscr{C}_0(X) = \mathscr{B}(\phi) \cap \mathscr{C}_0(X)$. For a nonempty closed set B, $B \cap \mathrm{Ker}(\phi)$ is compact if and only if $B \cap C$ is compact because by construction $\mathrm{Ker}(\phi) = C$. If $B \in \mathscr{C}_0(X)$ and A is a nonempty closed subset with $A \cap C = A \cap \mathrm{Ker}(\phi) = \emptyset$ then $\exists n$ with $A \cap V_n = \emptyset \Leftrightarrow \exists n$ with $A \subseteq X \backslash V_n \Leftrightarrow f$ is bounded above on $A \Leftrightarrow \inf_{a \in A} \phi(a) > 0$. The result now follows from Corollary 28.11. $\qquad\square$

A more refined question, that we do not address here, is this: given a bornology $\mathscr{B}$ on a metric space $\langle X, d \rangle$, when is there a d-uniformly continuous function ϕ such that $\mathscr{B} = \mathscr{B}(\phi)$? For its resolution, the reader may consult Theorems 4.22 and 4.23 of [49].

We now turn our attention to ϕ-spaces, that is, those metric spaces $\langle X, d \rangle$ for which X is a ϕ-subset, so that each subset of X is a ϕ-subset as well. Our first characterization of ϕ-spaces follows immediately from Theorem 28.10.

Theorem 28.19. *Let $\langle X, d \rangle$ be a metric space and suppose $\phi \in C(X, [0, \infty])$. Then X is a ϕ-space if and only if $\mathrm{Ker}(\phi)$ is compact and $\forall \delta > 0$, $\exists \varepsilon > 0$ such that $x \notin S_d(\mathrm{Ker}(\phi), \delta) \Rightarrow \phi(x) > \varepsilon$.*

Given a nonnegative continuous functional ϕ on $\langle X, d \rangle$, for $A \in \mathscr{P}_0(X)$, define $\underline{\phi}(A)$ and $\overline{\phi}(A)$ by the formulas
$$\underline{\phi}(A) := \inf_{a \in A} \phi(a), \qquad \overline{\phi}(A) = \sup_{a \in A} \phi(a).$$

Theorem 28.20. *Let $\langle X, d \rangle$ be a metric space and let $\phi \in C(X, [0, \infty])$. The following conditions are equivalent:*

(1) *the metric space $\langle X, d \rangle$ is a ϕ-space;*

(2) *whenever $\langle C_n \rangle$ is a decreasing sequence in $\mathscr{C}_0(X)$ with $\lim_{n \to \infty} \underline{\phi}(C_n) = 0$, then $\cap_{n=1}^{\infty} C_n \neq \emptyset$;*

(3) *whenever $\langle C_n \rangle$ is a decreasing sequence in $\mathscr{C}_0(X)$ with $\lim_{n \to \infty} \overline{\phi}(C_n) = 0$, then $\cap_{n=1}^{\infty} C_n \neq \emptyset$.*

Proof. (1) $\Rightarrow$ (2). Suppose $\langle X, d \rangle$ is a ϕ-space and $\langle C_n \rangle$ is a decreasing sequence in $\mathscr{C}_0(X)$ with $\underline{\phi}(C_n) \to 0$. For each $n \in \mathbb{N}$, pick $c_n \in C_n$ with $\phi(c_n) \leq \inf\{\phi(c) : c \in C_n\} + \frac{1}{n}$. As $\underline{\phi}(C_n) \to 0$, we have $\phi(c_n) \to 0$, so by (1) $\langle c_n \rangle$ must have a cluster point, say p. Then
$$p \in \cap_{n=1}^{\infty} \mathrm{cl}(\{c_j : j \geq n\}) \subseteq \cap_{n=1}^{\infty} \mathrm{cl}(\cup_{j=n}^{\infty} C_j) = \cap_{n=1}^{\infty} \mathrm{cl}(C_n) = \cap_{n=1}^{\infty} C_n.$$

(2) $\Rightarrow$ (3). This is trivial.

(3) $\Rightarrow$ (1). Let $\langle x_n \rangle$ be a sequence in $\langle X, d \rangle$ where $\lim_{n \to \infty} \phi(x_n) = 0$. For each $n \in \mathbb{N}$, put $C_n := \mathrm{cl}(\{x_j : j \geq n\})$. Fix $\varepsilon > 0$; $\exists n_0 \in \mathbb{N}$ such that $n \geq n_0 \Rightarrow \phi(x_n) < \varepsilon$. As a result, $\forall n \geq n_0$, $\sup\{\phi(c) : c \in C_n\} \leq \varepsilon \Rightarrow \lim_{n \to \infty} \overline{\phi}(C_n) = 0$. Hence $\cap_{n=1}^{\infty} \mathrm{cl}(\{x_j : j \geq n\}) \neq \emptyset$, and $\langle x_n \rangle$ has a cluster point. $\qquad\square$

Our last theorem is in the spirit of Cantor's intersection theorem [111, p. 413] that characterizes completeness of a metric space: $\langle X, d \rangle$ is complete if and only if whenever whenever $\langle C_n \rangle$ is a decreasing sequence in $\mathscr{C}_0(X)$ with $\lim_{n \to \infty} \mathrm{diam}_d(C_n) = 0$, then $\cap_{n=1}^{\infty} C_n \neq \emptyset$. Completeness of $\langle X, d \rangle$ can also be so characterized replacing $\mathrm{diam}_d(\cdot)$ by a *measure of noncompactness* functional $\alpha(\cdot)$ as discovered by Kuratowski [111, p. 412]. Let $A \subseteq X$; put
$$\alpha(A) := \inf \{\varepsilon > 0 : \exists F \in \mathscr{F}(X) \text{ with } A \subseteq S_d(F, \varepsilon)\}.$$

Actually, $\alpha(\cdot)$ would be more accurately called a measure of non-total boundedness, as $\alpha(A) = 0$ if and only if A is totally bounded. Notice that $\alpha(A)$ is finite if and only if $A \in \mathscr{B}_d(X)$, and that $A_1 \subseteq A_2 \Rightarrow \alpha(A_1) \leq \alpha(A_2)$. Here is the formal statement of Kuratowski's theorem.

Theorem 28.21. *A metric space $\langle X, d \rangle$ is complete if and only if whenever $\langle C_n \rangle$ is a decreasing sequence in $\mathscr{C}_0(X)$ with $\lim_{n \to \infty} \alpha(C_n) = 0$, then $\cap_{n=1}^{\infty} C_n \neq \emptyset$.*

In a complete space, the nonempty intersection in Cantor's Theorem must be a single point, whereas the intersection in Kuratowski's Theorem need not be; still it must be compact, as a closed subset of a complete metric space is complete, and a complete and totally bounded set is compact.

We leave the routine proof of the next proposition to the reader; it implies that the H_d-limit of a sequence of nonempty totally bounded subsets is totally bounded.

Proposition 28.22. *Let $\langle X, d \rangle$ be a metric space, and suppose $A, A_1, A_2, A_3, \ldots$ is a sequence in $\mathscr{P}_0(X)$ where $\lim_{n \to \infty} H_d(A_n, A) = 0$. Then $\lim_{n \to \infty} \alpha(A_n) = \alpha(A)$.*

Lemma 28.23. *Let $\langle X, d \rangle$ be a ϕ-space. Suppose $\langle C_n \rangle$ is a decreasing sequence in $\mathscr{C}_0(X)$ with $\lim_{n \to \infty} \overline{\phi}(C_n) = 0$. Then $C := \bigcap_{n=1}^{\infty} C_n$ is nonempty and compact and $\lim_{n \to \infty} H_d(C_n, C) = 0$.*

Proof. The set C is nonempty by Theorem 28.20. Choose an arbitrary sequence $x_1, x_2, x_3, \ldots$ in C. Since $\overline{\phi}$ is monotone and $\lim_{n \to \infty} \overline{\phi}(C_n) = 0$, we have $\overline{\phi}(C) = 0$. Hence $\forall n \in \mathbb{N}$, $\phi(x_n) = 0 \Rightarrow \langle x_n \rangle$ has a cluster point in C because X is a ϕ-space and C is closed. Thus, C is compact.

Now we show $\lim_{n \to \infty} H_d(C_n, C) = 0$. Suppose this does not hold; since our sequence of closed subsets is decreasing, $\exists \varepsilon > 0$ such that $\forall n \in \mathbb{N}$, $e_d(C_n, C) > \varepsilon$. Choose $c_n \in C_n$ with $d(c_n, C) > \varepsilon$. Since $\lim_{n \to \infty} \phi(c_n) = 0$, $\langle c_n \rangle$ must have a cluster point p which lies outside $S_d(C, \varepsilon)$. But since the subsequence that converges to p must lie in each C_n eventually, we also have $p \in C$, which is a contradiction. Thus, $\langle C_n \rangle$ converges to C in Hausdorff distance. $\qquad \square$

Theorem 28.24. *If $\langle X, d \rangle$ is complete and $\phi \in C(X, [0, \infty])$, then the following statements are equivalent:*

(1) *$\langle X, d \rangle$ is a ϕ-space;*

(2) *$\forall \varepsilon > 0$, $\exists \delta > 0$ such that $C \in \mathscr{C}_0(X)$ and $\overline{\phi}(C) < \delta \Rightarrow \alpha(C) < \varepsilon$.*

Proof. (2) $\Rightarrow$ (1). Let $\langle C_n \rangle$ be a decreasing sequence in $\mathscr{C}_0(X)$ with $\lim_{n \to \infty} \overline{\phi}(C_n) = 0$. Fix $\varepsilon > 0$; by condition (2), $\exists \delta > 0$ such that $\overline{\phi}(C_n) < \delta \Rightarrow \alpha(C_n) < \varepsilon$. Since $\lim_{n \to \infty} \overline{\phi}(C_n) = 0$, we have $\lim_{n \to \infty} \alpha(C_n) = 0$. Since $\langle X, d \rangle$ is complete, by Kuratowski's Theorem, $\bigcap_{n=1}^{\infty} C_n \neq \emptyset$. Hence, by Theorem 28.20, X is a ψ-space.

(1) $\Rightarrow$ (2). This direction does not require completeness. Assume (1) holds but (2) fails, i.e., $\exists \varepsilon > 0$ such that given $n \in \mathbb{N}$, $\exists B_n \in \mathscr{C}_0(X)$ with $\overline{\phi}(B_n) \leq \frac{1}{n}$ but $\alpha(B_n) \geq \varepsilon$. Let $C_n := \{x \in X : \phi(x) \leq \frac{1}{n}\}$ and put $C := \bigcap_{n=1}^{\infty} C_n$ which, by the last lemma, is nonempty and compact and $\lim_{n \to \infty} H_d(C_n, C) = 0$. Since $C_n \supseteq B_n$, by continuity of α with respect to Hausdorff distance, $\forall n \in \mathbb{N}$, $\alpha(C_n) \geq \varepsilon \Rightarrow \alpha(C) \geq \varepsilon$. But $\alpha(C) = 0$ as C is compact; thus we have a contradiction. $\qquad \square$

Corollary 28.25. *Let $\langle X, d \rangle$ be a complete metric space. Then X is a UC-space if and only if $\forall \varepsilon > 0$, $\exists \delta > 0$ such that whenever $C \in \mathscr{C}_0(X)$ and $\sup\{I(c) : c \in C\} < \delta$ then $\alpha(C) < \varepsilon$.*

We know from Theorem 28.19 that if a metric space $\langle X, d\rangle$ is to be a ϕ-space for a given functional ϕ, then $\mathrm{Ker}(\phi)$ must be compact. Conversely, given a compact subset K of a metric space $\langle X, d\rangle$, there evidently exists $\phi \in C(X, \mathbb{R})$ such that $\langle X, d\rangle$ is ϕ-space where $\mathrm{Ker}(\phi) = K$. In the case that $K = \emptyset$, take $\phi(x) \equiv 1$, and for $K \neq \emptyset$, let $\phi(x) = d(x, K)$.

A related and much less transparent question is this: suppose P is an hereditary property of open subsets of metrizable spaces. When does there exist a compatible metric d on a metrizable space X such that $\langle X, d\rangle$ is a ϕ_P-space? Note that the kernel of ϕ_P does not depend on the choice of the metric, for if d and ρ are compatible metrics, then at each $x \in X$,

$$\forall \alpha > 0, \ \neg P(S_d(x, \alpha)) \text{ if and only if } \forall \alpha > 0, \ \neg P(S_\rho(x, \alpha)).$$

However, whether or not X is a ϕ_P-subset does. Consider $\mathbb{N}$ with the discrete topology and $P(V) := V$ contains at most one point. Here, $\mathrm{Ker}(\phi_P) = \emptyset$. Equipping X with either the Euclidean metric or the discrete zero-one metric, $\mathbb{N}$ becomes a UC-space, that is, a ϕ_P space. But with $\rho(n, j) = |\frac{1}{n} - \frac{1}{j}|$, $\mathbb{N}$ fails to be a UC-space, as the isolation functional goes to zero along the sequence $\langle n \rangle$.

Theorem 28.26. *Let X be a metrizable topological space, and let P be an hereditary property of open sets. The following conditions are equivalent:*

(1) *X has a compatible metric d such that $\langle X, d\rangle$ is a ϕ_P-space;*

(2) *$\mathrm{Ker}(\phi_P)$ is compact.*

Proof. If d is an arbitrary compatible metric, let us write for the purposes of this proof ϕ_P^d for the functional induced by P with respect to the metric d. As we have noted just above, condition (2) is necessary for condition (1). For the sufficiency of (2) for (1), we use this technical fact about open covers: if X is metrizable and $\{\Omega_k : k \in \mathbb{N}\}$ is a family of open covers of X, then there exists a compatible metric d for X such that $\forall k \in \mathbb{N}$, $\{S_d(x, 1/k) : x \in X\}$ refines Ω_k [74, p. 196].

It is possible that while compact, $\mathrm{Ker}(\phi_P)$ is empty. Then each $x \in X$ has an open neighborhood V_x such that $P(V_x)$. By an attenuated form of the just-stated refinement result, there exists a compatible metric d such that $\{S_d(x, 1) : x \in X\}$ refines $\{V_x : x \in X\}$. Since P is hereditary, $\forall x$, $\phi_P^d(x) = \sup\{\alpha > 0 : P(S_d(x, \alpha))\} \geq 1$, and so $\langle X, d\rangle$ is a ϕ_P-space. Otherwise, $\mathrm{Ker}(\phi_P)$ is nonempty and compact and so there is a countable family of open neighborhoods $\{W_k : k \in \mathbb{N}\}$ of $\mathrm{Ker}(\phi_P)$ such that whenever V is open and $\mathrm{Ker}(\phi_P) \subseteq V$, $\exists k \in \mathbb{N}$ with $W_k \subseteq V$. Again, for each $x \notin \mathrm{Ker}(\phi_P)$, let V_x be an open neighborhood of x with $P(V_x)$. For each $k \in \mathbb{N}$, define an open cover Ω_k of X as follows:

$$\Omega_k := \{V_x : x \notin W_k\} \cup \{W_k\}.$$

Choose a compatible metric d such that for each k, $\{S_d(x, 1/k) : x \in X\}$ refines Ω_k. Now let $\langle x_n \rangle$ be a sequence in X satisfying $\lim_{n \to \infty} \phi_P^d(x_n) = 0$. For each $k \in \mathbb{N}$, W_k contains a tail of $\langle x_n \rangle$, specifically $x_n \in W_k$ whenever $\phi_P^d(x_n) < \frac{1}{k}$. Since $\{W_k : k \in \mathbb{N}\}$ forms a base for the neighborhoods of $\mathrm{Ker}(\phi_P)$, $\forall \varepsilon > 0$, $\exists n_\varepsilon \in \mathbb{N} \ \forall n \geq n_\varepsilon$, $x_n \in S_d(\mathrm{Ker}(\phi_P), \varepsilon)$. Since $\mathrm{Ker}(\phi_P)$ is compact, $\langle x_n \rangle$ has a cluster point and $\langle X, d\rangle$ is a ϕ_P-space in this second case, too. $\square$

Recall that an elementary proof of the following corollary was presented in Theorem 19.12.

Corollary 28.27. *Let $\langle X, \tau \rangle$ be a metrizable space. Then there is a compatible UC-metric for the topology if and only if X' is compact.*

29. Uniformly Paracompact Subsets

In this section we will study the family of those subsets A of a metric space $\langle X, d \rangle$ that are *uniformly paracompact*: for each open cover $\mathscr{V}$ of X there exists $\mu > 0$ and a locally finite open cover $\mathscr{U}$ refining $\mathscr{V}$ such that for each $a \in A$, $S_d(a, \mu)$ hits at most finitely many members of $\mathscr{U}$. Uniform paracompactness of the entire space and its relationship to other covering properties was clarified by Rice in the uniform space setting [133]. In the metric setting, the many parallels between uniformly paracompact spaces and UC-spaces were perhaps first pointed out in [20].

Our first task is to adapt Rice's arguments to subsets in the metric setting, following Beer and Di Maio [32]. The third condition of our next result is particularly tractable.

Theorem 29.1. *For a nonempty subset A of a metric space $\langle X, d \rangle$, the following conditions are equivalent:*

(1) *A is uniformly paracompact;*

(2) *whenever $\mathscr{U}$ is a locally finite open cover of X, there exists $\mu > 0$ such that for each $a \in A$, $S_d(a, \mu)$ hits at most finitely many members of $\mathscr{U}$;*

(3) *for each open cover $\mathscr{V}$ of X $\exists \mu > 0$ such that whenever $E \cap A \neq \emptyset$ and $\mathrm{diam}_d(E) < \mu$, $\exists \{V_1, V_2 \dots, V_n\} \subseteq \mathscr{V}$ with $E \subseteq \cup_{j=1}^{n} V_j$;*

(4) *whenever $\{V_i : i \in I\}$ is an open cover of X directed by inclusion, $\exists \mu > 0$ such that $\forall a \in A$, $\exists i \in I$ with $S_d(a, \mu) \subseteq V_i$.*

Proof. $(2) \Rightarrow (1)$. This is immediate from the paracompactness of metrizable spaces.

$(1) \Rightarrow (3)$. Choose by (1) $\mu > 0$ and $\mathscr{U}$ an open cover refining $\mathscr{V}$ such that for each $a \in A$, $S_d(a, \mu)$ hits at most finitely many elements of $\mathscr{U}$. Let E satisfy $\mathrm{diam}_d(E) < \mu$ and $E \cap A \neq \emptyset$. Fix $a_0 \in E \cap A$ and let $U_1, U_2, \ldots, U_n$ be those members of the cover that $S_d(a_0, \mu)$ hits. Choose for $j = 1, 2, \ldots, n$ $V_j \in \mathscr{V}$ with $U_j \subseteq V_j$. As $\mathscr{U}$ is a cover of X,

$$E \subseteq S_d(a_0, \mu) \subseteq \cup_{j=1}^{n} U_j \subseteq \cup_{j=1}^{n} V_j$$

as required.

$(3) \Rightarrow (4)$. Let $\{V_i : i \in I\}$ be an open cover of X directed by inclusion: whenever $\{i_1, i_2\} \subseteq I$, $\exists i_3 \in I$ with $V_{i_1} \cup V_{i_2} \subseteq V_{i_3}$. By (3) choose $\mu > 0$ such that if $E \cap A \neq \emptyset$ and $\mathrm{diam}_d(E) < 3\mu$, then E is contained in a finite union of members of the cover. Fix $a \in A$; since $\mathrm{diam}_d\left(S_d(a, \mu)\right) < 3\mu$, $\exists \{i_1, i_2, i_3, \ldots, i_n\} \subseteq I$ with $S_d(a, \mu) \subseteq \cup_{j=1}^{n} V_{i_j}$. But there exists $i_{n+1} \in I$ such that $\cup_{j=1}^{n} V_{i_j} \subseteq V_{i_{n+1}}$, and condition (4) follows.

$(4) \Rightarrow (2)$. Let $\mathscr{U}$ be a locally finite open cover of X. For each $x \in X$ choose $\delta_x > 0$ such that $S_d(x, \delta_x)$ hits only finitely many members of $\mathscr{U}$. Let $\mathscr{W}$ be the cover of X consisting of all finite unions of members of $\{S_d(x, \delta_x) : x \in X\}$. Since $\mathscr{W}$ is stable under finite unions, by (4) $\exists \mu > 0$

such that $\forall a \in A$, $S_d(a,\mu)$ is contained in a finite union of members of $\{S_d(x,\delta_x) : x \in X\}$, and so $S_d(a,\mu)$ hits only finitely many members of $\mathscr{U}$. $\qquad\square$

Corollary 29.2. *Let $\langle X, d \rangle$ be a metric space. The family of uniformly paracompact subsets of X forms a bornology with closed base that contains the UC-subsets of X.*

Proof. That the family of uniformly paracompact subsets contains the singletons, is hereditary, and is stable under finite unions are all immediate from condition (3) of Theorem 29.1. If A is uniformly paracompact and nonempty and $\mathscr{V}$ is an open cover of X, choose $\mu > 0$ such that $\operatorname{diam}_d(T) < 2\mu$ and $T \cap A \neq \emptyset \Rightarrow T$ has a finite subcover from $\mathscr{V}$. If $\operatorname{diam}_d(E) < \mu$ and $E \cap \operatorname{cl}(A) \neq \emptyset$ then $S_d(E, \mu/2) \cap A \neq \emptyset$ and the enlargement has diameter less than 2μ. Thus $S_d(E, \mu/2)$ has a finite subcover from $\mathscr{V}$ so that E does as well. That each UC-subset is uniformly paracompact follows from condition (2) of Theorem 18.3 in view of condition (3) of Theorem 29.1. $\qquad\square$

In the sequel, we denote the bornology of uniformly paracompact subsets by $\mathscr{B}_d^{up}$. We next show that $\mathscr{B}_d^{up}$ coincides with the bornology $\mathscr{B}(\nu)$ as considered in the last section, where ν is the measure of local compactness functional. Since $I(\cdot) \leq \nu(\cdot)$, it follows separately from this fact that each UC-subset is uniformly paracompact. Relative to the functional ν and consistent with the notation employed in [20, 31, 32], we introduce the notation

$$\operatorname{nlc}(X) := \{x \in X : \nu(x) = 0\} = \operatorname{Ker}(\nu),$$

as these are the points of non-local-compactness of the space. Concurrently, we show that a uniformly paracompact subset is one on which each continuous function is uniformly locally bounded in a sense that we now make precise.

Definition 29.3. Let $\langle X, d \rangle$ and $\langle Y, \rho \rangle$ be metric spaces and let $f \in Y^X$. We say f is *uniformly locally bounded* on $A \subseteq X$ if there exists $\delta > 0$ such that $\forall a \in A$, $f(S_d(a, \delta))$ is a metrically bounded subset of Y.

Clearly the family of subsets on which each member of a prescribed family of continuous function is uniformly locally bounded is a bornology containing $\mathscr{B}_d^{uc}(X)$, by strong uniform continuity of continuous functions on each UC-subset.

Theorem 29.4. *For a nonempty subset A of a metric space $\langle X, d \rangle$, the following conditions are equivalent:*

(1) *each sequence $\langle a_n \rangle$ in A such that $\lim_{n \to \infty} \nu(a_n) = 0$ clusters;*

(2) *$\operatorname{cl}(A) \cap \operatorname{nlc}(X)$ is compact but possibly empty, and $\forall \delta > 0$, $\exists \mu > 0$ such that whenever $a \in A \backslash S_d(\operatorname{cl}(A) \cap \operatorname{nlc}(X), \delta)$, we have $\nu(a) > \mu$;*

(3) *A is uniformly paracompact;*

(4) *whenever $f : \langle X, d \rangle \to \langle Y, \rho \rangle$ is continuous, f is uniformly locally bounded on A;*

(5) *whenever $f \in C(X, \mathbb{R})$, f is uniformly locally bounded on A.*

Proof. (1) $\Rightarrow$ (2). This is an immediate consequence of Theorem 28.10.

(2) $\Rightarrow$ (3). Suppose A satisfies condition (2). We show that A satisfies condition (3) of the Theorem 29.1 which is equivalent to uniform paracompactness. Let $\mathscr{V}$ be an open cover of X.

If $\mathrm{nlc}(X) \cap \mathrm{cl}(A) = \emptyset$, then $\inf\{\nu(a) : a \in A\} > 0$. Take $\mu > 0$ such that $\forall a \in A$, $S_d(a, \mu)$ has compact closure. Then whenever $\mathrm{diam}_d(E) < \mu$ and $E \cap A \ne \emptyset$, clearly E lies in a compact ball and thus in the union of finitely many members of the cover .

Otherwise, $\mathrm{nlc}(X) \cap \mathrm{cl}(A)$ is nonempty and compact. Let $\mathscr{V}_1$ be a finite subfamily of $\mathscr{V}$ such that $\mathrm{nlc}(X) \cap \mathrm{cl}(A) \subseteq \cup \mathscr{V}_1$. By compactness $\exists \varepsilon > 0$ with

$$S_d(\mathrm{nlc}(X) \cap \mathrm{cl}(A), \varepsilon) \subseteq \cup \mathscr{V}_1.$$

Clearly, $\forall x \in S_d(\mathrm{nlc}(X) \cap \mathrm{cl}(A), \varepsilon/2)$ we have $S_d(x, \frac{\varepsilon}{2}) \subseteq \cup \mathscr{V}_1$. By condition (2), $\exists \delta > 0$ such that $\forall a \in A$,

$$d(a, \mathrm{nlc}(X) \cap \mathrm{cl}(A)) > \frac{\varepsilon}{3} \Rightarrow \nu(a) > \delta.$$

Then for $\mu = \min\{\delta, \frac{\varepsilon}{2}\}$, we see that each open ball of radius μ about any point of A is contained in a finite union of members of $\mathscr{V}$.

(3) $\Rightarrow$ (4). Fix $y_0 \in Y$; then $\{f^{-1}(S_\rho(y_0, n)) : n \in \mathbb{N}\}$ is an open cover of X so for some $\mu > 0$ and each $a \in A$, $S_d(a, \mu)$ is contained in a finite union of these preimage sets.

(4) $\Rightarrow$ (5). This is trivial.

(5) $\Rightarrow$ (1). Suppose condition (1) fails: there exists a sequence $\langle a_n \rangle$ in A for which $\lim_{n \to \infty} \nu(a_n) = 0$ yet having no cluster point. Without loss of generality, we may assume that its terms are distinct. We consider two mutually exclusive and exhaustive cases for its set of terms A: (a) $\{a_n : n \in \mathbb{N}\}$ is a totally bounded set; (b) the set of terms is not totally bounded.

In case (a), by passing to a subsequence, we may assume that $\langle a_n \rangle$ is a Cauchy sequence. Let $f \in C(X, \mathbb{R})$ map each a_n to n. Then f fails to be uniformly locally bounded on $\{a_n : n \in \mathbb{N}\}$, in violation of condition (5). In case (b), by passing to a subsequence, we can find $\delta > 0$ such that $\{S_d(a_n, \delta) : n \in \mathbb{N}\}$ is a discrete family of balls and for each $n \in \mathbb{N}$, $C_n := \{x \in X : d(x, a_n) \le \frac{\delta}{n}\}$ is not compact. For each n, let $g_n : C_n \to \mathbb{R}$ be continuous and unbounded. By local finiteness, $\cup_{n=1}^\infty C_n$ is closed, so there is $g \in C(X, \mathbb{R})$ that extends each g_n. This, too, is not uniformly locally bounded on the set of terms, completing the proof. □

Corollary 29.5. *Each Cauchy sequence in a nonempty uniformly paracompact subset A of a metric space $\langle X, d \rangle$ clusters. As a result, $\mathrm{cl}(A)$ as a metric subspace of $\langle X, d \rangle$ is complete.*

Proof. Suppose $\langle a_n \rangle$ is a Cauchy sequence in A that does not cluster. Easily, $\lim_{n \to \infty} \nu(a_n) = 0$, in violation of condition (1). □

Corollary 29.6. *The bornology of uniformly paracompact subsets in a metric space is shielded from closed sets.*

Proof. The functional ν is arises as ϕ_P where P is an hereditary property of open subsets, namely $P(V) := V$ has compact closure. Apply Corollary 28.15 □

Corollary 29.7. *Let A be a subset of a metric space $\langle X, d \rangle$. Suppose each point of $\mathrm{cl}(A)$ has a relatively compact neighborhood in X. Then A is uniformly paracompact if and only if there exists $\mu > 0$ such that for each $a \in A, \{x \in X : d(x, a) \le \mu\}$ is compact.*

Proof. This follows from condition (2) of Theorem 29.4 as $\mathrm{cl}(A) \cap \mathrm{nlc}(X) = \emptyset$. □

Uniform paracompactness of a subset is not an intrinsic property of metric spaces. As with UC-ness, its defining properties all depend on points outside the set in question. The following result should come as no surprise.

Proposition 29.8. *Let A be a nonempty subset of a metric space $\langle X, d \rangle$. The following conditions are equivalent:*

(1) A is a uniformly paracompact subset of each metric space in which it is isometrically embedded;

(2) A is a UC-subset of each metric space in which it is isometrically embedded;

(3) $\langle A, d \rangle$ is a compact metric space.

Proof. We need only prove (1) $\Rightarrow$ (3). Suppose condition (3) fails; choose $\langle a_n \rangle$ in A that has no cluster point in A. Put $X = A \times \mathbb{Q}$, equipped with box metric $\rho((a_1, q_1), (a_2, q_2)) := \max \{d(a_1, a_2), |q_1 - q_2|\}$. Of course, $a \mapsto (a, 1)$ is an isometric embedding. Since no point in $A \times \{1\}$ has a compact neighborhood in X, $\forall n \in \mathbb{N}$, $\nu(a_n, 1) = 0$ with respect to ρ. Since the sequence $\langle (a_n, 1) \rangle$ fails to cluster, condition (1) fails. $\square$

Theorem 29.9. *Let A be a nonempty subset of a metric space $\langle X, d \rangle$. The following conditions are equivalent:*

(1) $A \in \mathscr{B}_d^{up}$;

(2) *whenever ρ is a metric equivalent to d on X, there exists $\delta > 0$ such that $\forall a \in A$, $S_d(a, \delta)$ is ρ-bounded.*

Proof. (1) $\Rightarrow$ (2). Suppose $A \in \mathscr{B}_d^{up}$; then the identity map $I_X : \langle X, d \rangle \to \langle X, \rho \rangle$ is uniformly locally bounded on A, and this is exactly condition (2).

(2) $\Rightarrow$ (1). We prove the contrapositive. Suppose A is not uniformly paracompact; then for some $f \in C(X, \mathbb{R})$, f fails to be uniformly locally bounded on A. Consider the equivalent metric ρ on X defined by

$$\rho(x, w) = d(x, w) + |f(x) - f(w)|.$$

For each $\delta > 0$, there exists $a \in A$ such that $f(S_d(a, \delta))$ is an unbounded subset of $\mathbb{R}$, and so $S_d(a, \delta) \notin \mathscr{B}_\rho(X)$. Thus, condition (2) fails if (1) fails. $\square$

In our study of UC-subsets, we saw that $A \in \mathscr{B}_d^{uc}$ if and only if each pseudo-Cauchy sequence in X with distinct terms such that $\lim_{n \to} d(x_n, A) = 0$ clusters. Since $\mathscr{B}_d^{up}$ also lies between the subsets with compact closure and the subsets with complete closure, one is motivated to look for a condition on sequences between pseudo-Cauchyness and Cauchyness that can be used to characterize uniformly paracompact subsets. We now introduce an appropriate condition.

Definition 29.10. A sequence $\langle x_n \rangle$ in a metric space $\langle X, d \rangle$ is called *cofinally Cauchy* if $\forall \varepsilon > 0$, there exists an infinite subset $\mathbb{N}_\varepsilon$ of $\mathbb{N}$ such that whenever $\{n, j\} \subseteq \mathbb{N}_\varepsilon$, we have $d(x_n, x_j) < \varepsilon$.

It is easy to generalize this property to nets in a uniform spaces, as was done by Howes [100] who introduced the language cofinally Cauchy to the literature: if $\langle x_\lambda \rangle_{\lambda \in \Lambda}$ is a net in a Hausdorff uniform space as determined by a diagonal uniformity with base $\mathscr{D}$, then the net is called *cofinally Cauchy* provided for each $D \in \mathscr{D}$ there exists a cofinal subset Λ_0 of the underlying directed set Λ such that whenever $\{\lambda_1, \lambda_2\} \subseteq \Lambda_0$, we have $(x_{\lambda_1}, x_{\lambda_2}) \in D$. Earlier an analagous property for filters was considered by Corson [68]. Naturally, Howes called the uniform space *cofinally complete* provided each cofinally Cauchy net clusters. Other authors required that the underlying directed set have additional structure, e.g., Burdick [60] studied cofinal Cauchyness for nets based on a well-ordered set.

Before looking at cofinally Cauchy sequences and uniform paracompactness, we digress to give a simple characterization of total boundedness in terms such sequences.

Proposition 29.11. *Let A be a nonempty subset of a metric space $\langle X, d \rangle$. The following conditions are equivalent:*

(1) *A is totally bounded;*

(2) *each sequence in A is cofinally Cauchy;*

(3) *each sequence in A is pseudo-Cauchy.*

Proof. Suppose A is totally bounded and $\langle a_n \rangle$ is a sequence in A. Let $\varepsilon > 0$, and choose a finite subset F of X such that $A \subseteq S_d(F, \frac{\varepsilon}{2})$. Then there exists an infinite subset $\mathbb{N}_\varepsilon$ of $\mathbb{N}$ and $x \in F$ with $d(a_n, x) < \frac{\varepsilon}{2}$ for each $n \in \mathbb{N}_\varepsilon$. As a result, whenever $\{n, j\} \subseteq \mathbb{N}_\varepsilon$, we have $d(a_n, a_j) < \varepsilon$. The implication $(2) \Rightarrow (3)$ is trivial. Finally if (1) fails, then for some $\varepsilon > 0$ there exists no finite subset F of A for which $A \subseteq S_d(F, \varepsilon)$. From this, starting with $a_1 \in A$, we can inductively find $a_2, a_3, a_4, \ldots$ in A with $a_{n+1} \notin S_d(\{a_1, a_2, \ldots, a_n\})$ for each $n \in \mathbb{N}$ and $\langle a_n \rangle$ so constructed is not pseudo-Cauchy. Thus, (3) fails. $\qquad\square$

When A is the entire metric space, we will see that clustering of each cofinally Cauchy sequence is necessary and sufficient for uniform paracompactness of the space. In view of condition (5) in Theorem 18.3, we might conjecture that if A is a nonempty subset of a metric space $\langle X, d \rangle$, then A is uniformly paracompact if and only if each cofinally Cauchy sequence $\langle x_n \rangle$ in X with distinct terms such that $\lim_{n \to \infty} d(x_n, A) = 0$ clusters. Unfortunately, this sequential condition fails to be sufficient!

Example 29.12. In the Hilbert space ℓ_2, let $A = \{e_n : n \in \mathbb{N}\}$, the standard orthonormal base for the space. Partition $\mathbb{N}$ into a countable family of infinite subsets: $\mathbb{N}_1 \cup \mathbb{N}_2 \cup \mathbb{N}_3 \cup \cdots$. Let $n_{(j,k)}$ be the kth member of $\mathbb{N}_j$ with respect to the usual ordering of the integers, and put $x_{(j,k)} := e_j + \frac{1}{3j} e_k$. Finally, let X be the following metric subspace of ℓ_2:

$$X := A \cup \{x_{(j,k)} : (j,k) \in \mathbb{N} \times \mathbb{N}\}.$$

Evidently, there is no cofinally Cauchy sequence with distinct terms in X satisfying $\lim_{n \to \infty} d(x_n, A) = 0$, so the condition on sequences with distinct terms holds vacuously. However, $A \notin \mathscr{B}_d^{up}$. To see this, note that since $\forall j \in \mathbb{N}$, $\langle x_{(j,k)} \rangle_{k \in \mathbb{N}}$ has no cluster point, we have $\lim_{j \to \infty} \nu(e_j) = 0$. But $\langle e_j \rangle$ has no cluster point, and so A is not a uniformly paracompact subset.

Still, we do have a characterization of uniform paracompactness for a nonempty subset A in terms of the clustering of cofinally Cauchy sequences, replacing the condition $\lim_{n \to} d(x_n, A) = 0$ by a weaker condition to force additional sequences to converge. Here, there is no need to require that the sequence have distinct terms.

Definition 29.13. Let $\langle X, d \rangle$ be a metric space and let A be a nonempty subset. We call a sequence $\langle x_n \rangle$ in X *asymptotically cofinally Cauchy with respect to A* if for each $\varepsilon > 0$, there exists an infinite subset $\mathbb{N}_\varepsilon$ of $\mathbb{N}$ such that $\sup\{d(x_n, A) : n \in \mathbb{N}_\varepsilon\} < \varepsilon$ and whenever $\{n, j\} \subseteq \mathbb{N}_\varepsilon$, we have $d(x_n, x_j) < \varepsilon$.

Theorem 29.14. *Let A be a nonempty subset of a metric space $\langle X, d \rangle$. Then A is uniformly paracompact if and only if each asymptotically cofinally Cauchy sequence with respect to A clusters.*

Proof. Suppose A is uniformly paracompact, and let $\langle x_n \rangle$ be an asymptotically cofinally Cauchy sequence with respect to A. For each $n \in \mathbb{N}$, let $\mathbb{M}_n$ be an infinite subset of $\mathbb{N}$ such that whenever $\{k, j\} \subseteq \mathbb{M}_n$, we have $d(x_k, x_j) < \frac{1}{2n}$ and $\forall j \in \mathbb{M}_n$, $d(x_j, A) < \frac{1}{2n}$. If for some $n \in \mathbb{N}$ the subsequence $\langle x_j \rangle_{j \in \mathbb{M}_n}$ clusters, we are done. Otherwise, for each $n \in \mathbb{N}$ we can find $a_n \in A$ such that for each $j \in \mathbb{M}_n$ we have $d(x_j, a_n) < \frac{1}{n}$ and so $\nu(a_n) \leq \frac{1}{n}$. By uniform paracompactness, $\langle a_n \rangle$ clusters. Since each set $\mathbb{M}_n$ is infinite, we can find a strictly increasing sequence of positive integers $\langle j_n \rangle$ such that for each $n \in \mathbb{N}$, $j_n \in \mathbb{M}_n$. As a result, $\langle x_{j_n} \rangle$ clusters as well, and therefore $\langle x_n \rangle$ clusters.

Conversely, suppose each sequence that is asymptotically cofinally Cauchy with respect to A clusters. Let $\langle a_n \rangle$ be a sequence in A with $\lim_{n \to \infty} \nu(a_n) = 0$. Since $S_d(a_n, \nu(a_n) + \frac{1}{n})$ fails to be relatively compact for each $n \in \mathbb{N}$, there exists a sequence $\langle x_{(n,j)} \rangle_{j \in \mathbb{N}}$ in $S_d(a_n, \nu(a_n) + \frac{1}{n})$ that fails to have a cluster point. Let $g : \mathbb{N} \to \mathbb{N} \times \mathbb{N}$ be a bijection. By our construction, $\langle x_{g(n)} \rangle$ is asymptotically cofinally Cauchy with respect to A, and each cluster point of $\langle x_{g(n)} \rangle$ will also be a cluster point of $\langle a_n \rangle$. $\qquad\square$

30. Uniformly Paracompact Spaces and Uniformly Locally Lipschitz Functions

We now put together our results of the last section to characterize uniformly paracompact metric spaces, also called cofinally complete spaces, in a large number of ways. The equivalence of all the conditions save the last two follows from the analysis of the last section while the last two follow from Theorem 28.20.

Theorem 30.1. *Let $\langle X, d \rangle$ be a metric space; the following conditions are equivalent:*

(1) *X is a uniformly paracompact space;*

(2) *each cofinally Cauchy sequence in X clusters;*

(3) *for each locally finite open cover of X, there exists $\mu > 0$ such that for each $x \in X$, $S_d(x, \mu)$ hits at most finitely many members of the cover;*

(4) *for each open cover $\mathcal{V}$ of X, there exists $\mu > 0$ such that for each $x \in X$, $S_d(x, \mu)$ has a finite subcover from $\mathcal{V}$;*

(5) *for each open cover $\mathcal{V}$ of X directed by inclusion, there exists $\mu > 0$ such that $\{S_d(x, \mu) : x \in X\}$ refines $\mathcal{V}$;*

(6) *each sequence $\langle x_n \rangle$ in X such that $\lim_{n \to \infty} \nu(x_n) = 0$ clusters;*

(7) *$\mathrm{nlc}(X)$ is compact, and $\forall \delta > 0$, $\exists \mu > 0$ such that $x \in X \backslash S_d(\mathrm{nlc}(X), \delta) \Rightarrow \nu(x) > \mu$;*

(8) *whenever $f : \langle X, d \rangle \to \langle Y, \rho \rangle$ is continuous, f is uniformly locally bounded on X;*

(9) *whenever $f \in C(X, \mathbb{R})$, f is uniformly locally bounded on X;*

(10) *whenever ρ is a metric equivalent to d on X, there exists $\delta > 0$ such that $\forall x \in X$, $S_d(x, \delta)$ is ρ-bounded;*

(11) *whenever $\langle A_n \rangle$ is a decreasing sequence in $\mathscr{C}_0(X)$ with $\lim_{n \to \infty} \inf\{\nu(x) : x \in A_n\} = 0$, then $\cap_{n=1}^{\infty} A_n \neq \emptyset$;*

(12) *whenever $\langle A_n \rangle$ is a decreasing sequence in $\mathscr{C}_0(X)$ with $\lim_{n \to \infty} \sup\{\nu(x) : x \in A_n\} = 0$, then $\cap_{n=1}^{\infty} A_n \neq \emptyset$.*

By condition (7), if $\mathrm{nlc}(X) = \emptyset$ in a uniformly paracompact space, then $\inf\{\nu(x) : x \in X\}$ is positive. Less formally, a locally compact space is uniformly paracompact if and only if it uniformly locally compact. The open unit ball as a metric subspace of Euclidean space is an example of a locally compact space that is not uniformly paracompact. Also note that a normed linear space is uniformly paracompact if and only if it is finite dimensional, as in an infinite dimensional space we have $\mathrm{nlc}(X) = X$.

Several historical remarks are in order. The measure of local compactness functional ν was intoduced in [20] in a deliberate attempt to find a signature functional for the uniformly paracompact metric spaces, in the same way that the isolation functional is the signature functional for UC-spaces. In this article, the equivalence of conditions (2), (6), (8), (9), (10), (11) and (12) was established. Although the equivalence of uniform paracompactness for a uniform space and the clustering of each cofinally Cauchy net had been known for some time in the uniform setting (see, e.g., [60, p. 438]), the equivalence of conditions (1) and (6) in the metric setting somehow came later - [32] is a source, but it may not be the first. In any case, this means that in a metric space, if each cofinally Cauchy sequence clusters, then each cofinally Cauchy net with respect to the induced metric uniformity also clusters, and so there is no confusion possible in the meaning of cofinal completeness for metric spaces. The equivalence of conditions (1) and (7) may be due to Hohti [98, Theorem 2.1.1].

From condition (2) of Theorem 30.1, we most easily get these two corollaries.

Corollary 30.2. *Each uniformly paracompact metric space is complete.*

Corollary 30.3. *Each closed metric subspace of a uniformly paracompact metric space is uniformly paracompact.*

The next theorem also appears in the paper of Hohti [98, Theorem 2.2.1].

Theorem 30.4. *Let $\langle X, d_1 \rangle$ and $\langle Y, d_2 \rangle$ be metric spaces, and equip $X \times Y$ with the box metric ρ. Then the product is uniformly paracompact if and only if one of the following conditions holds:*

(1) *either $\langle X, d_1 \rangle$ or $\langle Y, d_2 \rangle$ is compact and the other is uniformly paracompact, or*

(2) *both $\langle X, d_1 \rangle$ and $\langle Y, d_2 \rangle$ are locally compact and uniformly paracompact.*

Proof. We denote the measure of local compactness functionals for the factors by ν_X and ν_Y and use ν for the product. For sufficiency, suppose first that one of the spaces, say $\langle X, d_1 \rangle$, is compact while $\langle Y, d_2 \rangle$ is just uniformly paracompact. Now if $(x, y) \in X \times Y$ is arbitrary and $S_{d_2}(y, \varepsilon)$ is relatively compact, then $S_\rho((x, y), \varepsilon)$ is relatively compact. Thus, if $\langle (x_n, y_n) \rangle$ is a sequence in $X \times Y$ along which ν tends to zero, then $\lim_{n \to \infty} \nu_Y(y_n) = 0$. As a result, $\langle y_n \rangle$ has a convergent subsequence, and passing to a second subsequence, so does $\langle x_n \rangle$. We conclude that $\langle (x_n, y_n) \rangle$ clusters.

Now suppose that X and Y are both locally compact and uniformly paracompact. As $\mathrm{nlc}(X) = \emptyset = \mathrm{nlc}(Y)$ we conclude by condition (7) of Theorem 30.1 that $\inf\{\nu_X(x) : x \in X\} > 0$ and $\inf\{\nu_Y(y) : y \in Y\} > 0$. As a result, $\inf\{\nu((x, y)) : (x, y) \in X \times Y\}$ is positive, so the product is uniformly paracompact.

For the converse, assume $X \times Y$ is uniformly paracompact. From the immediately preceding corollary, this is also true for both factors. Suppose neither factor is compact and at least one, say $\langle X, d_1 \rangle$, is not locally compact, that is, $\mathrm{nlc}(X) \neq \emptyset$. Pick $x_0 \in \mathrm{nlc}(X)$ and a sequence $\langle y_n \rangle$ in Y that does not cluster. Now for each $n \in \mathbb{N}$ we have $(x_0, y_n) \in \mathrm{nlc}(X \times Y)$, but the sequence $\langle (x_0, y_n) \rangle$ fails to cluster in $X \times Y$. By condition (6) of Theorem 30.1, the product is not uniformly paracompact, and a contradiction ensues. □

Uniform paracompactness of $\langle \mathscr{C}_0(X), H_d \rangle$ was characterized by Beer and Di Maio [31, Theorem 3.9] who in the process of coming up with the next result, displayed a rather complex formula for the values of the local compactness functional in the hyperspace [31, Theorem 3.7]. As the analysis is quite technical, the details are omitted.

Theorem 30.5. *Let $\langle X, d \rangle$ be a metric space. The following conditions are equivalent:*

(1) *X is a point of local compactness of $\langle \mathscr{C}_0(X), H_d \rangle$;*

(2) *$\inf\{\nu(A) : A \in \mathscr{C}_0(X)\} > 0$, where ν is determined by Hausdorff distance;*

(3) *$\langle \mathscr{C}_0(X), H_d \rangle$ is uniformly paracompact.*

The second condition of the last theorem says that the hyperspace is uniformly locally compact. Actually, the equivalence of the first two conditions was established much earlier in the uniform setting by Burdick [59, pp. 29–31] who also presented an internal condition on X characteristic of (1) and (2). In the metric setting, his condition may be stated as follows: $\langle X, d \rangle$ is uniformly locally compact and $\exists \varepsilon > 0 \ \forall \delta > 0$ there exists a compact subset C_δ such that $S_d(x, \varepsilon) \subseteq S_d(x, \delta)$ whenever $x \notin C_\delta$.

Theorem 28.24 tells us what must be added to completeness of a metric space to produce uniform paracompactness for the space.

Theorem 30.6. *Let $\langle X, d \rangle$ be a complete metric space and let $\alpha(\cdot)$ be the measure of noncompactness functional for $\mathscr{P}_0(X)$. Then X is uniformly paracompact if and only if $\forall \varepsilon > 0$, $\exists \delta > 0$ such that whenever $C \in \mathscr{C}_0(X)$ and $\sup\{\nu(c) : c \in C\} < \delta$ then $\alpha(C) < \varepsilon$.*

We now specify what precisely must be added to uniform paracompactness to achieve the UC-property for a metric space.

Theorem 30.7. *Let $\langle X, d \rangle$ be a uniformly paracompact metric space. Then the space is a UC-space if and only if each sequence $\langle x_n \rangle$ in X with $\lim_{n \to \infty} I(x_n) = 0$ has a Cauchy subsequence.*

Proof. Suppose $\langle X, d \rangle$ is a UC-space, and let $\langle x_n \rangle$ satisfy $\lim_{n \to \infty} I(x_n) = 0$. By the definition of UC-space $\langle x_n \rangle$ clusters which means that it has a convergent and therefore a Cauchy subsequence. For the converse, we rely on condition (5) of Theorem 19.1. Let $\langle x_n \rangle$ be a pseudo-Cauchy sequence with distinct terms. Clearly, $\langle x_n \rangle$ has a subsequence along which the isolation function tends to zero which by assumption has a Cauchy subsequence. Since this secondary subsequence is cofinally Cauchy and X is a uniformly paracompact space, it follows from condition (2) of Theorem 30.1 that this subsequence clusters. Thus, X is a UC-space. $\qquad \square$

We are able to characterize the existence of a uniformly paracompact metric for a metrizable space using Theorem 28.26. This was first achieved by Romaguera [137] using a very different approach.

Theorem 30.8. *Let $\langle X, \tau \rangle$ be a metrizable space. Then there is a compatible metric making the space uniformly paracompact if and only if $nlc(X)$ is compact.*

Finally, we characterize those bornologies in a metrizable space that are bornologies of uniformly paracompact subsets with respect to some compatible metric. The proof is almost identical to the one given for Theorem 18.8 and is left to the reader.

Theorem 30.9. *Let $\langle X, \tau \rangle$ be a metrizable space and let $\mathscr{B}$ be a bornology on X. Then $\mathscr{B} = \mathscr{B}_d^{up}$ for some compatible metric d if and only if there is a star-development $\langle \mathscr{U}_n \rangle$ for X such that*

$(*)$ $\qquad \mathscr{B} = \{E \in \mathscr{P}(X) :$ *whenever $\mathscr{V}$ is an open cover of $X, \exists n \in \mathbb{N}$ such that*

$\qquad \qquad \forall U \in \mathscr{U}_n, \ U \cap E \neq \emptyset \Rightarrow U$ *has a finite subcover from $\mathscr{V}\}$*

Uniform paracompactness plays a fundamental role in answering these four questions about the uniformly locally Lipschitz functions:

- When does the class of uniformly locally Lipschitz functions on $\langle X, d \rangle$ agree with the formally larger class of locally Lipschitz functions?

- When does the class of uniformly locally Lipschitz functions on $\langle X, d \rangle$ agree with the intermediate class of Cauchy-Lipschitz functions?

- When can we uniformly approximate an arbitrary Cauchy continuous real-valued function on $\langle X, d \rangle$ by uniformly locally Lipschitz functions?

- When are the nonvanishing real-valued uniformly locally Lipschitz functions stable under reciprocation?

Recall that the locally Lipschitz functions (i) agree with the Cauchy-Lipschitz functions exactly when the domain is a complete metric space, and (ii) they agree with the Lipschitz in the small functions exactly when $\langle X, d \rangle$ is a UC-space. Further the Cauchy-Lipschitz functions and the uniformly locally Lipschitz functions agree with the Lipschitz in the small functions if and only if the domain has UC-completion.

Theorem 30.10. *Let $\langle X, d \rangle$ be a metric space. The following conditions are equivalent:*

(1) *$\langle X, d \rangle$ is a uniformly paracompact metric space;*

(2) *each locally Lipschitz function on X with values in a metric space $\langle Y, d \rangle$ is uniformly locally Lipschitz;*

(3) *each real-valued locally Lipschitz function on X is uniformly locally Lipschitz;*

(4) *the real-valued uniformly locally Lipschitz functions are uniformly dense in $C(X, \mathbb{R})$.*

Proof. $(1) \Rightarrow (2)$. Let $f : X \to Y$ be locally Lipschitz. Fix $y_0 \in Y$; for each $n \in \mathbb{N}$, let V_n be the following open subset of X:

$$V_n := \left\{ x \in X : \exists \varepsilon > 0 \; \forall a, b \in S_d\left(x, \frac{1}{n} + \varepsilon\right), \; \rho(f(a), f(b)) < n \cdot d(a, b) \text{ and } f(x) \in S_\rho(y_0, n) \right\}.$$

Clearly, $\{V_n : n \in \mathbb{N}\}$ is an open cover of X such that $\forall n \in \mathbb{N}, V_n \subseteq V_{n+1}$. By condition (5) of Theorem 30.1, there exist $\mu > 0$ such that each open ball of radius μ is contained in some single member of the cover. Fix $x \in X$ and choose $k \in \mathbb{N}$ such that $S_d(x, \mu) \subseteq V_k$. It remains to show that f is Lipschitz on $S_d(x, \mu)$.

To this end, suppose $\{a, b\} \subseteq S_d(x, \mu)$. If $d(a, b) < \frac{1}{k}$, then by definition of V_k, we have $\rho(f(a), f(b)) < kd(a, b)$. On the other if $d(a, b) \geq \frac{1}{k}$, then as $\{f(a), f(b)\} \subseteq S_\rho(y_0, k)$, we have

$$\frac{\rho(f(a), f(b))}{d(a, b)} < \frac{2k}{1/k} = 2k^2.$$

Since $2k^2 > k$, it will be a Lipschitz constant for f restricted to $S_d(x, \mu)$.

$(2) \Rightarrow (3)$. This is obvious.

$(3) \Rightarrow (4)$. Condition (4) is an immediate consequence of (3) and Theorem 26.3.

$(4) \Rightarrow (1)$. If we can uniformly approximate each real-valued continuous function by uniformly locally Lipschitz functions, then each real-valued continuous function on X is uniformly locally bounded. Invoke condition (9) of Theorem 30.1 to conclude that $\langle X, d \rangle$ is uniformly paracompact. □

We note that Theorem 27.8 immediately falls out of Theorem 27.11 combined with Theorem 30.10 which we just proved.

We now turn to the second and third questions. They have a common answer as given by the next result, which provides a companion to Theorem 30.10.

Theorem 30.11. *Let $\langle X, d \rangle$ be a metric space. The following conditions are equivalent:*

(1) *the completion $\langle \hat{X}, \hat{d} \rangle$ of $\langle X, d \rangle$ is uniformly paracompact;*

(2) *each Cauchy-Lipschitz function on X with values in an arbitrary metric space $\langle Y, \rho \rangle$ is uniformly locally Lipschitz;*

(3) *each real-valued Cauchy-Lipschitz function on X is uniformly locally Lipschitz;*

(4) *the uniformly locally Lipschitz real-valued functions on X are uniformly dense in $CC(X, \mathbb{R})$.*

Proof. $(1) \Rightarrow (2)$. Let $f : X \to Y$ be Cauchy-Lipschitz. As we can view f as a Cauchy-Lipschitz function into the completion $\langle \hat{Y}, \hat{\rho} \rangle$, by Proposition 27.2, f has a Cauchy-Lipschitz extension $\hat{f}$ from the completion $\langle \hat{X}, \hat{d} \rangle$ to $\langle \hat{Y}, \hat{\rho} \rangle$. By Theorem 30.10, as $\hat{f}$ is locally Lipschitz and $\langle \hat{X}, \hat{d} \rangle$ is uniformly paracompact, $\hat{f}$ is uniformly locally Lipschitz, and so its restriction to X is uniformly locally Lipschitz as well.

$(2) \Rightarrow (3)$. This is trivial.

$(3) \Rightarrow (4)$. This follows from Theorem 26.8: on a general metric space, the real-valued Cauchy-Lipschitz functions are uniformly dense in $CC(X, \mathbb{R})$.

$(4) \Rightarrow (1)$. Suppose (4) holds and let $\hat{f} \in C(\hat{X}, \mathbb{R})$ be arbitrary. By condition (9) of Theorem 30.1, it suffices to show that $\hat{f}$ is uniformly locally bounded. Put $f := \hat{f}|_X$; by the completeness of $\langle \hat{X}, \hat{d} \rangle$, the function $\hat{f}$ is Cauchy continuous, so $f \in CC(X, \mathbb{R})$. By condition (4) we can find a uniformly locally Lipschitz function $g : X \to \mathbb{R}$ whose uniform distance from f is finite. Choose $\delta > 0$ such that g restricted to each open ball of radius δ in X is Lipschitz. Now for each $\hat{x} \in \hat{X}$ we can find $x \in X$ with

$$S_{\hat{d}}(\hat{x}, \frac{\delta}{2}) \subseteq \mathrm{cl}_{\hat{X}}(S_d(x, \delta)).$$

Since f restricted to $S_d(x, \delta)$ is bounded, by continuity, $\hat{f}$ restricted to $S_{\hat{d}}(\hat{x}, \frac{\delta}{2})$ is also bounded. We have shown that $\hat{f}$ is uniformly locally bounded. □

We state the following result from [33] that answers our fourth question without proof.

Theorem 30.12. *Let $\langle X, d \rangle$ be a metric space. The following conditions are equivalent.*

(1) *the metric space is uniformly paracompact;*

(2) *the reciprocal of each nonvanishing uniformly locally Lipschitz real-valued function on X is uniformly locally Lipschitz;*

(3) *the reciprocal of each nonvanishing member of $Lip(X, \mathbb{R})$ is uniformly locally Lipschitz.*

We leave it as an easy exercise to show that the real-valued uniformly locally Lipschitz functions are always stable under pointwise product.

31. Bornological Convergence of Nets of Closed Subsets

In the following sections, we look at bornological convergence of nets of closed subsets of a metric space as first considered abstractly by Lechicki, Levi and Spakowski [113]. In the literature, this has been studied for nets of arbitrary subsets and with respect to a general family of truncating subsets rather than with respect to a truncating cover. Further, the two halves of the convergence can be considered separately and a surprising lack of symmetry is in evidence.

Definition 31.1. Let $\mathscr{B}$ be a cover of a metric space $\langle X, d \rangle$ and let $A \in \mathscr{C}(X)$. A net $\langle A_\lambda \rangle$ of closed subsets of X is declared $(\mathscr{B}, d)$-*convergent to* A provided whenever $B \in \mathscr{B}$ and $\varepsilon > 0$, then eventually both

$$A \cap B \subseteq S_d(A_\lambda, \varepsilon) \ \text{ and } \ A_\lambda \cap B \subseteq S_d(A, \varepsilon).$$

If $\langle A_\lambda \rangle$ is convergent to A in this sense, we will write $A \in (\mathscr{B}, d) - \lim A_\lambda$, as limits need not be unique! No finer convergence results if in the definition we replace $\mathscr{B}$ by $\downarrow \sum(\mathscr{B})$, and so we may assume without loss of generality that $\mathscr{B}$ is a bornology. For this reason, we call our notion of set convergence *bornological convergence*. The larger the bornology, the finer the convergence. It is also clear that no coarser convergence occurs if we replace the bornology $\mathscr{B}$ by a base $\mathscr{B}_0$ in the definition. Notice that a net of closed subsets is $(\mathscr{B}, d)$-convergent to the empty set if and only if it is eventually outside each member of $\mathscr{B}$.

The finest bornological convergence is obtained when we use $\mathscr{B} = \mathscr{P}(X)$. Since $\{X\}$ is a base for the trivial bornology, $A \in (\mathscr{P}(X), d) - \lim A_\lambda$ iff both

$$A \subseteq S_d(A_\lambda, \varepsilon) \ \text{ and } \ A_\lambda \subseteq S_d(A, \varepsilon),$$

so that for nets of nonempty closed sets, this agrees with convergence in Hausdorff distance. Here, a net $\langle A_\lambda \rangle$ in $\mathscr{C}(X)$ is convergent to $\emptyset$ if and only if $A_\lambda = \emptyset$ eventually.

Example 31.2. Let $X = \{0\} \cup \{\frac{1}{n} : n \in \mathbb{N}\}$ equipped with the usual metric d of the line. A base for the bornology of finite subsets is $\{B_j : j \in \mathbb{N}\}$ where for each $j, B_j = \{0\} \cup \{\frac{1}{j}, \frac{1}{j-1}, \ldots, 1\}$. For each $n \in \mathbb{N}$, let $A_n = \{\frac{1}{n+1}, 1\}$. Obviously, $\langle A_n \rangle$ converges in Hausdorff distance to $\{0, 1\}$ so that $\{0, 1\} = (\mathscr{F}(X), d) - \lim A_n$. But $\{1\} = (\mathscr{F}(X), d) - \lim A_n$ as well, because for each $j, A_n \cap B_j = \{1\}$ for all n sufficiently large. This shows that limits need not be unique, even for sequences.

Example 31.3. In the plane $\mathbb{R}^2$ equipped with the Euclidean metric d, if L is the horizontal axis and L_n is the line through the origin with slope $\frac{1}{n}$, then $\langle L_n \rangle$ ought to be convergent to L with respect to any reasonable convergence notion. But clearly convergence in Hausdorff distance fails this criterion, as for each $n, H_d(L_n, L) = \infty$. On the other hand we do have $(\mathscr{B}_d(X), d)$-convergence of $\langle L_n \rangle$ to L.

To see this, we confirm convergence by working with a convenient base for the bornology: $\{[-j,j] \times [-j,j] : j \in \mathbb{N}\}$. For a fixed $j \in \mathbb{N}$ and $\varepsilon \in (0,1)$, whenever $n > \frac{j}{\varepsilon}$, we have

$$L_n \cap [-j,j] \times [-j,j] \subseteq S_d(L, \varepsilon) \text{ and } L \cap [-j,j] \times [-j,j] \subseteq S_d(L_n, \varepsilon),$$

as required

Bornological convergence as we have defined it extends convergence in the underlying metric space.

Proposition 31.4. *Let $\langle a_\lambda \rangle$ be a net in $\langle X, d \rangle$ and let $a \in X$. Suppose $\mathscr{B}$ is a bornology on X. Then $\{a\} \in (\mathscr{B}, d) - \lim \{a_\lambda\}$ if and only if $\lim_\lambda d(a_\lambda, a) = 0$.*

Proof. Necessity comes from taking $B = \{a\}$; given $\varepsilon > 0$, eventually

$$\{a\} = \{a\} \cap B \subseteq S_d(\{a_\lambda\}, \varepsilon) = S_d(a_\lambda, \varepsilon).$$

that is, $d(a, a_\lambda) < \varepsilon$ eventually. For sufficiency, notice that $\{a\} = H_d - \lim \{a_\lambda\}$, and thus the net is $(\mathscr{P}(X), d)$-convergent to $\{a\}$ and so it is $(\mathscr{B}, d)$-convergent with respect to any smaller bornology. $\qquad\square$

Given $B \in \mathscr{B}$, a bornology on $\langle X, d \rangle$, we put $[B, d, \varepsilon] := \{(A_1, A_2) \in \mathscr{C}(X) \times \mathscr{C}(X) : A_1 \cap B \subseteq S_d(A_2, \varepsilon) \text{ and } A_2 \cap B \subseteq S_d(A_1, \varepsilon)\}$. Such objects look like they might form a base for a uniformity on the closed subsets compatible with $(\mathscr{B}, d)$-convergence; indeed, this is sometimes the case but not always! However, in general we have $A \in (\mathscr{B}, d) - \lim A_\lambda$ if and only if for each $B \in \mathscr{B}$ (or in a given base for $\mathscr{B}$) and each $\varepsilon > 0$, we have eventually $(A, A_\lambda) \in [B, d, \varepsilon]$. We will show in a subsequent section that all sets of the form $[B, d, \varepsilon]$ form a base for a uniformity if and only if $\mathscr{B}$ is stable under small enlargements.

Replacing the given metric by an equivalent metric can produce a different bornological convergence with respect to the same bornology $\mathscr{B}$ on X.

Example 31.5. The map $\phi : (0, \infty) \to (0, 1)$ be defined by $\phi(x) = \frac{x}{1+x}$ is a homeomorphism. As a result, if d is the usual metric on $(0, \infty)$ and $\rho = d \circ \phi$, we have equivalent metrics on the ray. Put $A = \{2k : k \in \mathbb{N}\}$ and let $A_n = A \cup \{2k + 1 : k \geq n\}$. Note that $H_d(A_n, A) = 1$ for each n while $\lim_{n \to \infty} H_\rho(A_n, A) = 0$. Thus, $(\mathscr{P}(X), d)$-convergence fails while $(\mathscr{P}(X), \rho)$-convergence holds.

The proof of the next result is easy and is left to the reader.

Proposition 31.6. *Let d and ρ be uniformly equivalent metrics on a set X and let $\mathscr{B}$ be a bornology on X. Then $(\mathscr{B}, d)$-convergence on $\mathscr{C}(X)$ agrees with $(\mathscr{B}, \rho)$-convergence.*

Our next result speaks to uniqueness of limits.

Theorem 31.7. *Let $\mathscr{B}$ be a bornology on a metric space $\langle X, d \rangle$. The following conditions are equivalent.*

(1) *$\mathscr{B}$ is local;*

(2) *$(\mathscr{B}, d)$-limits for nets in $\mathscr{C}(X)$ are unique;*

(3) *$(\mathscr{B}, d)$-limits for nets in $\mathscr{F}(X)$ are unique.*

Proof. Only the implications (1) $\Rightarrow$ (2) and (3) $\Rightarrow$ (1) require proof.

(1) $\Rightarrow$ (2). Assume $\mathscr{B}$ is local but that the net of closed sets $\langle A_\lambda \rangle$ is convergent to different closed limits A and C. Without loss of generality, we may take $c \in C \backslash A$. Choose $\delta < \frac{1}{2} d(c, A)$ such that $S_d(c, \delta) \in \mathscr{B}$. Eventually, $C \cap \{c\} \subseteq S_d(A_\lambda, \delta)$ so that eventually, $A_\lambda \cap S_d(c, \delta) \neq \emptyset$. But by the choice of δ, for all λ, $A_\lambda \cap S_d(c, \delta) \cap S_d(A, \delta) = \emptyset$. This contradicts the $(\mathscr{B}, d)$-convergence of $\langle A_\lambda \rangle$ to A.

(3) $\Rightarrow$ (1). Suppose some p in X has no neighborhood in $\mathscr{B}$. Remembering that X has at least two points, choose $q \neq p$, and for each $B \in \mathscr{B}$ and each $n \in \mathbb{N}$ choose $x_{n,B} \notin B$ with $d(x_{n,B}, p) < \frac{1}{n}$. Directing $\mathscr{B} \times \mathbb{N}$ by $(B_1, n_1) \preceq (B_2, n_2)$ provided $B_1 \subseteq B_2$ and $n_1 \leq n_2$, the net $(B, n) \mapsto \{x_{n,B}, q\}$ is $(\mathscr{B}, d)$-convergent to both $\{p, q\}$ and $\{q\}$. $\quad\square$

In the case our bornology is local, we can write $A = (\mathscr{B}, d) - \lim A_\lambda$ in lieu of $A \in (\mathscr{B}, d) - \lim A_\lambda$. From the last result, we see that $(\mathscr{K}(X), d)$-convergence produces unique limits if and only if the space is locally compact, while $(\mathscr{F}(X), d)$-convergence produces unique limits if and only if $X' = \emptyset$.

Proposition 31.6 above is far from a characterization. Let $\mathscr{B}$ and $\mathscr{S}$ be bornologies on X. We next intend to give necessary and sufficient conditions on two pairs $(\mathscr{S}, \rho)$ and $(\mathscr{B}, d)$ as established by Beer and Levi [42] so that the same convergence ensues for nets of closed sets. This was initially accomplished for $\mathscr{S} = \mathscr{B}_\rho(X)$ and $\mathscr{B} = \mathscr{B}_d(X)$ in [30]. By Proposition 31.4, there is no hope of getting the same convergence unless the metrics are equivalent. As a main step, we first give necessary and sufficient conditions for one bornological convergence to be finer than the other.

Let X be a metrizable space and let $\mathscr{S}$ be a bornology on X. Given compatible metrics d and ρ, we say ρ is *uniformly stronger than* d *on* $\mathscr{S}$ provided the identity map $I_X : \langle X, \rho \rangle \to \langle X, d \rangle$ is strongly uniformly continuous on $\mathscr{S}$. If this is also true when the metrics are reversed, we say that d and ρ are *uniformly equivalent with respect to* $\mathscr{S}$.

Theorem 31.8. *Let d and ρ be compatible metrics for a metrizable topological space X, and let $\mathscr{B}$ and $\mathscr{S}$ be two bornologies on X. The following conditions are equivalent:*

   (1) *$(\mathscr{S}, \rho)$-convergence ensures $(\mathscr{B}, d)$-convergence for nets in $\mathscr{C}(X)$;*

   (2) *$(\mathscr{S}, \rho)$-convergence ensures $(\mathscr{B}, d)$-convergence for nets in $\mathscr{C}_0(X)$;*

   (3) *$\mathscr{B} \subseteq \mathscr{S}$ and ρ is uniformly stronger than d on $\mathscr{B}$.*

Proof. (1) $\Rightarrow$ (2). This is trivial.

(2) $\Rightarrow$ (3). Suppose first that $\mathscr{B} \nsubseteq \mathscr{S}$. Then $\exists B_0 \in \mathscr{B}$ such that $\forall S \in \mathscr{S}, B_0 \backslash S \neq \emptyset$. For each $S \in \mathscr{S}$, pick $x_S \in B_0 \backslash S$. Clearly, $\forall x \in X$, we have $\{x\} \in (\mathscr{S}, \rho)$-lim $\{x_S, x\}$. By (2), we have $\{x\} \in (\mathscr{B}, d)$-lim $\{x_S, x\}$. But since $\langle x_S \rangle$ is a net in B_0, we conclude that for all x, lim $d(x_S, x) = 0$. This contradicts our groundrule that X is not a singleton, and we now have shown that $\mathscr{B} \subseteq \mathscr{S}$.

It remains to show that ρ is uniformly stronger than d on $\mathscr{B}$. If not, then by definition $I_X : \langle X, \rho \rangle \to \langle X, d \rangle$ is not strongly uniformly continuous on $\mathscr{B}$. Thus for some nonempty $B \in \mathscr{B}$ and $\varepsilon > 0$ there exists sequences $\langle b_n \rangle$ in B and $\langle w_n \rangle$ in X such that $\forall n \in \mathbb{N}$

$$\rho(b_n, w_n) \leq \frac{1}{n} \text{ but } d(b_n, w_n) > \varepsilon.$$

Notice that since d and ρ are equivalent metrics, neither sequence can have a cluster point. By the Efremovic lemma, by passing to a subsequence we may assume $\forall n, k \in \mathbb{N}$ that $d(b_n, w_k) \geq \varepsilon/4$. For each $k \in \mathbb{N}$ write

$$A_k := \{w_n : n \in \mathbb{N}\} \cup \{b_n : n \geq k\} \in \mathscr{C}_0(X).$$

Clearly the sequence $\langle A_k \rangle$ converges to the closed set $\{w_n : n \in \mathbb{N}\}$ in Hausdorff distance H_ρ and thus is $(\mathscr{S}, \rho)$-convergent, too. But for all k, $(A_k, \{w_n : n \in \mathbb{N}\}) \notin [B, d, \varepsilon/4]$ and so $(\mathscr{B}, d)$-convergence fails. This contradiction shows ρ is uniformly stronger than d on $\mathscr{T}$.

(3) $\Rightarrow$ (1). Suppose $\langle A_\lambda \rangle$ is $(\mathscr{S}, \rho)$-convergent to A. Fix $\varepsilon > 0$ and $B \in \mathscr{B} \subseteq \mathscr{S}$. By strong uniform continuity of the identity on B, there exists $\delta > 0$ such that if $\{x, w\} \cap B \neq \emptyset$ and $\rho(x, w) < \delta$, then $d(x, w) < \varepsilon$. This yields $[B, \rho, \delta] \subseteq [B, d, \varepsilon]$ from which $(\mathscr{B}, d)$-convergence immediately follows as $[B, \rho, \delta]$ contains (A, A_λ) eventually. $\qquad \square$

Corollary 31.9. *Let d and ρ be compatible metrics for a metrizable topological space X, and let $\mathscr{B}$ and $\mathscr{S}$ be two bornologies on X. Then $(\mathscr{S}, \rho)$-convergence coincides with $(\mathscr{B}, d)$-convergence either on $\mathscr{C}(X)$ or on $\mathscr{C}_0(X)$ if and only if $\mathscr{B} = \mathscr{S}$ and ρ is uniformly equivalent to d on $\mathscr{B}$.*

The next corollary is well-known [17, p. 92].

Corollary 31.10. *Let d and ρ be compatible metrics for a metrizable topological space X. Then H_d-convergence agrees with H_ρ-convergence on $\mathscr{C}_0(X)$ if and only if d and ρ are uniformly equivalent.*

Proof. The identity map is strongly uniformly continuous on $\mathscr{P}(X)$ if and only if it is strongly uniformly continuous on X, that is, it is globally uniformly continuous. This means that the identity map is biuniformly continuous if and only if $(\mathscr{P}(X), d)$-convergence agrees with $(\mathscr{P}(X), \rho)$-convergence on $\mathscr{C}_0(X)$. So restricted, we get convergence in Hausdorff distance. $\qquad \square$

32. Attouch-Wets Convergence

When $\mathscr{B} = \mathscr{B}_d(X)$, the bornological convergence that we get is called *Attouch-Wets convergence* [8, 17, 30]. This convergence has also been called *bounded Hausdorff convergence* [131]. From Theorem 31.7, we see that Attouch-Wets convergence produces unique limits, and if $A = (\mathscr{B}_d(X), d) - \lim A_\lambda$, we will just write $A = AW_d - \lim A_\lambda$. All sets of the form $[S_d(p, n), d, \frac{1}{n}]$ where p is a fixed but arbitrary point of X are easily seen directly to be a countable base for a uniformity compatible with the convergence; the induced metrizable topology on $\mathscr{C}(X)$ is called the *Attouch-Wets topology*.

If we replace d by $\rho := \min\{d, 1\}$, then $X \in \mathscr{B}_\rho(X)$, so that AW_ρ-convergence for nets in $\mathscr{C}(X)$ becomes convergence in Hausdorff distance with respect to H_ρ and thus with respect to H_d because d and ρ are uniformly equivalent. Attouch-Wets convergence for nets of closed sets as determined by metrics d_1 and d_2 coincide if and only if $\mathscr{B}_{d_1}(X) = \mathscr{B}_{d_2}(X)$ and the identity map is strongly uniformly continuous on bounded subsets in both directions [30].

Notably, Attouch-Wets convergence of a net of nonempty closed subsets $\langle A_\lambda \rangle$ to a nonempty closed set A amounts to the uniform convergence of $\langle d(\cdot, A_\lambda) \rangle$ to $d(\cdot, A)$ on bounded subsets of X [17, 30]. We first give a generally applicable proposition.

Proposition 32.1. *Let $\langle A_\lambda \rangle$ be a net of nonempty closed sets in a metric space $\langle X, d \rangle$ and let $A \in \mathscr{C}_0(X)$. Suppose $\langle d(\cdot, A_\lambda) \rangle$ converges uniformly to $d(\cdot, A)$ on a bornology $\mathscr{B}$. Then $A \in (\mathscr{B}, d) - \lim A_\lambda$.*

Proof. Given $B \in \mathscr{B}$ and $\varepsilon > 0$, we must show that eventually, $A_\lambda \cap B \subseteq S_d(A, \varepsilon)$ and $A \cap B \subseteq S_d(A_\lambda, \varepsilon)$. We just verify the first property. Choose $\lambda_0 \in \Lambda$ such that $\lambda \succeq \lambda_0 \Rightarrow \sup_{b \in B} |d(b, A) - d(b, A_\lambda)| < \varepsilon$. Fix $\lambda \succeq \lambda_0$ and let $x \in A_\lambda \cap B$. Since $d(x, A_\lambda) = 0$ and $d(x, A) - d(x, A_\lambda) < \varepsilon$, we have $d(x, A) < \varepsilon$ which means that $x \in S_d(A, \varepsilon)$. This shows that $A_\lambda \cap B \subseteq S_d(A, \varepsilon)$. $\square$

Theorem 32.2. *Let $\langle A_\lambda \rangle$ be a net of nonempty closed subsets of a metric space $\langle X, d \rangle$ and let $A \in \mathscr{C}_0(X)$. Then $A = AW_d - \lim A_\lambda$ if and only if $\langle d(\cdot, A_\lambda) \rangle$ converges uniformly to $d(\cdot, A)$ on $\mathscr{B}_d(X)$.*

Proof. Sufficiency follows from Proposition 32.1. For necessity, let $A \in \mathscr{C}_0(X), B \in \mathscr{B}_d(X) \cap \mathscr{P}_0(X)$, and $\varepsilon > 0$ be arbitrary. We produce a second metrically bounded set B_1 and $\delta > 0$ such that

$$(A, C) \in [B_1, d, \delta] \Rightarrow \sup_{x \in B} |d(x, A) - d(x, C)| < \varepsilon.$$

Fix $p \in X$ and $n \in \mathbb{N}$ with $B \subseteq S_d(p, n)$ and $\frac{1}{n} < \varepsilon$. Choose an integer $n_0 > n$ such that $A \cap S_d(p, n_0 - 1) \neq \emptyset$. Let $k = 2n + n_0 + 1$; we claim that $B_1 = S_d(p, k)$ and $\delta = \frac{1}{k}$ do the job.

Fix $C \in \mathscr{C}_0(X)$ with $(C,A) \in [S_d(p,k), d, \frac{1}{k}]$. For each $x \in B \subseteq S_d(p,k)$ take $c_x \in C$ with $d(x,c_x) < d(x,C) + \frac{1}{k}$. Since $A \cap S_d(p, n_0 - 1) \neq \emptyset \Rightarrow C \cap S_d(p, n_0) \neq \emptyset$, the triangle inequality yields

$$d(p, c_x) \leq d(p,x) + d(x, c_x) < d(p,x) + d(x,C) + 1 < n + (n + n_0) + 1 = k.$$

Since $C \cap S_d(p,k) \subseteq S_d(A, \frac{1}{k})$, there exists $a_x \in A$ with $d(a_x, c_x) < \frac{1}{k}$. From this, we get

$$d(x,A) \leq d(x, a_x) \leq d(x, c_x) + d(c_x, a_x) \leq d(x,C) + \frac{1}{k} + \frac{1}{k} < d(x,C) + \frac{2}{3}\varepsilon.$$

Similarly we have $d(x,C) \leq d(x,A) + \frac{2}{3}\varepsilon$ for each $x \in B$, so that $\sup_{x \in B} |d(x,A) - d(x,C)| < \varepsilon$. $\square$

As a result of Theorem 32.2, the following metric is compatible with AW_d-convergence for nonempty closed subsets:

$$\rho_{AW_d}(A_1, A_2) := \sum_{n=1}^{\infty} 2^{-n} \min\{1, \ \sup\{|d(x,A_1) - d(x,A_2)| : x \in S_d(p,n)\}\}.$$

The point $p \in X$ is a fixed but arbitrary point of X; changing p produces a uniformly equivalent metric. It can be shown that this metric is complete provided that the underlying metric d is complete. One way to see this is to observe that $C(X, \mathbb{R})$ equipped with the natural metric for uniform convergence on bounded subsets

$$\rho(f,g) := \sum_{n=1}^{\infty} 2^{-n} \min\{1, \ \sup\{|f(x) - g(x)| : x \in S_d(p,n)\}\}$$

is a complete metric space and that distance functions are a closed subset of $C(X, \mathbb{R})$ in the topology [17, Lemma 3.1.1].

Example 32.3. If $\langle X, d \rangle$ is an unbounded metric space, the uniformity determined by the metric ρ_{AW_d} on $\mathscr{C}_0(X)$ is strictly finer than the uniformity having as a base all sets of the form $[B, d, \varepsilon]$ where $B \in \mathscr{B}_d(X)$ and $\varepsilon > 0$. To see this, given $B \in \mathscr{B}$ and $\varepsilon > 0$ we produce nonempty closed subsets A_1 and A_2 with $(A_1, A_2) \in [B, d, \varepsilon]$ yet $\rho_{AW_d}(A_1, A_2) = 1$. Fix $p \in X$ and choose $\{a_1, a_2\} \subseteq B^c$ with $d(a_1, p) > d(a_2, p) + 1$. As $\{a_1\} \cap B \subseteq S_d(\{a_2\}, \varepsilon)$ and $\{a_2\} \cap B \subseteq S_d(\{a_1\}, \varepsilon)$, we have $(\{a_1\}, \{a_2\}) \in [B, d, \varepsilon]$. But for each positive integer n,

$$\sup\{|d(x, \{a_1\}) - d(x, \{a_2\})| : x \in S_d(p,n)\} \geq |d(p, \{a_1\}) - d(p, \{a_2\})| > 1.$$

This shows that $\rho_{AW_d}(\{a_1\}, \{a_2\}) = 1$.

Attouch-Wets convergence imposes itself on linear analysis in a fundamental way. Let $\langle X, ||\cdot||_X \rangle$ and $\langle Y, ||\cdot||_Y \rangle$ be real normed linear spaces, with $\mathbf{B}(X,Y)$ as the space of continuous linear transformations from X to Y, equipped with the usual operator norm $||\cdot||_{\mathrm{op}}$. As shown by Penot and Zalinescu [131], a sequence $T, T_1, T_2, T_3, \ldots$ in $\mathbf{B}(X,Y)$ satisfies $\lim_{n \to \infty} ||T - T_n||_{\mathrm{op}} = 0$ if and only if $\langle \mathrm{Gr}(T_n) \rangle$ is Attouch-Wets convergent with respect to the box metric on $X \times Y$ to $\mathrm{Gr}(T)$. This was proved for continuous linear functionals much earlier [15]. We will establish a more comprehensive result along these lines in the next section.

Attouch-Wets convergence is also of importance with respect to convex duality. Let $f : X \to (-\infty, \infty]$ be a lower semicontinuous convex function, that is, a function whose epigraph is a closed convex set. Assuming $\mathrm{dom}(f) \neq \emptyset$, there will be a continuous affine functional $x \mapsto y(x) - \alpha$ where $y \in X^*$ that f fails to majorize. Nevertheless, it is a consequence of the separation theorem that f is the supremum of the continuous affine functions that it does majorize [76, Proposition

3.1]. By the *Fenchel conjugate* f^{conj} we mean the convex function on X^* whose epigraph consists of all $(y, \alpha) \in X^* \times \mathbb{R}$ such that f majorizes $x \mapsto y(x) - \alpha$. The actual formula for the Fenchel conjugate is given by

$$f^{\text{conj}}(y) := \sup\{y(x) - f(x) : x \in X\}.$$

As a supremum of weak-star continuous functions on X^*, the conjugate is not only norm lower semicontinuous on X^* but is moreover weak-star lower semicontinuous. The map $f \mapsto f^{\text{conj}}$ is one-to-one; in fact [76, Proposition 4.1]

$$f(x) = \sup\{y(x) - f^{\text{conj}}(y) : y \in X^*\}.$$

In qualitative terms, the conjugate of f^* when restricted to the isometric image of X in X^{**} is the original function f.

The result we wish to call to the reader's attention linking Attouch-Wets convergence to Fenchel conjugacy as established by the author in [14] is this: if $f, f_1, f_2, f_3, \ldots$ is a sequence of lower semicontinuous convex functions each with nonempty effective domain, then then $\langle \text{epi}(f_n) \rangle$ is Attouch-Wets convergent to $\text{epi}(f)$ if and only if $\langle \text{epi}(f_n^{\text{conj}}) \rangle$ is Attouch-Wets convergent to $\text{epi}(f^{\text{conj}})$.

Providing additional details would only be useful to the reader with a background in convex analysis, and as this is not expected, we keep our discussion short.

33. Topologies of Uniform Convergence for B(X,Y) and Convergence of Graphs

Let $\langle X, ||\cdot||_X \rangle$ and $\langle Y, ||\cdot||_Y \rangle$ be real normed linear spaces. In this section, and following [22], we explain how convergence of sequences in $\mathbf{B}(X,Y)$ with respect to a topology of uniform convergence on a cover of X by norm bounded subsets of X is reflected in the bornological convergence of the associated sequence of graphs with respect to a cover of $X \times Y$ where the product is equipped with the box norm as determined by $||\cdot||_X$ and $||\cdot||_Y$, that is, $||(x,y)|| = \max\{||x||_X, ||y||_Y\}$. The results are valid for any norm equivalent to the box norm, as equivalent norms determine uniformly equivalent metrics. In particular, we will see that if $T, T_1, T_2, T_3, \ldots$ is a sequence in $\mathbf{B}(X,Y)$, then $\lim_{n\to\infty}||T_n - T||_{\mathrm{op}} = 0$ if and only if $\langle \mathrm{Gr}(T_n) \rangle$ is Attouch-Wets convergent to $\mathrm{Gr}(T)$. This means that for each bounded subset E of $X \times Y$ we have eventually both

$$\mathrm{Gr}(T_n) \cap E \subseteq \mathrm{Gr}(T) + \varepsilon U_{X\times Y},$$

$$\mathrm{Gr}(T) \cap E \subseteq \mathrm{Gr}(T_n) + \varepsilon U_{X\times Y}.$$

Given a cover $\mathscr{A}$ of X by norm bounded subsets, the topology of uniform convergence on $\mathscr{A}$ for $\mathbf{B}(X,Y)$ can be easily shown to be a Hausdorff locally convex topology; the requirement that the sets be norm bounded of course is included to guarantee that each neighborhood of the zero transformation be absorbing with respect to the vector space $\mathbf{B}(X,Y)$. The strongest such topology is obtained when $\mathscr{A} = \{nU_X : n \in \mathbb{N}\}$ and is the operator norm topology. The weakest such topology is obtained when $\mathscr{A} = \mathscr{F}(X)$, and is sometimes oddly called the *strong operator topology* in the literature [75, p. 475]. This of course is the *weak-star topology* in the case of X^*, i.e., $\mathbf{B}(X,\mathbb{R})$. By the *bounded weak-star topology* for X^*, we mean the strongest topology that coincides with the weak-star topology on each norm bounded subset of X^*. By the Banach-Dieudonné theorem [110, p. 272], this agrees with the topology of uniform convergence on $\mathscr{T}\mathscr{B}_d(X)$ where d is the metric associated with $||\cdot||_X$. This topology has also been called the equicontinuous weak-star topology.

In the theory of locally convex spaces, topologies of uniform convergence are often presented in terms of families of absolutely convex sets, in view of the importance of the Mackey-Arens theorem [135, pg. 62]. For example, the topology of uniform convergence on the weakly compact absolutely convex subsets of $\langle X, ||\cdot||_X \rangle$ for X^* is called the *Mackey topology*. This is the finest locally convex topology on X^* for which X is its continuous dual as a locally convex space. When X is a Banach space, by the Krein-Šmulian Theorem [75, pg. 434] and Lemma 33.12 below, the topology of uniform convergence on weakly compact absolutely convex subsets for $\mathbf{B}(X,Y)$ agrees with the formally stronger topology of uniform convergence on weakly compact subsets whatever the target space $\langle Y, ||\cdot||_Y \rangle$ may be.

The next result attempts to reconcile some of the above approaches and indicates the flexibility we have in describing our cover with respect to which our topology of uniform convergence on $\mathbf{B}(X,Y)$ is defined.

Proposition 33.1. *Let $\mathscr{A}$ be a family of norm bounded subsets of $\langle X, ||\cdot||_X\rangle$ that is a cover of X. Then there a bornology of norm bounded subsets $\mathscr{B}$ containing $\mathscr{A}$ having a base $\mathscr{B}_0$ consisting of norm bounded absolutely convex sets that is closed under homothetic images such that $\mathscr{T}_{\mathscr{A}} = \mathscr{T}_{\mathscr{B}} = \mathscr{T}_{\mathscr{B}_0}$ on $\mathbf{B}(X, Y)$.*

Proof. Put $\mathscr{D} = \mathrm{born}(\mathscr{A})$, and consider this family of subsets of X:

$$\mathscr{B}_0 := \{\alpha\mathrm{aco}(D) : \alpha > 0 \text{ and } D \in \mathscr{D}\}.$$

Note $\mathscr{B}_0$ is a family of norm bounded absolutely convex sets closed under homothetic images. In view of the description of the absolutely convex hull of a set in terms of linear combinations of elements of the set, clearly $\mathscr{T}_{\mathscr{A}} = \mathscr{T}_{\mathscr{D}} = \mathscr{T}_{\mathscr{B}_0}$.

Put $\mathscr{B} := \downarrow \mathscr{B}_0$. Evidently $\mathscr{B}$ is an hereditary cover of X such that $\mathscr{T}_{\mathscr{A}} = \mathscr{T}_{\mathscr{B}} = \mathscr{T}_{\mathscr{B}_0}$. We still must show that $\mathscr{B}$ is stable under finite unions. To this end, suppose $E_1 \subseteq \alpha_1\mathrm{aco}(D_1)$ and $E_2 \subseteq \alpha_2\mathrm{aco}(D_2)$ where $\{D_1, D_2\} \subseteq \mathscr{D}$ and α_1 and α_2 are positive. Then

$$E_1 \cup E_2 \subseteq (\alpha_1 + \alpha_2)(\mathrm{aco}(D_1) \cup \mathrm{aco}(D_2)) \subseteq (\alpha_1 + \alpha_2)\mathrm{aco}(D_1 \cup D_2) \in \mathscr{B}_0$$

as required.

Finally, note that for each $A \in \mathscr{A}$, $A \subseteq \mathrm{aco}(A) \in \mathscr{B}_0$ and this shows $\mathscr{A} \subseteq \mathscr{B}$. $\square$

By our last result, we saw that uniform convergence in $\mathbf{B}(X, Y)$ with respect to a certain cover of norm bounded subsets can be determined by looking at a related cover of (absolutely) convex sets. We do not have this luxury in the case of bornological convergence, even for sequences of compact convex sets.

Example 33.2. Consider the plane $\mathbb{R}^2$ equipped with the box norm and with the bornology of finite subsets $\mathscr{F}(\mathbb{R}^2)$. Let ρ denote the metric determined by the box norm. Let $\langle (x_n, y_n)\rangle$ be a sequence of distinct norm one elements. For each $n \in \mathbb{N}$, let $A_n = \mathrm{co}(\{(x_n, y_n), (0, 0)\})$. While $\{(0, 0)\} \in (\mathscr{F}(\mathbb{R}^2), \rho) - \lim A_n$, bornological convergence to this single point fails with respect to the bornology generated by all polytopes, i.e., convex hulls of members of $\mathscr{F}(\mathbb{R}^2)$, which of course coincides with the bornology of norm bounded sets in this setting. However, bornological convergence with respect to this larger bornology might occur to a line segment.

A nonempty subset A of $\langle X, ||\cdot||\rangle$ is called *starshaped with respect to* $a_0 \in A$ if $\forall a \in A$, $\mathrm{co}(\{a, a_0\}) \subseteq A$. The set of points with respect to which a nonempty subset A of X is starshaped is easily seen to be convex [148]. The smallest set containing $A \subseteq X$ that is starshaped with respect to the origin 0_X is

$$\mathrm{star}(A; 0_X) := \cup_{a \in A} \mathrm{co}(\{a, 0_X\}) = \cup_{\alpha \in [0,1]} \alpha A.$$

Proposition 33.3. *Let $\langle X, ||\cdot||\rangle$ be a normed linear space and let $A \subseteq X$ be a nonempty subset. If A is norm (resp. weakly) compact and $\alpha > 0$, then $\alpha\mathrm{star}(A; 0_X)$ is norm (resp. weakly) compact.*

Proof. The product $[0, \alpha] \times A$ is compact and $\alpha\mathrm{star}(A; \theta)$ is the image of $[0, \alpha] \times A$ under the continuous function $(\alpha, a) \mapsto \alpha a$, whether the normed linear space is equipped with the norm or weak topology. $\square$

Proposition 33.4. *Let $\langle X, ||\cdot||\rangle$ be a normed linear space and let $A \subseteq X$ be a nonempty totally bounded subset. Then $\alpha\mathrm{star}(A; 0_X)$ is totally bounded.*

Proof. Equip $[0, \alpha] \times A$ with the box metric ρ. Since a uniformly continuous function between metric spaces sends totally bounded subsets to totally bounded subsets and $[0, \alpha] \times A$ is totally bounded, we need only show $\phi(t, a) = ta$ is uniformly continuous on the product. Actually, we will show that it is Lipschitz. Pick $n \in \mathbb{N}$ such that $A \subseteq nU_X$. Then if (t_1, a_1) and (t_2, a_2) are points of the domain we compute

$$\|\phi(t_1, a_1) - \phi(t_2, a_2)\| = \|t_1 a_1 - t_1 a_2 + t_1 a_2 - t_2 a_2\| \leq |t_1| \cdot \|a_1 - a_2\| + |t_1 - t_2| \cdot \|a_2\|$$

$$\leq \alpha \|a_1 - a_2\| + n|t_1 - t_2| \leq (\alpha + n)\rho((t_1, a_1), (t_2, a_2)).$$

We may conclude that $\alpha \mathrm{star}(A; 0_X)$ - the range of ϕ - is totally bounded. $\qquad \square$

The convex hull (and thus the closed convex hull) of a totally bounded subset A of a general normed linear space is again totally bounded (if we can $\frac{\varepsilon}{2}$-approximate A by a finite set F, then we can ε-approximate $\mathrm{co}(A)$ by a fixed finite subset of $\mathrm{co}(F)$, which itself is totally bounded). However, the closed convex hull of a norm compact subset need not be norm compact, and the closed convex hull of a weakly compact set need not be weakly compact (notice that the weak closure and norm closure coincide by convexity). However in view of Mazur's theorem [75, p. 416] and the Krein-Šmulian theorem [75, p. 434], we do get compactness when the norm of X is complete.

Let $\langle X, \|\cdot\|_X \rangle$ and $\langle Y, \|\cdot\|_Y \rangle$ be normed linear spaces. For the purposes of our main theorem which comes next [22, Theorem 3.1], we let ρ denote the metric on $X \times Y$ determined by the box norm on the product. If $\mathscr{A}$ is a family of nonempty subsets of X and $\mathscr{S}$ is a family of nonempty subsets of Y, we write $\mathscr{A} \otimes \mathscr{S}$ for the associated family of subsets of $X \times Y$ consisting of all products of the form $A \times S$ where $A \in \mathscr{A}$ and $S \in \mathscr{S}$.

Theorem 33.5. *Let $\langle X, \|\cdot\|_X \rangle$ and $\langle Y, \|\cdot\|_Y \rangle$ be normed linear spaces. Let $\mathscr{B}_0$ be a cover of nonempty norm bounded subsets of X each starshaped with respect to the origin that is closed under multiplication by positive scalars, and suppose $\mathscr{B}_0 \subseteq \mathscr{B} \subseteq \downarrow \mathscr{B}_0$. Suppose $T, T_1, T_2, T_3, \ldots$ is a sequence in $\mathbf{B}(X, Y)$. The following conditions are equivalent:*

(1) $Gr(T) = (\mathscr{B} \otimes \{Y\}, \rho)\text{-}lim\ Gr(T_n)$;

(2) $Gr(T) = (\mathscr{B} \otimes \{\alpha U_Y : \alpha > 0\}, \rho)\text{-}lim\ Gr(T_n)$;

(3) $Gr(T) = (\mathscr{B}_0 \otimes \{\alpha U_Y : \alpha > 0\}, \rho)\text{-}lim\ Gr(T_n)$;

(4) $T = \mathscr{T}_{\mathscr{B}_0}\text{-}lim\ T_n$;

(5) $T = \mathscr{T}_{\mathscr{B}}\text{-}lim\ T_n$.

Proof. Since $\mathscr{B}_0$ refines $\mathscr{B}$ and $\{\alpha U_Y : \alpha > 0\}$ refines $\{Y\}$, the implication string $(1) \Rightarrow (2) \Rightarrow (3)$ is obvious. Clearly, (4) and (5) are equivalent. We prove $(3) \Rightarrow (4)$ and $(5) \Rightarrow (1)$.

Assume (3) holds and fix $B \in \mathscr{B}_0$. By scaling we may assume without loss of generality that $B \subseteq \frac{1}{2}U_X$ because by assumption B is norm bounded and uniform convergence on any homothetic image of B implies uniform convergence on B. We first claim that $\cup_{n=1}^{\infty} T_n(B)$ is a norm bounded subset of Y. If $\cup_{n=1}^{\infty} T_n(B)$ fails to be bounded, then since linear transformations preserve starshapedness with respect to the origin, there exists for infinitely many $n \in \mathbb{N}$ $x_n \in B$ with

$$\|T_n(x_n)\|_Y = 2(\|T\|_{\mathrm{op}} + \frac{1}{2}).$$

Choose by graph convergence $k_0 \in \mathbb{N}$ such that $\forall n > k_0$ we have

$$\mathrm{Gr}(T_n) \cap (B \times 2(\|T\|_{\mathrm{op}} + \tfrac{1}{2})U_Y) \subseteq \mathrm{Gr}(T) + \tfrac{1}{2}(U_{X \times Y}).$$

Now pick $n_1 > k_0$ and $x_{n_1} \in B$ with $\|T_{n_1}(x_{n_1})\|_Y = 2(\|T\|_{\mathrm{op}} + \tfrac{1}{2})$, and then pick $x \in X$ with

$(\Diamond)$ $\qquad\qquad \|(x_{n_1}, T_{n_1}(x_{n_1})) - (x, T(x))\|_{\mathrm{box}} \leq \tfrac{1}{2}.$

Since $x_{n_1} \in B \subseteq \tfrac{1}{2}U_X$, we conclude $x \in U_X$, and we obtain

$$\|T_{n_1}(x_{n_1}) - T(x)\|_Y \geq | \|T_{n_1}(x_{n_1})\|_Y - \|T(x)\|_Y | \geq 2(\|T\|_{\mathrm{op}} + \tfrac{1}{2}) - \|T\|_{\mathrm{op}} > \tfrac{1}{2}.$$

and this contradicts $(\Diamond)$.

The claim established, next choose $\alpha > 0$ where $\cup_{n=1}^{\infty} T_n(B) \subset \alpha U_Y$ and let $\varepsilon > 0$. To show that $\langle T_n \rangle$ converges uniformly to T on B, we consider two mutually exclusive and exhaustive cases: (i) $\forall x \in X$, $T(x) = 0$; and (ii) $\|T\|_{\mathrm{op}} > 0$.

In case (i) choose $k \in \mathbb{N}$ such that $\forall n > k$

$$\mathrm{Gr}(T_n) \cap (B \times \alpha U_Y) \subseteq \mathrm{Gr}(T) + \varepsilon U_{X \times Y} = X \times \varepsilon U_Y.$$

By the choice of α, for each $x \in B$ and for all $n \in \mathbb{N}$ we have $(x, T_n(x)) \in B \times \alpha U_Y$ and so for $n > k$ we obtain $\|T_n(x) - T(x)\|_Y = \|T_n(x)\|_Y \leq \varepsilon$ as required.

In case (ii), put $\delta := \varepsilon(2\|T\|_{\mathrm{op}} + 2)^{-1}$ and choose $k \in \mathbb{N}$ so large that $\forall n > k$, we have

$(\heartsuit)$ $\qquad\qquad \mathrm{Gr}(T_n) \cap (B \times \alpha U_Y) \subseteq \mathrm{Gr}(T) + \delta U_{X \times Y}.$

Let $x \in B$ and $n > k$ be arbitrary; by the choice of δ and $(\heartsuit)$, we can find $w \in X$ such that $\|w - x\|_X < \tfrac{\varepsilon}{2}\|T\|_{\mathrm{op}}^{-1}$ and $\|T(w) - T_n(x)\|_Y < \tfrac{\varepsilon}{2}$. As a result,

$$\|T(w - x)\|_Y \leq \|T\|_{\mathrm{op}} \cdot \frac{\varepsilon}{2\|T\|_{\mathrm{op}}} = \frac{\varepsilon}{2},$$

and it follows that

$$\|T_n(x) - T(x)\|_Y \leq \|T_n(x) - T(w)\|_Y + \|T(w) - T(x)\|_Y < \frac{\varepsilon}{2} + \frac{\varepsilon}{2} = \varepsilon.$$

In either case we conclude that $\sup\{\|T(x) - T_n(x)\|_Y : x \in B\} \leq \varepsilon$, and since $B \in \mathscr{B}_0$ was arbitrary, we have $T = \mathscr{T}_{\mathscr{B}_0}\text{-lim}\, T_n$. Thus (4) holds

The proof of (5) $\Rightarrow$ (1) is very easy. Suppose $T = \mathscr{T}_{\mathscr{B}}\text{-lim}\, T_n$ and $\varepsilon > 0$ and $B \in \mathscr{B}$ are arbitrary. Choose k so large that $\forall n > k$, $\sup_{x \in B}\|T_n(x) - T(x)\|_Y < \varepsilon$. Then for all such n we actually have $\mathrm{Gr}(T) \cap (B \times Y) \subseteq \mathrm{Gr}(T_n) + (\{0_X\} \times \varepsilon U_Y)$ and $\mathrm{Gr}(T_n) \cap (B \times Y) \subseteq \mathrm{Gr}(T) + (\{0_X\} \times \varepsilon U_Y)$, and so both

$$\mathrm{Gr}(T) \cap (B \times Y) \subseteq \mathrm{Gr}(T_n) + \varepsilon U_{X \times Y},$$

and

$$\mathrm{Gr}(T_n) \cap (B \times Y) \subseteq \mathrm{Gr}(T) + \varepsilon U_{X \times Y}.$$

This completes the proof. $\qquad\qquad\qquad\qquad\qquad\qquad\qquad\qquad\qquad\qquad\qquad\quad \square$

We note that while $\tau_{\mathscr{B}}^s$-convergence implies two-sided bornological graph convergence, the proof of Theorem 3.1 shows that $\tau_{\mathscr{B}}^s$-convergence follows just from the upper half of bornological graph convergence. For example, condition (1) can be replaced by

$$\forall B \in \mathscr{B}\ \forall \varepsilon > 0\ \exists n_0 \in \mathbb{N}\ \forall n > n_0\ \mathrm{Gr}(T_n) \cap (B \times Y) \subseteq \mathrm{Gr}(T) + \varepsilon U_{X \times Y}.$$

From this perspective, the equivalence of conditions (1) and (4) is a special case of Theorem 6.18 of [45].

Our first corollary to Theorem 33.5 gives the Penot-Zalinescu theorem where $\mathscr{B}_0 = \{nU_X : n \in \mathbb{N}\}$ and $\mathscr{B}$ is the family of metrically bounded subsets of X, since convergence in the operator norm topology is equivalent to uniform convergence on metrically bounded subsets of X.

Corollary 33.6. *Let $\langle X, ||\cdot||_X \rangle$ and $\langle Y, ||\cdot||_Y \rangle$ be normed linear spaces and let $T, T_1, T_2, T_3, \ldots$ be a sequence in $\mathbf{B}(X,Y)$. Then $\langle T_n \rangle$ is convergent to T in the operator norm topology if and only if $\langle Gr(T_n) \rangle$ is Attouch-Wets convergent to $Gr(T)$ with respect to the box norm on $X \times Y$.*

In our next corollary we use the fact that uniform convergence of linear transformations on finite sets is equivalent to uniform convergence on polytopes, and the fact that each polytope is a subset of a polytope containing the origin.

Corollary 33.7. *Let $\langle X, ||\cdot||_X \rangle$ and $\langle Y, ||\cdot||_Y \rangle$ be normed linear spaces and let $\{T, T_1, T_2, \ldots\} \subseteq \mathbf{B}(X,Y)$. Then $\langle T_n \rangle$ converges pointwise to T if and only if for each polytope P in X and $\alpha > 0$ and $\varepsilon > 0$, we have eventually both*

$$Gr(T_n) \cap (P \times \alpha U_Y) \subseteq Gr(T) + \varepsilon U_{X \times Y},$$

$$Gr(T) \cap (P \times \alpha U_Y) \subseteq Gr(T_n) + \varepsilon U_{X \times Y}.$$

Our next corollary for continuous linear functionals is a consequence of the Banach-Dieudonné theorem and Proposition 33.4.

Corollary 33.8. *Let $\langle X, ||\cdot|| \rangle$ a normed linear space and let $\{f, f_1, f_2, \ldots\}$ be a sequence in X^*. Then $\langle f_n \rangle$ converges to f in the bounded weak-star topology if and only if for each totally bounded subset B of X and $\alpha > 0$ and $\varepsilon > 0$, we have eventually both*

$$Gr(f_n) \cap (B \times [-\alpha, \alpha]) \subseteq Gr(f) + \varepsilon(U_X \times [-1, 1]),$$

$$Gr(f) \cap (B \times [-\alpha, \alpha]) \subseteq Gr(T_n) + \varepsilon(U_X \times [-1, 1]).$$

Our next two corollaries use Proposition 33.3.

Corollary 33.9. *Let $\langle X, ||\cdot||_X \rangle$ and $\langle Y, ||\cdot||_Y \rangle$ be normed linear spaces and let $\{T, T_1, T_2, \ldots\} \subseteq \mathbf{B}(X,Y)$. Then $\langle T_n \rangle$ converges to T uniformly on norm compact subsets of X if and only if for each norm compact subset K of X and $\alpha > 0$ and $\varepsilon > 0$, we have eventually both*

$$Gr(T_n) \cap (K \times \alpha U_Y) \subseteq Gr(T) + \varepsilon U_{X \times Y},$$

$$Gr(T) \cap (K \times \alpha U_Y) \subseteq Gr(T_n) + \varepsilon U_{X \times Y}.$$

Corollary 33.10. *Let $\langle X, ||\cdot||_X \rangle$ and $\langle Y, ||\cdot||_Y \rangle$ be normed linear spaces and let $\{T, T_1, T_2, \ldots\} \subseteq \mathbf{B}(X,Y)$. Then $\langle T_n \rangle$ converges to T uniformly on weakly compact subsets of X if and only if for each weakly compact subset K of X and $\alpha > 0$ and $\varepsilon > 0$, we have eventually both*

$$Gr(T_n) \cap (K \times \alpha U_Y) \subseteq Gr(T) + \varepsilon U_{X \times Y},$$

$$Gr(T) \cap (K \times \alpha U_Y) \subseteq Gr(T_n) + \varepsilon U_{X \times Y}.$$

The last corollary can be stated in a more attractive way when $Y = W^*$ for some Banach space W. Without assuming completeness of W, by Alaoglu's theorem [75], αU_Y is weak-star compact for each $\alpha > 0$, and with completeness, each weak-star compact subset is norm bounded [143, p. 174].

Proposition 33.11. *Let $\langle X, ||\cdot||_X \rangle$ be a normed linear space and let $\langle W^*, ||\cdot||_{op} \rangle$ be the dual of a Banach space W. Suppose $\{T, T_1, T_2, \ldots\} \subseteq \mathbf{B}(X, W^*)$. Then $\langle T_n \rangle$ converges to T uniformly on weakly compact compact subsets of X if and only if for each weakly compact subset K of X and each weak-star compact subset C of W^* and $\varepsilon > 0$, we have eventually both*

$$Gr(T_n) \cap (K \times C) \subseteq Gr(T) + \varepsilon U_{X \times W^*},$$
$$Gr(T) \cap (K \times C) \subseteq Gr(T_n) + \varepsilon U_{X \times W^*}.$$

We close this section with a description of convergence of sequences in the Mackey topology for continuous linear functionals in terms of graph convergence that uses convex subsets for truncating families. We need the following lemma.

Lemma 33.12. *Let A be a nonempty convex weakly compact subset of $\langle X, ||\cdot|| \rangle$. Then $aco(A)$ is weakly compact.*

Proof. Recall that $aco(A) = co(A \cup -A)$. Since both A and $-A$ are convex,

$$aco(A) = \{ta + (1-t)b : a \in A, b \in -A, t \in [0,1]\} = \{ta_1 + (t-1)a_2 : a_1, a_2 \in A \text{ and } t \in [0,1]\}.$$

Thus $aco(A)$ is the continuous image of $[0,1] \times A \times A$ (where the last two factors are equipped with the weak topology) and so this absolutely convex set is weakly compact. □

Proposition 33.13. *Let $\langle X, ||\cdot|| \rangle$ be a normed linear space, and let $\{f, f_1, f_2, \ldots\} \subseteq X^*$. The following are equivalent:*

(1) $\langle f_n \rangle$ is convergent to f uniformly on absolutely convex weakly compact subsets of X;

(2) whenever C is a weakly compact convex subset of $X \times \mathbb{R}$ and $\varepsilon > 0$, then eventually both

$$Gr(f_n) \cap C \subseteq Gr(f) + \varepsilon(U_X \times [-1,1]) \text{ and } Gr(f) \cap C \subseteq Gr(f_n) + \varepsilon(U_X \times [-1,1]).$$

Proof. To apply Theorem 33.5, we take $\mathscr{B} =$ the weakly compact convex subsets of X while $\mathscr{B}_0 =$ the weakly compact absolutely convex subsets of X. From our Lemma 33.12, $\mathscr{B}_0 \subseteq \mathscr{B} \subseteq \downarrow \mathscr{B}_0$. Each element of $\mathscr{B} \otimes \{[-\alpha, \alpha] : \alpha > 0\}$ is a weakly compact convex subset of the product while if E is a weakly compact convex subset $X \times \mathbb{R}$, then $E \subseteq \pi_X(E) \times \pi_{\mathbb{R}}(E)$; the projection onto X is weakly compact and convex while the projection onto $\mathbb{R}$ is contained in $[-\alpha, \alpha]$ for some positive α. We have shown that $E \in \downarrow \mathscr{B} \otimes \{[-\alpha, \alpha] : \alpha > 0\}$. □

34. Bornological Convergence and Uniform Convergence of Distance Functionals

We know from Proposition 32.1 that if $\mathscr{B}$ is a bornology on $\langle X, d \rangle$ and $\langle A_\lambda \rangle$ is a net in $\mathscr{C}_0(X)$ for which $\langle d(\cdot, A_\lambda) \rangle$ is uniformly convergent on $\mathscr{B}$ to $d(\cdot, A)$ (where $A \in \mathscr{C}_0(X)$), then $\langle A_\lambda \rangle$ is $(\mathscr{B}, d)$-convergent to A. In this section, we see exactly when the converse holds. We start with a simple example showing that the converse can fail when our bornology is $\mathscr{F}(X)$.

Example 34.1. Consider $\mathbb{N}$ equipped with the following metric: $d(1, n) = 2$ for $n > 1$ and $d(j, n) = 1$ whenever $1 < j < n$. The bornology $\mathscr{F}(\mathbb{N})$ is clearly stable under small enlargements; in fact, for any finite set A, we have $A = S_d(A, 1)$. The sequence $\langle \{1, n+1\} \rangle$ is $(\mathscr{F}(\mathbb{N}), d)$-convergent to $\{1\}$. However, $d(2, \{1, n+1\}) = 1$ eventually while $d(2, \{1\}) = 2$.

Even when $\mathscr{B}$ is stable under arbitrary enlargements, uniform convergence of distance functionals may be properly stronger than $(\mathscr{B}, d)$-convergence [48].

Example 34.2. In the plane $\mathbb{R}^2$ equipped with the usual metric d, let $\mathscr{B}$ be the bornology with countable base $\{\mathbb{R} \times [-n, n] : n \in \mathbb{N}\}$. This bornology is not only stable under arbitrary enlargements, but it is a bornology of metrically bounded subsets with respect to the metric

$$\rho((x_1, y_1), (x_2, y_2)) = \min\{1, d((x_1, y_1), (x_2, y_2))\} + |y_1 - y_2|$$

which is uniformly equivalent to d. Let $(x_n, y_n) = (n, n-1)$ for $n \in \mathbb{N}$. Put $A_n = \{(0, 0), (x_n, y_n)\}$ and $A = \{(0, 0)\}$; since $\langle (x_n, y_n) \rangle$ converges to infinity with respect to $\mathscr{B}$, $A = (\mathscr{B}, d) - \lim A_n$. On the other hand, uniform convergence of $\langle d(\cdot, A_n) \rangle$ to $d(\cdot, A)$ fails on the horizontal axis, as $d((n, 0), A_n) = n - 1$ while $d((n, 0), A) = n$.

The topology of uniform convergence of distance functionals for nonempty closed sets on finite subsets of $\langle X, d \rangle$ is called the *Wijsman topology* in the literature, and the corresponding convergence notion for nets of nonempty closed subsets is called *Wijsman convergence* [16, 17, 69, 70, 41, 112, 154]. Of course, Wijsman convergence of a net of nonempty closed subsets $\langle A_\lambda \rangle$ to $A \in \mathscr{C}_0(X)$ is nothing more than pointwise convergence of $\langle d(\cdot, A_\lambda) \rangle$ to $d(\cdot, A)$. When this occurs, we write $A = W_d - \lim A_\lambda$.

Here are some important facts about the Wijsman topology on $\mathscr{C}_0(X)$, all of which are recorded in [17]:

- the Wijsman topology is metrizable if and only if $\langle X, d \rangle$ is separable; in this case it is second countable as well; [112];

- if $\langle X, d \rangle$ is a separable complete metric space, then the Wijsman topology on $\mathscr{C}_0(X)$ is completely metrizable [16]. More generally, each Wijsman topology on a separable completely metrizable space is completely metrizable [69];

- the topology on $\mathscr{C}_0(X)$ generated by all Wijsman topologies as determined by the family $\mathbf{D}(X)$ of metrics compatible with the topology of X coincides with the Vietoris topology [41];

- when $\langle X, d \rangle$ is separable, the sigma algebra on $\mathscr{C}_0(X)$ generated by all sets of the form $\{A \in \mathscr{C}_0(X) : A \cap V \neq \emptyset\}$ where V runs over the open subsets of X coincides with the Borel field as determined by the Wijsman topology [95].

Necessary and sufficient conditions for two equivalent metrics on X to determine the same Wijsman topologies on $\mathscr{C}_0(X)$ were provided by Costantini, Levi and Zieminska [17, 70].

The next result gives a necessary and sufficient condition on the structure of a bornology $\mathscr{B}$ on $\langle X, d \rangle$ so that bornological convergence of nets of nonempty closed subsets ensures their Wijsman convergence to the same limit [43, Theorem 1]. Note that the condition ensures that the bornology is local, so that a bornological limit of a net of closed sets (if it exists) must be unique.

Theorem 34.3. *Let $\mathscr{B}$ be a bornology on a metric space $\langle X, d \rangle$. The following conditions are equivalent:*

(1) *each closed ball that is a proper subset of X lies in $\mathscr{B}$;*

(2) *$(\mathscr{B}, d)$-convergence of nets in $\mathscr{C}_0(X)$ to a nonempty subset ensures their Wijsman convergence;*

(3) *$(\mathscr{B}, d)$-convergence of nets of finite subsets to a nonempty finite limit ensures their Wijsman convergence.*

Proof. $(1) \Rightarrow (2)$. Suppose $A = (\mathscr{B}, d)\text{-}\lim A_\lambda$ in $\mathscr{C}_0(X)$ and $p \in X$ is arbitrary. Let $\varepsilon > 0$ and choose $a \in A$ with $d(p, a) < d(p, A) + \frac{\varepsilon}{2}$. Since eventually $\{a\} \cap A \subseteq S_d(A_\lambda, \frac{\varepsilon}{2})$, we eventually have $d(p, A_\lambda) < d(p, A) + \varepsilon$. It follows that without the assumption on closed balls,

$$\limsup d(p, A_\lambda) \leq d(p, A).$$

Now set $\alpha := \liminf d(p, A_\lambda)$. To show $A = W_d - \lim A_\lambda$, it remains to verify

$$\forall \varepsilon > 0, \ d(p, A) \leq \alpha + \varepsilon.$$

Fix $\varepsilon > 0$; if $\{x : d(p, x) \leq \alpha + \varepsilon/2\} = X$, then $d(p, A) \leq \alpha + \varepsilon/2 \leq \alpha + \varepsilon$. Otherwise, by condition (1), $B_0 := \{x : d(p, x) \leq \alpha + \varepsilon/2\} \in \mathscr{B}$, and so for some index λ_0 and all $\lambda \succeq \lambda_0$, we have

$$A_\lambda \cap B_0 \subseteq S_d(A, \frac{\varepsilon}{2}).$$

By the definition of α, we have $d(p, A_\lambda) < \alpha + \frac{\varepsilon}{2}$ frequently. Thus we can find $\lambda_1 \succeq \lambda_0$ and $a_{\lambda_1} \in A_{\lambda_1}$ with $d(p, a_{\lambda_1}) < \alpha + \frac{\varepsilon}{2}$. As $a_{\lambda_1} \in A_{\lambda_1} \cap B_0$, there exists $a \in A$ with $d(a_{\lambda_1}, a) < \varepsilon/2$, and the triangle inequality yields

$$d(p, A) \leq d(p, a) < (\alpha + \frac{\varepsilon}{2}) + \frac{\varepsilon}{2} = \alpha + \varepsilon.$$

$(2) \Rightarrow (3)$. This is obvious.

(3) $\Rightarrow$ (1). Suppose (1) fails : $\exists p \in X \ \exists \alpha > 0$ such that $\{x : d(p,x) \leq \alpha\} \neq X$ and $\{x : d(p,x) \leq \alpha\} \notin \mathscr{B}$. Pick $q \in X$ with $d(p,q) > \alpha$, and $\forall B \in \mathscr{B}$ pick $x_B \in X \backslash B$ with $d(p,x_B) \leq \alpha$. The net $B \mapsto \{x_B, q\}$ is $(\mathscr{B}, d)$-convergent to $\{q\}$ but

$$\lim \sup \, d(p, \{x_B, q\}) \leq \alpha < d(p, \{q\}).$$

Thus, W_d-convergence of the net to $\{q\}$ fails. $\qquad\qquad\qquad\qquad\qquad\qquad\qquad\square$

Example 34.2 shows that $(\mathscr{B}, d)$-convergence in $\mathscr{C}_0(X)$ can ensure Wijsman convergence without ensuring $\mathscr{T}_{\mathscr{B}}$-convergence of distance functionals.

Let $\mathscr{B}$ be a bornology on $\langle X, d \rangle$. We return to the main objective of this section which is to give necessary and sufficient conditions for $(\mathscr{B}, d)$-convergence to ensure $\mathscr{T}_{\mathscr{B}}$-convergence on $\mathscr{C}_0(X)$, under the identification $A \leftrightarrow d(\cdot, A)$. As we mentioned when we introduced Hausdorff distance, when A and B are nonempty subsets of $\langle X, d \rangle$ we have [17, p. 29].

$$H_d(A, B) := \sup_{x \in X} |d(x, A) - d(x, B)|.$$

Thus, when $\mathscr{B} = \mathscr{P}(X)$, the property holds. For this reason, we now confine our attention to bornologies $\mathscr{B}$ not containing X. It turns out that $(\mathscr{B}, d)$-convergence ensures uniform convergence of distance functionals provided it does for two very special types of nets of closed sets, as discovered by Beer, Naimpally and Rodríguez-López [48].

Theorem 34.4. *Let $\mathscr{B}$ be a bornology on a metric space $\langle X, d \rangle$ not containing X as a member. Then $(\mathscr{B}, d)$-convergence in $\mathscr{C}_0(X)$ ensures uniform convergence of distance functionals on nonempty members of $\mathscr{B}$ if only if the following two conditions hold:*

(1) *whenever $\langle x_\lambda \rangle_{\lambda \in \Lambda}$ converges to infinity with respect to $\mathscr{B}$, $\forall p \in X, \langle d(\cdot, \{p, x_\lambda\}) \rangle_{\lambda \in \Lambda}$ converges uniformly to $d(\cdot, p)$ on $\mathscr{B}$, and*

(2) *whenever $A \in \mathscr{C}_0(X)$ and $\mathscr{B}$ is directed by inclusion, then $\langle d(\cdot, A \cap B) \rangle_{B \in \mathscr{B}}$ converges uniformly to $d(\cdot, A)$ on $\mathscr{B}$.*

Proof. Condition (1) is obviously necessary because for each $p \in X, \{p\} = (\mathscr{B}, d) - \lim \{p, x_\lambda\}$. Also note that with $\mathscr{B}$ directed by inclusion, eventually $\text{cl}(A \cap B)$ is a nonempty closed set. As $A = (\mathscr{B}, d) - \lim \text{cl}(A \cap B)$ and $d(\cdot, \text{cl}(A \cap B)) = d(\cdot, A \cap B)$, condition (2) is also necessary.

Conversely, we will show that if condition (1) holds, then whenever $A = (\mathscr{B}, d) - \lim A_\lambda$, for each nonempty $B \in \mathscr{B}$ and for each $\varepsilon > 0$, eventually

$$\sup_{b \in B} \, d(b, A) - d(b, A_\lambda) < \varepsilon,$$

and if condition (2) holds, then eventually

$$\sup_{b \in B} \, d(b, A_\lambda) - d(b, A) < \varepsilon.$$

Suppose $A, A_\lambda \ (\lambda \in \Lambda)$ are nonempty closed sets with $A = (\mathscr{B}, d) - \lim A_\lambda$, yet for some nonempty $B_0 \in \mathscr{B}$ and $\varepsilon > 0$ cofinally $\sup_{b \in B_0} \, d(b, A) - d(b, A_\lambda) \geq \varepsilon$. We can find a cofinal subset Λ_0 of Λ and a net $\langle b_\lambda \rangle_{\lambda \in \Lambda_0}$ in B_0 such that

($\spadesuit$) $\qquad\qquad \forall \lambda \in \Lambda_0, \ d(b_\lambda, A) - d(b_\lambda, A_\lambda) > \frac{\varepsilon}{2}.$

For each $\lambda \in \Lambda_0$ pick $a_\lambda \in A_\lambda$ with $d(b_\lambda, A) - d(b_\lambda, a_\lambda) > \varepsilon/2$. We claim that for all $(\lambda_0, B) \in \Lambda_0 \times \mathscr{B}$, there exists $\gamma = \gamma(\lambda_0, B) \in \Lambda_0$ with $\gamma \succeq \lambda_0$ and $a_\gamma \notin B$. Otherwise, residually in Λ_0 we have

$a_\lambda \in B$, and by $(\mathscr{B}, d)$-convergence, we have for $\lambda \in \Lambda_0$ sufficiently large, $a_\lambda \in A_\lambda \cap B \subseteq S_d(A, \frac{\varepsilon}{3})$, from which

$$d(b_\lambda, A) - d(b_\lambda, a_\lambda) < d(b_\lambda, a_\lambda) + \frac{\varepsilon}{3} - d(b_\lambda, a_\lambda) = \frac{\varepsilon}{3},$$

a contradiction to ($\spadesuit$). The claim established, $(\lambda_0, B) \mapsto a_{\gamma(\lambda_0, B)}$ is a net in X defined on $\Lambda_0 \times \mathscr{B}$ equipped with the natural product direction convergent to infinity with respect to $\mathscr{B}$. Let $p \in A$ be arbitrary; we compute for each value of γ

$$d(b_\gamma, p) - d(b_\gamma, \{p, a_\gamma\}) \geq d(b_\gamma, A) - d(b_\gamma, a_\gamma) > \frac{\varepsilon}{2},$$

and this violates condition (1).

Using condition (2), we now show directly that given nonempty $B_1 \in \mathscr{B}$ and $\varepsilon > 0$, eventually we have $\sup_{b \in B_1} d(b, A_\lambda) - d(b, A) < \varepsilon$. To this end, we claim there exists $\widetilde{B_1} \in \mathscr{B}$ such that for each $b \in B_1$,

($\heartsuit$) $\{a \in A \cap \widetilde{B_1} : d(b, a) < d(b, A) + \frac{\varepsilon}{2}\} \neq \emptyset.$

If this fails, then for each $B \in \mathscr{B}$ with $A \cap B \neq \emptyset$ we can find $b_B \in B_1$ with $d(b_B, A \cap B) - d(b_B, A) \geq \frac{\varepsilon}{2}$. This means that $\langle d(\cdot, A \cap B) \rangle_{B \in \mathscr{B}}$ cannot converge uniformly to $d(\cdot, A)$ on B_1, in violation of (2). Thus $\widetilde{B_1}$ exists with the asserted properties. For each $b \in B_1$, pick $a_b \in A \cap \widetilde{B_1}$ with $d(b, a_b) < d(b, A) + \frac{\varepsilon}{2}$. As $\mathscr{B}$ is hereditary, $\{a_b : b \in B_1\} \in \mathscr{B}$ as well.

Returning to the main argument, choose λ_0 such that $\lambda \succeq \lambda_0$ implies

$$\{a_b : b \in B_1\} = A \cap \{a_b : b \in B_1\} \subseteq S_d(A_\lambda, \frac{\varepsilon}{2}).$$

Fix $\lambda \succeq \lambda_0$ and $b \in B_1$. By ($\heartsuit$), $d(b, a_b) < d(b, A) + \frac{\varepsilon}{2}$. Choosing $a_\lambda \in A_\lambda$ with $d(a_\lambda, a_b) < \frac{\varepsilon}{2}$, we compute

$$d(b, A_\lambda) \leq d(b, a_\lambda) \leq d(b, a_b) + \frac{\varepsilon}{2} < d(b, A) + \varepsilon$$

as required. $\square$

Example 34.1 shows that condition (1) can fail. We next give an equally simple example showing that condition (2) can also fail.

Example 34.5. Our metric space will be the set of integers $\mathbb{Z}$ equipped with bornology $\mathscr{B}$ generated by $\{\mathbb{N}\} \cup \mathscr{F}(\mathbb{Z})$ and the metric d defined by

$$d(i, j) = \begin{cases} 0 & \text{if } i = j \\ 1 & \text{if } i = -j \\ 2 & \text{otherwise.} \end{cases}$$

Let $A = \{-j : j \in \mathbb{N}\}$. Clearly, $\forall n \in \mathbb{N}$, $d(n, A) = d(n, -n) = 1$. On the other hand, whenever $B \in \mathscr{B}$, there exists $n \in \mathbb{N}$ with $-n \notin B \cap A$ and so $d(n, A \cap B) \geq 2$. Thus $\langle d(\cdot, A \cap B) \rangle_{B \in \mathscr{B}}$ cannot converge to $d(\cdot, A)$ uniformly on $\mathbb{N}$.

We introduce a third convergence notion for sequences of nonempty closed subsets relative to a bornology $\mathscr{B}$ on $\langle X, d \rangle$ intermediate in strength between $(\mathscr{B}, d)$-convergence and uniform convergence of distance functions on nonempty members of $\mathscr{B}$. We shall need an elementary lemma whose proof is left to the reader (note that uniform convergence cannot be replaced by pointwise convergence in its statement).

Lemma 34.6. *Let S be a nonempty set without any assumed structure and let $\langle f_\lambda \rangle$ be a net of real-valued functions each defined on S that converges uniformly to a real-valued function $f : S \to \mathbb{R}$. Then $\inf \{f(s) : s \in S\} = \lim_\lambda \inf \{f_\lambda(s) : s \in S\}$ and $\sup \{f(s) : s \in S\} = \lim_\lambda \sup \{f_\lambda(s) : s \in S\}$.*

Proposition 34.7. *Let $\mathscr{B}$ be a bornology on a metric space $\langle X, d \rangle$, and let $\{A\} \cup \{A_\lambda : \lambda \in \Lambda\} \subseteq \mathscr{C}_0(X)$ where Λ is a directed set.*

(1) *if $\langle d(\cdot, A_\lambda) \rangle_{\lambda \in \Lambda}$ is uniformly convergent to $d(\cdot, A)$ on each nonempty member of $\mathscr{B}$, then for all nonempty $B \in \mathscr{B}$, we have $e_d(B, A) = \lim e_d(B, A_\lambda)$ and $D_d(B, A) = \lim D_d(B, A_\lambda)$;*

(2) *If for all nonempty $B \in \mathscr{B}$, we have $e_d(B, A) = \lim e_d(B, A_\lambda)$ and $D_d(B, A) = \lim D_d(B, A_\lambda)$, then $A \in (\mathscr{B}, d) - \lim A_\lambda$.*

Proof. Statement (1) follows immediately from the last lemma where $f_\lambda = d(\cdot, A_\lambda)$ and $f = d(\cdot, A)$ and S runs over the nonempty members of $\mathscr{B}$. For statement (2), fix $B_0 \neq \emptyset \in \mathscr{B}$ and $\varepsilon > 0$; we will show assuming $e_d(B, A) = \lim e_d(B, A_\lambda)$ and $D_d(B, A) = \lim D_d(B, A_\lambda)$ where B runs over the nonempty members of $\mathscr{B}$ that

(i) eventually $A \cap B_0 \subseteq S_d(A_\lambda, \varepsilon)$,

(ii) eventually $A_\lambda \cap B_0 \subseteq S_d(A, \varepsilon)$.

For (i), if $A \cap B_0 = \emptyset$, there is nothing to show. Otherwise, $A \cap B_0 \neq \emptyset$ and $e_d(A \cap B_0, A) = 0$. As $A \cap B_0 \in \mathscr{B}$,

$$\lim e_d(A \cap B_0, A_\lambda) = e_d(A \cap B_0, A),$$

given $\varepsilon > 0$, we have eventually $e_d(A \cap B_0, A_\lambda) < \varepsilon$. In particular, we have eventually $A \cap B_0 \subseteq S_d(A_\lambda, \varepsilon)$. For (ii), suppose to the contrary that frequently $A_\lambda \cap B_0 \not\subseteq S_d(A, \varepsilon)$. There exists a cofinal set of indices $\Lambda_0 \subseteq \Lambda$ such that

$$\forall \lambda \in \Lambda_0 \; \exists x_\lambda \in B_0 \cap A_\lambda \text{ with } d(x_\lambda, A) \geq \varepsilon.$$

Let $B_1 = \{x_\lambda : \lambda \in \Lambda_0\} \in \mathscr{B}$; frequently, $D_d(B_1, A_\lambda) = 0$ whereas $D_d(B_1, A) \geq \varepsilon$. This contradicts $D_d(B_1, A) = \lim D_d(B_1, A_\lambda)$. $\qquad\square$

Example 34.1 shows that we can have $(\mathscr{F}(X), d)$-convergence of a sequence $\langle A_n \rangle$ of nonempty closed subsets to a nonempty closed subset A without having either (i) $e_d(F, A) = \lim e_d(F, A_n)$ for each nonempty finite subset or (ii) $D_d(F, A) = \lim D_d(F, A_n)$ for each nonempty finite subset; simply let $F = \{2\}$. More generally, if we have $(\mathscr{B}, d)$-convergence of a net of nonempty closed subsets in a metric $\langle X, d \rangle$ to $A \in \mathscr{C}_0(X)$ without Wijsman convergence, then neither convergence of associated gap and excess functions whose left argument is an appropriate singleton will occur.

In view of the next proposition, Example 34.2 provides an example of a sequence of nonempty closed subsets $A, A_1, A_2, \ldots$ relative to a bornology $\mathscr{B}$ in a metric space such that

$$\lim e_d(B, A_n) = e_d(B, , A) \text{ and } \lim D_d(B, A_n) = D_d(B, , A) \quad (B \in \mathscr{B}, \; B \neq \emptyset)$$

while $\langle d(\cdot, A_n) \rangle$ fails to converge uniformly to $d(\cdot, A)$ on $\{(x, 0) : x \in \mathbb{R}\} \in \mathscr{B}$.

Proposition 34.8. *Let $\langle X, d \rangle$ be a metric space, and let $\mathscr{B}$ be a bornology on X such that for each $B \in \mathscr{B}$ and each $\alpha > 0$ we have $S_d(B, \alpha) \in \mathscr{B}$. Then if $\langle A_\lambda \rangle$ is a net of nonempty closed* *subsets and $A \in \mathscr{C}_0(X)$ satisfies $A = (\mathscr{B}, d) - \lim A_\lambda$, then for all nonempty $B \in \mathscr{B}$, we have $e_d(B, A) = \lim e_d(B, A_\lambda)$ and $D_d(B, A) = \lim D_d(B, A_\lambda)$.*

Proof. The condition that $\mathscr{B}$ be stable under arbitrary enlargements of course implies the bornology is local, and so A is the unique $(\mathscr{B}, d)$-limit of the net of closed sets. We just prove the statement involving excesses, leaving the easier proof of the statement involving gaps to the reader.

Fix $B \in \mathscr{B}, B \neq \emptyset$. It suffices to show for all $\alpha \in (0, \infty)$ (i) if $\alpha < e_d(B, A)$, then eventually $\alpha < e_d(B, A_\lambda)$, and (ii) if $e_d(B, A) < \alpha$ then $e_d(B, A_\lambda) < \alpha$ eventually. For (i), choose $b \in B$ and $\varepsilon > 0$ such that $d(b, A) > \alpha + \varepsilon$. We claim that eventually $A_\lambda \cap S_d(b, \alpha + \frac{\varepsilon}{2}) = \emptyset$ from which eventually

$$e_d(B, A_\lambda) \geq d(b, A_\lambda) \geq \alpha + \frac{\varepsilon}{2} > \alpha.$$

If the claim fails, then for some large enough λ, since $S_d(b, \alpha + \frac{\varepsilon}{2}) \in \mathscr{B}$, we have

$$\emptyset \neq A_\lambda \cap S_d(b, \alpha + \frac{\varepsilon}{2}) \subseteq S_d(A, \frac{\varepsilon}{2})$$

so that $A \cap S_d(b, \alpha + \varepsilon) \neq \emptyset$ and this contradicts $d(b, A) > \alpha + \varepsilon$.

For (ii), we prove the contrapositive. Suppose frequently we have $e_d(B, A_\lambda) \geq \alpha$. Let $\varepsilon > 0$ be arbitrary. Then frequently, $A_\lambda \cap S_d(B, \alpha - \varepsilon) = \emptyset$ which means that frequently

$$(\heartsuit) \qquad\qquad S_d(A_\lambda, \tfrac{\varepsilon}{2}) \cap S_d(B, \alpha - \tfrac{\varepsilon}{2}) = \emptyset.$$

Thus, there exist λ such both $A \cap S_d(B, \alpha - \frac{\varepsilon}{2}) \subseteq S_d(A_\lambda, \frac{\varepsilon}{2})$ and $(\heartsuit)$ hold so that $A \cap S_d(B, \alpha - \frac{\varepsilon}{2}) = \emptyset$. We conclude that $e_d(A, B) \geq \alpha - \frac{\varepsilon}{2}$, and $e_d(A, B) \geq \alpha$ follows. $\qquad \square$

Rather technical necessary and sufficient conditions for $(\mathscr{B}, d)$-convergence of nets in $\mathscr{P}_0(X)$ rather than $\mathscr{C}_0(X)$ to ensure convergence of the associated nets of gap and excess functionals with fixed left argument running over the nonememnpty members of $\mathscr{B}$ are presented in [43, Theorem 5].

35. Bornological Convergence with Respect to the Compact Bornology

Bornological convergence with respect to the compact bornology, which is coarser than Attouch-Wets convergence, is also compatible with a topology on $\mathscr{C}(X)$. Convergence in this sense has a tangible visual interpretation.

Proposition 35.1. *Let A be a closed subset of a metric space $\langle X, d \rangle$ and let $\langle A_\lambda \rangle$ be a net of closed subsets. Then $A \in (\mathscr{K}(X), d) - \lim A_\lambda$ if and only if both of the following two conditions are satisfied:*

- *whenever V is an open subset of X, $A \cap V \neq \emptyset \Rightarrow A_\lambda \cap V \neq \emptyset$ eventually;*

- *whenever K is a compact subset of X, $A \cap K = \emptyset \Rightarrow A_\lambda \cap K = \emptyset$ eventually.*

Proof. Suppose we have bornological convergence, and V is an open subset with $A \cap V \neq \emptyset$. Choose $a_0 \in A \cap V$ and $\delta > 0$ with $S_d(a_0, \delta) \subseteq V$. Eventually, $A \cap \{a_0\} \subseteq S_d(A_\lambda, \delta)$ and for all such λ, we have $A_\lambda \cap V \neq \emptyset$. Suppose now that K is nonempty and compact and $A \cap K = \emptyset$. There exists $\delta > 0$ such that $S_d(A, \delta) \cap K = \emptyset$. Eventually, $A_\lambda \cap K \subseteq S_d(A, \delta)$ and so for all such λ, $A_\lambda \cap K = \emptyset$ as well.

Conversely, suppose the two bulleted conditions hold. It suffices to show that for each nonempty compact subset K and each $\delta > 0$, we eventually have (1) $A_\lambda \cap K \subseteq S_d(A, \delta)$ and (2) $A \cap K \subseteq S_d(A_\lambda, \delta)$. For (1), put $K_0 = K \backslash S_d(A, \delta)$. If K_0 is empty, there is nothing to prove. Otherwise, since K_0 is nonempty, compact and disjoint from A, it is disjoint from A_λ eventually, which means that $A_\lambda \cap K \subseteq S_d(A, \delta)$ for all such λ. For (2), if $A \cap K$ is empty, there is nothing to prove. Otherwise let $\{x_1, x_2, \ldots, x_n\}$ be a finite subset of $A \cap K$ such that $A \cap K \subseteq \cup_{i=1}^n S_d(x_i, \delta/2)$. Now eventually A_λ hits each ball $S_d(x_i, \delta/2)$ because A hits each, and it follows that $A \cap K \subseteq S_d(A_\lambda, \delta)$ for all such λ. $\qquad \square$

Thus, $(\mathscr{K}(X), d)$-convergence in $\mathscr{C}(X)$ is convergence with respect to the topology on $\mathscr{C}(X)$ generated by all sets of the form $\{A \in \mathscr{C}(X) : A \cap V \neq \emptyset\}$ where V is open, plus all sets of the form $\{A \in \mathscr{C}(X) : A \cap K = \emptyset\}$ where K is compact. We call this topology the *Fell topology* [17, 79]. Since limits are unique if and only if the bornology $\mathscr{K}(X)$ is local, i.e., X is locally compact, the Fell topology is Hausdorff precisely when the underlying space is locally compact [155, p. 86]. When the metric space is in addition separable, then by Vaughan's Theorem, we can find an equivalent metric ρ such that $\mathscr{B}_\rho(X) = \mathscr{K}(X)$. This means that $(\mathscr{K}(X), d)$-convergence is Attouch-Wets convergence under an equivalent remetrization. As we will presently see, whenever $(\mathscr{B}, d)$-convergence is metrizable, this must be true.

The Fell topology makes sense on the closed subsets of a arbitrary Hausdorff space. Remarkably, it is compact with no additional assumptions whatsoever!

Theorem 35.2. *Let $\langle X, \tau \rangle$ be a Hausdorff space. Then the Fell topology on $\mathscr{C}(X)$ is compact.*

Proof. We use the Alexander subbase theorem [155, p. 129]: if each open cover of a topological space consisting of members from a fixed subbase for a topology has a finite subcover, then the space is compact. For convenience, we introduce some notation: for V open in X, put $V^- := \{A \in \mathscr{C}(X) : A \cap V \neq \emptyset\}$, and for K compact, put $K^\natural := \{A \in \mathscr{C}(X) : A \cap K = \emptyset\}$. Let $\{K_i^\natural : i \in I\} \cup \{V_j^- : j \in J\}$ be an open cover of $\mathscr{C}(X)$. If $\{\emptyset\}^\natural = \mathscr{C}(X)$ is an element of the cover, we are done. Otherwise, $I \neq \emptyset$, else $\emptyset$ lies in no member of the cover, and $J \neq \emptyset$, else X lies in no member of the cover. We claim that for some $i_0 \in I$, $K_{i_0} \subseteq \cup_{j \in J} V_j$. Otherwise, for each $i \in I$ choosing $x_i \in K_i$ that lies in no V_j, we see that $\mathrm{cl}(\{x_i : i \in I\})$ lies in no member of the cover. By compactness of K_{i_0}, there exists $\{j_1, j_2, \ldots, j_n\} \subseteq J$ such that $K_{i_0} \subseteq \cup_{k=1}^n V_{j_k}$, from which it follows that $\{K_{i_0}^\natural, V_{j_1}^-, \ldots, V_{j_n}^-\}$ is a finite subcover of $\mathscr{C}(X)$. $\square$

We now show that convergence in this sense for sequences of closed subsets of a metric space amounts to their classical Kuratowski-Painlevé convergence (see [111, pp. 335–340] or [17, pp. 148–149]).

Theorem 35.3. *Let $A, A_1, A_2, A_3, \ldots$ be a sequence of closed subsets in a metric space $\langle X, d \rangle$. Then $A \in (\mathscr{K}(X), d) - \lim A_n$ if and only if the following two conditions hold:*

(1) *for each $a \in A$, each ball about a hits A_n for all but finitely many n;*

(2) *for each $p \in X$, if each ball about p hits A_n infinitely often, then $p \in A$.*

Proof. For necessity, first pick $a \in A$ and let $\varepsilon > 0$. Since $\{a\}$ is compact, eventually $\{a\} \cap A \subseteq S_d(A_n, \varepsilon)$ and this yields (1). For (2), we can find $n_1 < n_2 < n_3 < \cdots$ and $x_k \in A_{n_k}$ such that $\lim_{k \to \infty} x_k = p$. Put $E_k = \{p\} \cup \{x_k, x_{k+1}, x_{k+2}, \ldots\}$, a compact set. Fixing k for the moment, by convergence in the Fell topology, since A_n hits E_k frequently, we have $A \cap E_k \neq \emptyset$. Since A is closed and $k \in \mathbb{N}$ is arbitrary, we conclude that $p \in A$.

For sufficiency, suppose (1) and (2) both hold. We intend to verify convergence in the Fell topology. By (1) it is easy to see that if V is open and $A \cap V \neq \emptyset$, then $A_n \cap V \neq \emptyset$ for all n sufficiently large. On the other hand suppose K is nonempty and compact with $A \cap K = \emptyset$. For each $p \in K$, by (2) there exists $n_p \in \mathbb{N}$ and an open neighborhood V_p of p such $V_p \cap A_n = \emptyset$ whenever $n \geq n_p$. Taking a finite cover $\{V_{p_1}, V_{p_2}, \ldots, V_{p_n}\}$ of K, we see that for all n sufficiently large, we have

$$A_n \cap K \subseteq A_n \cap (\cup_{i=1}^n V_{p_i}) = \emptyset$$

as required. $\square$

The reader should verify that Kuratowski-Painlevé convergence of $\langle A_n \rangle$ to A in a metric space implies $A = \cap_{n=1}^\infty \mathrm{cl}(\cup_{j=n}^\infty A_j)$, so that bornological limits of sequences with respect to the compact bornology are unique for sequences in a metric space, even when local compactness is absent.

It is a classical result that in a separable metric space $\langle X, d \rangle$, each sequence in $\mathscr{C}(X)$ has a Kuratowski-Painlevé convergent subsequence to a closed set [111, pp. 340–341].

Kuratowski-Painlevé convergence of nets of closed sets in a Hausdorff space $\langle X, \tau \rangle$ can be defined in the natural way, and provided the space is locally compact it is equivalent to convergence with respect to the Fell topology [17, 109]. It can be shown that without local compactness, Kuratowski-Painlevé convergence of nets is not topological, i.e., there is no topology on $\mathscr{C}(X)$ compatible with the convergence [109, 125].

We state a result parallel to Proposition 35.1, leaving the similar proof to the reader (see [27, Theorem 4.5]).

Proposition 35.4. *Let A be a compact subset of a metric space $\langle X, d \rangle$ and let $\langle A_\lambda \rangle$ be a net of compact subsets. Let $\mathscr{B}$ be a bornology on X with closed base. Then $A \in (\mathscr{B}, d) - \lim A_\lambda$ if and only if both of the following two conditions are satisfied:*

(1) *whenever V is an open subset of X, $A \cap V \neq \emptyset \Rightarrow A_\lambda \cap V \neq \emptyset$ eventually;*

(2) *whenever $F \in \mathscr{B} \cap \mathscr{C}(X)$, $A \cap F = \emptyset \Rightarrow A_\lambda \cap F = \emptyset$ eventually.*

Proposition 35.4 of course shows that if $\mathscr{B}$ is a bornology with closed base, then $(\mathscr{B}, d)$-convergence of nets of compact subsets is topological. A subbase for the topology consists of all sets of the form $\{K : K \text{ is compact and } K \cap V \neq \emptyset\}$ where V runs over the open subsets of X plus all sets of the form $\{K : K \text{ is compact and } K \cap F = \emptyset\}$ where F runs over $\mathscr{B} \cap \mathscr{C}(X)$.

In the case that $\mathscr{B} = \mathscr{P}(X)$, we see that bornological convergence of a net of compact sets - that is, its convergence in Hausdorff distance - agrees with its convergence in the Vietoris topology.

While $(\mathscr{B}, d)$-convergence of a net of compact subsets implies convergence in the above topology without the assumption that $\mathscr{B}$ has a closed base, bornological convergence may be properly stronger without this assumption. The counterexample we offer is delicate.

Example 35.5. Let $X = \mathbb{R}$ equipped with the usual metric d. Let $\mathscr{B}$ be the principal bornology determined by the single subset

$$B := \{k - \frac{1}{1+n} : (k, n) \in \mathbb{N} \times \mathbb{N}\}.$$

Since $\operatorname{cl}(B) \backslash B$ is infinite, $\operatorname{cl}(B) \notin \mathscr{B}$, and so $\mathscr{B}$ does not have a closed base.

Let E be the set of sequences in $\mathbb{N}$ and let $\Lambda = \mathbb{N} \times E$ directed by $(i, \langle n_k \rangle) \succeq (j, \langle m_k \rangle)$ provided $i \geq j$ and $\forall k, n_k \geq m_k$. Put $A = \{0\}$, and for $\lambda = (i, \langle n_k \rangle)$, put

$$A_\lambda = \{0, 1 + i - \frac{1}{1 + n_{1+i}}\}.$$

Each A_λ is a two-element set and is thus compact. We show that A and $\langle A_\lambda \rangle_{\lambda \in \Lambda}$ satisfy conditions (1) and (2) of Proposition 35.4. If V is open and $A \cap V \neq \emptyset$ then for all λ we have $A_\lambda \cap V \neq \emptyset$ as $A \subseteq A_\lambda$. Now suppose F is a (nonempty) closed member of $\mathscr{B}$ such that $A \cap F = \emptyset$. Since $F \in \mathscr{B}$, there exists $j \in \mathbb{N}$ such that $\forall k > j$, $k \notin F$. Since F is closed, it follows that $\forall k > j$ there exists $m_k \in \mathbb{N}$ such that $\forall m \geq m_k$, $k - \frac{1}{1+m} \notin F$. Put $m_1 = m_2 = \cdots = m_j = 1$. It follows that if $\lambda = (i, \langle n_k \rangle)$ where $(i, \langle n_k \rangle) \succeq (j, \langle m_k \rangle)$ we have

$$A_\lambda \cap F = \{0, 1 + i - \frac{1}{1 + n_{1+i}}\} \cap F = \emptyset$$

because $1 + i > j$ and $n_{1+i} \geq m_{1+i}$. This shows that condition (2) is satisfied.

However $\langle A_\lambda \rangle_{\lambda \in \Lambda}$ is not $(\mathscr{B}, d)$-convergent to A because $\forall \lambda \in \Lambda$, $A_\lambda \cap B$ is not contained $S_d(A, 1)$.

36. When is Bornological Convergence Topological?

In this section we give a simple answer to the question: when is bornological convergence on $\mathscr{C}(X)$ topological, that is, when is there a topology τ on $\mathscr{C}(X)$ such that $(\mathscr{B}, d)$-convergence of nets in $\mathscr{C}(X)$ is convergence in the topology? As shown by Beer, Costantini and Levi [29], it turns out that $\mathscr{B}$ being shielded from closed sets is exactly what is required.

For necessity, we will show that if $\mathscr{B}$ fails to be shielded from closed sets, then the standard iterated limit condition for the convergence to be topological (see [107, pp. 74–75] or [109, pp. 29–30]) fails. Given a bornology $\mathscr{B}$ on $\langle X, d \rangle$, we now write $\widetilde{\mathscr{B}}$ for the subfamily of $\mathscr{B}$ each member of which has a shield in $\mathscr{B}$. Evidently, this is a bornology as well.

We will use Proposition 12.9 that characterizes when a superset of a nonempty closed set is a shield for it.

Proposition 36.1. *Let $\mathscr{B}$ be a bornology on $\langle X, d \rangle$ for which $(\mathscr{B}, d)$-convergence is topological on $\mathscr{C}_0(X)$. Then $\mathscr{B}$ is shielded from closed sets.*

Proof. Suppose $B_0 \in \mathscr{B}$ has no shield in $\mathscr{B}$. As each singleton shields itself from closed sets, B_0 contains distinct points b_1 and b_2. Put $\alpha = \frac{1}{2} d(b_1, b_2)$; since $\widetilde{\mathscr{B}}$ is stable under finite unions, either $B_1 := B_0 \cap (X \backslash S_d(b_1, \alpha)) \notin \widetilde{\mathscr{B}}$ or $B_2 := B_0 \cap (X \backslash S_d(b_2, \alpha)) \notin \widetilde{\mathscr{B}}$. Assume without loss of generality that the former is true.

We now consider two cases:

(1) $\mathrm{cl}(B_1) \notin \mathscr{B}$;

(2) $\mathrm{cl}(B_1) \in \mathscr{B}$.

In case (1), for each $B \in \mathscr{B}$, pick $a_B \in \mathrm{cl}(B_1) \backslash B$ and let $\langle x_{n,B} \rangle$ be a sequence in B_1 convergent to a_B. Since $\langle \{x_{n,B}, b_1\} \rangle$ converges to $\{a_B, b_1\}$ in Hausdorff distance for each $B \in \mathscr{B}$, we have

$$\{a_B, b_1\} \in (\mathscr{B}, d) - \lim_{n \to \infty} \{x_{n,B}, b_1\}.$$

Evidently, the net $B \mapsto \{a_B, b_1\}$ is $(\mathscr{B}, d)$-convergent to $\{b_1\}$, but the iterated limit condition fails because $\forall n \in \mathbb{N} \ \forall B \in \mathscr{B}$, $\{x_{n,B}, b_1\} \cap B_1$ fails to be contained in $S_d(b_1, \alpha)$.

In case (2) put $\mathscr{G} := \{B \in \mathscr{B} : \mathrm{cl}(B_1) \subseteq B\}$, directed by inclusion (note that $\mathrm{cl}(B_1) \in \mathscr{G}$). Since $\widetilde{\mathscr{B}}$ is hereditary, $\mathrm{cl}(B_1)$ has no shield in $\mathscr{G}$. By Proposition 12.9, for each $B \in \mathscr{G}$, there exists a sequence $\langle y_{n,B} \rangle$ in $\mathrm{cl}(B_1)$ having no cluster point and an asymptotic sequence $\langle x_{n,B} \rangle$ in $X \backslash B$ with $\lim_{n \to \infty} d(x_{n,B}, y_{n,B}) = 0$. Now define closed subsets $E_{n,B}$ by

$$E_{n,B} := \{b_1\} \cup \{x_{n,B} : n \in \mathbb{N}\} \cup \{y_{k,B} : k \geq n\} \quad (B \in \mathscr{G}, \ n \in \mathbb{N}).$$

For each $B \in \mathscr{G}$, the sequence $n \mapsto E_{n,B}$ is convergent in Hausdorff distance and thus is $(\mathscr{B}, d)$-convergent to the closed set $E_B = \{b_1\} \cup \{x_{n,B} : n \in \mathbb{N}\}$. Further, the net $B \mapsto E_B$ based on $\mathscr{G}$

directed by inclusion is $(\mathscr{B}, d)$-convergent to $\{b_1\}$. The iterated limit condition clearly fails for $(n, B) \mapsto E_{n,B}$ as before. $\qquad\square$

Unfortunately, the analysis required to obtain sufficiency is not simple. As we have noted in a previous section, since $\mathscr{B}$ has a closed base, we have $A = (\mathscr{B}, d) - \lim A_\lambda$ provided for each $B \in \mathscr{B} \cap \mathscr{C}_0(X)$ and each $\varepsilon > 0$, eventually $(A, A_\lambda) \in [B, d, \varepsilon]$. Consistent with our notation for uniformities, we now put for $A \in \mathscr{C}(X)$, $B \in \mathscr{B} \cap \mathscr{C}_0(X)$ and $\varepsilon > 0$,

$$[B, d, \varepsilon][A] := \{C \in \mathscr{C}(X) : A \cap B \subseteq S_d(C, \varepsilon) \text{ and } C \cap B \subseteq S_d(A, \varepsilon)\}.$$

It is clear that $(\mathscr{B}, d)$-convergence is topological provided at each $A \in \mathscr{C}(X)$, $\{[B, d, \varepsilon][A] : B \in \mathscr{B} \cap \mathscr{C}_0(X), \varepsilon > 0\}$ is a local base for a topology.

To obtain this fact, we view $[B, d, \varepsilon][A]$ as an intersection of two families:

$$[B, d, \varepsilon]^+[A] := \{C \in \mathscr{C}(X) : C \cap B \subseteq S_d(A, \varepsilon)\},$$
$$[B, d, \varepsilon]^-[A] := \{C \in \mathscr{C}(X) : A \cap B \subseteq S_d(C, \varepsilon)\}.$$

We intend to show separately that all sets of the form $[B, d, \varepsilon]^+[A]$ (resp. $[B, d, \varepsilon]^-[A]$) form a local base at A for a topology on $\mathscr{C}(X)$ provided $\mathscr{B}$ is shielded from closed sets. That accomplished, all sets of the form $[B, d, \varepsilon]^+[A] \cap [B, d, \varepsilon]^-[A]$ will also form a local base for a topology, as required.

We of course rely on the standard conditions which we now recall (see, e.g., [74, 155]). Suppose Y is a nonempty set and $\{\mathscr{U}_y : y \in Y\}$ is a collection of families of subsets of Y such that $\forall y \in Y \ \forall U \in \mathscr{U}_y$, we have $y \in U$. Then $\mathscr{U}_y$ is a local base for a topology on Y at each $y \in Y$ if and only if the following conditions are satisfied.

- whenever $U_1, U_2 \in \mathscr{U}_y$, there exists $U_3 \in \mathscr{U}_y$ with $U_3 \subseteq U_1 \cap U_2$;

- whenever $U \in \mathscr{U}_y$, there is some $U_0 \in \mathscr{U}_y$ such that whenever $w \in U_0$ there exists $W \in \mathscr{U}_w$ with $W \subseteq U$.

Clearly, only the second condition is at issue for each of the potential local bases under consideration, as we have both

$$[B_1 \cup B_2, d, \min\{\varepsilon_1, \varepsilon_2\}]^+[A] \subseteq [B_1, d, \varepsilon_1]^+[A] \cap [B_2, d, \varepsilon_2]^+[A]$$

and

$$[B_1 \cup B_2, d, \min\{\varepsilon_1, \varepsilon_2\}]^-[A] \subseteq [B_1, d, \varepsilon_1]^-[A] \cap [B_2, d, \varepsilon_2]^-[A].$$

Proposition 36.2. *Let $\mathscr{B}$ be a bornology on a metric space $\langle X, d \rangle$ that is shielded from closed sets. Then all sets of the form $[B, d, \varepsilon]^+[A]$ where $B \in \mathscr{B} \cap \mathscr{C}_0(X)$ and $\varepsilon > 0$ form a local base at $A \in \mathscr{C}(X)$ for a topology on $\mathscr{C}(X)$.*

Proof. We show that given $B \in \mathscr{B} \cap \mathscr{C}_0(X)$, $\varepsilon > 0$, and $A \in \mathscr{C}(X)$, $\exists B_1 \in \mathscr{B} \cap \mathscr{C}_0(X)$ and $\delta > 0$ such that

$$\forall D \in [B_1, d, \delta]^+[A], \ \exists \mu > 0 \text{ with } [B_1, d, \mu]^+[D] \subseteq [B, d, \varepsilon]^+(A).$$

Let $B_1 \in \mathscr{B}$ be a closed shield for B; we claim that $\delta = \frac{\varepsilon}{2}$ does the job. To see this, fix a closed $D \in [B_1, d, \frac{\varepsilon}{2}]^+[A]$. Since $D \cap (X \backslash S_d(A, \varepsilon/2)) \cap B_1 = \emptyset$ and B_1 is a shield for B, we can find $\alpha > 0$ such that

$$S_d(D \cap (X \backslash S_d(A, \varepsilon/2)), \alpha) \cap B = \emptyset.$$

Choose $0 < \mu < \min\{\frac{\varepsilon}{2}, \alpha\}$. We will show that $[B_1, d, \mu]^+[D] \subseteq [B, d, \varepsilon]^+[A]$.

For this purpose, suppose C is closed and $C \in [B_1, d, \mu]^+[D]$. Let $c \in C \cap B$ be arbitrary. As $c \in C \cap B_1, c \in S_d(D, \mu)$, and so $\exists x \in D$ with $d(c, x) < \mu$. As $c \in B$, we have $c \notin S_d(D \cap (X \backslash S_d(A, \frac{\varepsilon}{2})), \alpha)$. Since $\mu < \alpha$, we conclude that $x \in S_d(A, \frac{\varepsilon}{2})$. We compute

$$c \in S_d(x, \mu) \subseteq S_d(S_d(A, \frac{\varepsilon}{2}), \mu) \subseteq S_d(A, \mu + \frac{\varepsilon}{2}) \subseteq S_d(A, \varepsilon).$$

This shows that $C \cap B \subseteq S_d(A, \varepsilon)$ as required. $\square$

While the proof of the last proposition is a little technical, it is far less subtle than the proof of the companion result with respect to sets of the form $[B, d, \varepsilon]^-[A]$. We need a preliminary result with respect to shields for nonempty closed sets.

Proposition 36.3. *Let C be a nonempty closed subset of $\langle X, d \rangle$ and let A be a superset. The following conditions are equivalent.*

(1) *A is a shield for C;*

(2) *C contains a nonempty compact set K such that $\forall \varepsilon > 0, \exists \delta > 0$ such that $S_d(C \backslash S_d(K, \varepsilon), \delta) \subseteq A$.*

Proof. (1) $\Rightarrow$ (2). If A contains $S_d(C, \delta_0)$ for some $\delta_0 > 0$ let K be any point of C and let $\delta = \delta_0$ whatever $\varepsilon > 0$ may be. Otherwise, A contains no enlargement of C. It follows easily from Proposition 12.9 that $K = C \cap \mathrm{cl}(X \backslash A)$ is nonempty and compact. It follows from the same result that $S_d(K, \varepsilon)$ contains $\{c \in C : d(c, X \backslash A) \leq 1/n_0\}$ for sufficiently large $n_0 \in \mathbb{N}$, else $K \cap \{x : d(x, K) \geq \varepsilon\}$ would be nonempty. Taking $\delta = 1/n_0$ does the job, because if $c \in C$ and $c \notin S_d(K, \varepsilon)$, then $d(c, X \backslash A) > 1/n_0$.

(2) $\Rightarrow$ (1). Suppose (2) holds. Let $T \in \mathscr{C}_0(X)$ be disjoint from A. Choose $\varepsilon > 0$ such that $S_d(K, 2\varepsilon) \cap T = \emptyset$. Choose $\delta > 0$ corresponding to ε. With $\sigma = \min\{\varepsilon, \delta\}$, we have $D_d(T, C) \geq \sigma > 0$. $\square$

Proposition 36.4. *Let $\mathscr{B}$ be a bornology on $\langle X, d \rangle$ that is shielded from closed sets. Suppose $B \in \mathscr{B}$ is nonempty and closed and $\varepsilon > 0$. Then there exists $\delta > 0$ such that whenever $M \in \mathscr{P}_0(X)$ with $B \subseteq S_d(M, \delta)$, there exist $B_M \in \mathscr{B}$ with $B_M \subseteq M$ such that $B \subseteq S_d(B_M, \varepsilon)$.*

Proof. Pick $B_0 \in \mathscr{B}$ that shields B from closed sets and let $\varepsilon > 0$. By Proposition 36.3, let K be a nonempty compact subset of B such that for some $\delta > 0$,

$$S_d(B \backslash S_d(K, \frac{\varepsilon}{3}), \delta) \subseteq B_0.$$

Without loss of generality we may assume that $\delta < \frac{\varepsilon}{3}$.

Suppose now that $M \in \mathscr{P}_0(X)$ satisfies $B \subseteq S_d(M, \delta)$. Put $T := M \cap S_d(B \backslash S_d(K, \frac{\varepsilon}{3}), \delta)$. By our choice of δ, evidently $T \subseteq B_0$. Also note that each point of $B \backslash S_d(K, \frac{\varepsilon}{3})$ has distance less than δ from some point m of M and so m must lie in T.

Let E be a finite subset of K with $K \subseteq S_d(E, \frac{\varepsilon}{3})$, and then choose a finite subset F of M with $H_d(E, F) < \delta$. Since $\delta < \frac{\varepsilon}{3}$ we obtain

$$S_d(K, \frac{\varepsilon}{3}) \subseteq S_d(E, \frac{2\varepsilon}{3}) \subseteq S_d(F, \varepsilon).$$

With the above estimates in mind, we compute

$$B = (B \backslash S_d(K, \frac{\varepsilon}{3})) \cup (B \cap S_d(K, \frac{\varepsilon}{3})) \subseteq S_d(T, \delta) \cup S_d(F, \varepsilon) = S_d(T \cup F, \varepsilon).$$

By construction, $B_M := T \cup F \subseteq M$ and finally, $B_M \in \mathscr{B}$ because $T \subseteq B_0 \in \mathscr{B}$ and $F \in \mathscr{B}$. $\square$

Proposition 36.5. *Let $\mathscr{B}$ be a bornology in a metric space $\langle X, d \rangle$ that is shielded from closed sets. Then all sets of the form $[B, d, \varepsilon]^-[A]$ where $B \in \mathscr{B} \cap \mathscr{C}_0(X)$ and $\varepsilon > 0$ form a local base at $A \in \mathscr{C}(X)$ for a topology on $\mathscr{C}(X)$.*

Proof. We show that given $B \in \mathscr{B} \cap \mathscr{C}_0(X)$, $\varepsilon > 0$, and $A \in \mathscr{C}(X)$, $\exists \delta > 0$ such that

$$\forall D \in [B, d, \delta]^-[A], \ \exists T \in \mathscr{B} \cap \mathscr{C}_0(X) \text{ with } [T, d, \frac{\varepsilon}{2}]^-[D] \subseteq [B, d, \varepsilon]^-(A).$$

If $A \cap B = \emptyset$, then $[B, d, \varepsilon]^-(A) = \mathscr{C}(X)$ and there is nothing to show. Otherwise, by the last result, given $\varepsilon > 0$, there exists $\delta > 0$ such that if $A \cap B \subseteq S_d(M, \delta)$ whatever M may be, then there exists $B_M \subseteq M$ in the bornology with $A \cap B \subseteq S_d(B_M, \frac{\varepsilon}{2})$. Let $D \in [B, d, \delta]^-[A]$. Then we have $A \cap B \subseteq S_d(B_D, \frac{\varepsilon}{2})$ where $B_D \in \mathscr{B}$ and $B_D \subset D$. We claim $[cl(B_D), d, \frac{\varepsilon}{2})]^-[D] \subseteq [B, d, \varepsilon]^-[A]$. Let $C \in [cl(B_D), d, \frac{\varepsilon}{2})]^-[D]$. Since D is a closed superset of B_D, we get

$$A \cap B \subseteq S_d(cl(B_D), \frac{\varepsilon}{2}) = S_d(cl(B_D) \cap D, \frac{\varepsilon}{2}) \subseteq S_d(C, \varepsilon)$$

as required. $\square$

Putting together Propositions 36.1, 36.2, and 36.5, we can state

Theorem 36.6. *Let $\mathscr{B}$ be a bornology on a metric space $\langle X, d \rangle$. The following conditions are equivalent.*

(1) *$(\mathscr{B}, d)$-convergence is topological on $\mathscr{C}(X)$;*

(2) *$(\mathscr{B}, d)$-convergence is topological on $\mathscr{C}_0(X)$;*

(3) *$\mathscr{B}$ is shielded from closed sets.*

When our bornology is shielded from closed sets, we will denote the topology of $(\mathscr{B}, d)$-convergence on $\mathscr{C}(X)$ by $\tau_{\mathscr{B},d}$.

By Theorem 35.4, $(\mathscr{B}, d)$-convergence is topological on the compact subsets of X provided the bornology $\mathscr{B}$ has a closed base. The hyperspace topology is generated by all sets of the form $\{K : K \text{ is compact and } K \cap V \neq \emptyset\}$ where $V \in \tau_d$ plus all sets of the form $\{K : K \text{ is compact and } K \cap B = \emptyset\}$ where $B \in \mathscr{B} \cap \mathscr{C}(X)$. On the other hand the argument in case (1) in the proof of Proposition 36.1 shows that $\mathscr{B}$ having a closed base is necessary for the convergence to be topological even restricted to the nonempty finite subsets of X. Thus, $(\mathscr{B}, d)$-convergence of nets of compact subsets is topological if and only if the bornology $\mathscr{B}$ has a closed base.

37. Uniformizability and Metrizability

Our first result of this section is distilled from [29, Theorem 5.18 and Theorem 5.19].

Theorem 37.1. *Let $\mathscr{B}$ be a bornology that is shielded from closed sets on a metric space $\langle X, d \rangle$. The following statements are equivalent.*

(1) *$\mathscr{B}$ is local;*

(2) *the topology of $(\mathscr{B}, d)$-convergence on $\mathscr{C}(X)$ is Hausdorff;*

(3) *$\mathscr{B}$ is has an open base;*

(4) *$\mathscr{B}$ is stable under small enlargements;*

(5) *all sets of the form $[B, d, \varepsilon]$ where $B \in \mathscr{B}$ and $\varepsilon > 0$ form a base for a uniformity compatible with the topology of $(\mathscr{B}, d)$-convergence on $\mathscr{C}(X)$;*

(6) *the topology of $(\mathscr{B}, d)$-convergence on $\mathscr{C}(X)$ is completely regular;*

(7) *the topology of $(\mathscr{B}, d)$-convergence on $\mathscr{C}(X)$ is regular.*

Proof. By Theorem 31.7 and [155, p. 86], the topology is Hausdorff if and only if the bornology is in addition local. By Theorem 12.15, a local bornology that is shielded from closed sets is already stable under small enlargements. Condition (3) is intermediate in strength between condition (1) and condition (4). Thus, conditions (1) through (4) are equivalent. Clearly $(5) \Rightarrow (6) \Rightarrow (7)$ holds. To complete the proof, we establish $(4) \Rightarrow (5)$ and $(7) \Rightarrow (2)$.

$(4) \Rightarrow (5)$. Only the composition condition needs to be verified. Let $B \in \mathscr{B}$ and $\varepsilon > 0$ be arbitrary, and pick $\delta < \frac{\varepsilon}{2}$ such that $S_d(B, \delta) \in \mathscr{B}$. Suppose A, C and D are closed sets with $(A, C) \in [S_d(B, \delta), d, \delta]$ and $(C, D) \in [S_d(B, \delta), d, \delta]$. We would like to show $A \cap B \subseteq S_d(D, \varepsilon)$ and $D \cap B \subseteq S_d(A, \varepsilon)$ so that $(A, D) \in [B, d, \varepsilon]$. We just verify the first inclusion.

Let $x \in A \cap B$. Since $A \cap B \subseteq A \cap S_d(B, \delta) \subseteq S_d(C, \delta)$, there exists $c \in S_d(B, \delta) \cap C$ with $d(x, c) < \frac{\varepsilon}{2}$. Since $(C, D) \in [S_d(B, \delta), d, \delta]$, there exists $w \in D$ with $d(c, w) < \delta$. By the triangle inequality, $d(x, w) < 2\delta < \varepsilon$ and since $x \in A \cap B$ was arbitrary, $A \cap B \subseteq S_d(D, \varepsilon)$.

$(7) \Rightarrow (2)$. Without the regularity assumption, the topology $\tau_{\mathscr{B}, d}$ on $\mathscr{C}(X)$ contains

$$\{C \in \mathscr{C}(X) : \exists \varepsilon > 0 \text{ with } S_d(C, \varepsilon) \cap B = \emptyset\} \quad (B \in \mathscr{B}),$$

, and

$$\{C \in \mathscr{C}(X) : C \cap V \neq \emptyset\} \quad (V \text{ open in } X).$$

For the first assertion, if $S_d(C, \varepsilon) \cap B = \emptyset$, then any member of $[B, d, \varepsilon][C]$ has the same property. For the second, if $p \in C \cap V$, $\exists \varepsilon > 0$ with $S_d(p, \varepsilon) \subseteq V$. Then each set in $[\{p\}, d, \varepsilon][C]$ hits V as well.

From these two facts, we next show that $\tau_{\mathscr{B},d}$ satisfies the T_1 separation axiom. To this end, suppose C_1 and C_2 are distinct closed sets, and without loss of generality pick $c_1 \in C_1$ and $\delta > 0$ such that $S_d(c_1, \delta) \cap C_2 = \emptyset$. Then $\{C \in \mathscr{C}(X) : C \cap S_d(c_1, \delta) \neq \emptyset\}$ is a $\tau_{\mathscr{B},d}$-neighborhood of C_1 that does not contain C_2 while $\{C \in \mathscr{C}(X) : \exists \varepsilon > 0 \text{ with } S_d(C, \varepsilon) \cap \{c_1\} = \emptyset\}$ is a $\tau_{\mathscr{B},d}$-neighborhood of C_2 that does not contain C_1. Since each regular T_1 topology is Hausdorff, condition (2) follows from condition (7).   □

One wonders from our last proof whether all sets of the from $\{C \in \mathscr{C}(X) : \exists \varepsilon > 0 \text{ with } S_d(C, \varepsilon) \cap B = \emptyset\}$ where B runs over $\mathscr{B}$ plus all sets of the form $\{C \in \mathscr{C}(X) : C \cap V \neq \emptyset\}$ where V runs over τ_d generate $\tau_{\mathscr{B},d}$ when $\mathscr{B}$ is stable under small enlargements. Our expectations are heightened by the fact that for a bornology $\mathscr{B}$ with closed base, the family $\{K : K$ is compact and $K \cap B = \emptyset\}$ where B runs over $\mathscr{B} \cap \mathscr{C}(X)$ coincides with the family $\{K : K$ is compact and for some $\varepsilon > 0$ $S_d(K, \varepsilon) \cap B = \emptyset\}$ where B runs over $\mathscr{B}$. Alas, this family may generate a properly coarser topology. In the real line, let $\{q_n : n \in \mathbb{N}\}$ list the rationals without repetition and let $\mathscr{B} = \mathscr{P}(\mathbb{R})$. Then with $A_n = \{q_1, q_2, \ldots, q_n\}$, we have $\langle A_n \rangle$ convergent to $\mathbb{R}$ in the hyperspace topology so generated but $\langle A_n \rangle$ fails to converge to $\mathbb{R}$ in Hausdorff distance.

Assuming the bornology is stable under under small enlargement which by Theorem 37.1 is equivalent to $\langle \mathscr{C}(X), \tau_{\mathscr{B},d} \rangle$ being a Tychonoff space, let us denote the uniformity of condition (5) by $\Upsilon_{\mathscr{B},d}$. We finally turn to metrizability of the hyperspace topology as considered in [42, Theorem 3.18].

Theorem 37.2. *Let $\mathscr{B}$ be a bornology in a metric space $\langle X, d \rangle$ that is stable under small enlargements. The following conditions are equivalent:*

(1) *$\mathscr{B}$ has a countable base;*

(2) *there is an equivalent metric ρ for X such that $(\mathscr{B}, d)$-convergence is Attouch-Wets convergence with respect to the metric ρ;*

(3) *the uniformity $\Upsilon_{\mathscr{B},d}$ has a countable base;*

(4) *the hyperspace $\langle \mathscr{C}(X), \tau_{\mathscr{B},d} \rangle$ is metrizable;*

(5) *the hyperspace $\langle \mathscr{C}_0(X), \tau_{\mathscr{B},d} \rangle$ is first countable.*

Proof. $(1) \Rightarrow (2)$. By condition (1) we know immediately from Hu's Theorem that $\mathscr{B}$ is a metric bornology, but we must be careful in the construction of the appropriate metric so that it fulfills the requirements of Corollary 31.9.

If $\mathscr{B} = \mathscr{P}(X)$, then $(\mathscr{B}, d)$-convergence is convergence in Hausdorff distance H_d which is $(\mathscr{B}_\rho(X), \rho)$-convergence where ρ is the uniformly equivalent bounded metric defined by $\rho = \min\{d, 1\}$. Otherwise, for each $B \in \mathscr{B}$, $X \backslash B$ is nonempty. Let $\langle B_n \rangle$ be an increasing sequence of open sets that is cofinal in $\mathscr{B}$. Set $A_1 = B_1$ and choose $\varepsilon_1 > 0$ with $S_d(A_1, \varepsilon_1) \in \mathscr{B}$. Now let $A_2 = B_2 \cup S_d(A_1, \varepsilon_1) \in \mathscr{B}$. Having produced open $A_1, A_2, \ldots, A_n$ in $\mathscr{B}$ with $A_{j-1} \subseteq A_j$ for $j = 2, 3, \ldots n$, choose $\varepsilon_n > 0$ with $S_d(A_n, \varepsilon_n) \in \mathscr{B}$ and set $A_{n+1} = B_{n+1} \cup S_d(A_n, \varepsilon_n)$. The construction produces a new sequence $\langle A_n \rangle$ with these properties:

(a) $\forall n \in \mathbb{N}$, A_n is open;

(b) $\forall n \in \mathbb{N}$, $S_d(A_n, \varepsilon_n) \subseteq A_{n+1}$;

(c) $\forall n \in \mathbb{N}$, $X \backslash A_n \neq \emptyset$;

(d) $\{A_n : n \in \mathbb{N}\}$ is cofinal in $\mathscr{B}$.

For each $n \in \mathbb{N}$ define $f_n : \langle X, d \rangle \to \mathbb{R}$ by

$$f_n(x) = \min \left\{ 1, \frac{1}{\varepsilon_n} d(x, A_n) \right\}.$$

Each f_n is Lipschitz, $f_n(x) = 0$ if $x \in A_n$, and by property (b) above $f_n(x) = 1$ if $x \notin A_{n+1}$. Condition (d) says that $\{A_n : n \in \mathbb{N}\}$ is a cover of X, and we may assert by (b) and (d)

(e) $\forall x \in X \ \exists \delta > 0 \ \exists n_0 \in \mathbb{N}$ such that $d(x, w) < \delta \Rightarrow \forall n \geqslant n_0 \ f_n(w) = 0$.

By (e), $f := f_1 + f_2 + f_3 + \cdots$ defines a finite-valued nonnegative continuous function. Note also that if $E \subseteq X$, then $f|_E$ is bounded if and only if $E \subseteq A_n$ for some n because (i) if $x \in A_n$, then $f(x) \leq n - 1$, and (ii) if $x \notin A_{n+1}$, then $f(x) \geq n$.

Next define an equivalent metric ρ on X by

$$\rho(x, w) := \min \{d(x, w), 1\} + |f(x) - f(w)|.$$

By condition (d), $f|_E$ is bounded if and only if $E \in \mathscr{B}$ which gives $\mathscr{B}_\rho(X) = \mathscr{B}$. Since $\min\{d, 1\} \leq \rho$ and d and $\min\{d, 1\}$ are uniformly equivalent, to show that d and ρ are uniformly equivalent with respect to $\mathscr{B}$, it suffices to show that $f \in \mathscr{B}_f$ with respect to the metric d. To see this, fix $B \in \mathscr{B}$ and choose n with $B \subseteq A_n$. Now f restricted to the ε_n-enlargement of B is uniformly continuous as $f = f_1 + f_2 + f_2 + \cdots + f_n$ so restricted. Thus f is strongly uniformly continuous on $\mathscr{B}$, and we are done.

$(2) \Rightarrow (3)$. Fix $x_0 \in X$; clearly $\{[S_\rho(x_0, n), d, \frac{1}{n}] : n \in \mathbb{N}\}$ is a countable base for $\Upsilon_{\mathscr{B}, d}$ because by Corollary 31.9, $\mathscr{B} = \mathscr{B}_\rho(X)$.

$(3) \Rightarrow (4)$. A uniformity with a countable base is a uniformity determined by a compatible pseudometric (see, e.g., [155, p. 257]), and the hyperspace is already Hausdorff because $\mathscr{B}$ is local.

$(4) \Rightarrow (5)$. This is obvious.

$(5) \Rightarrow (1)$. Let $x_0 \in X$ be arbitrary, and let $\{\mathscr{C}_0(X) \cap [B_n, d, \varepsilon_n][\{x_0\}] : n \in \mathbb{N}\}$ be a countable local base for $\langle \mathscr{C}_0(X), \tau_{\mathscr{B}, d} \rangle$ at $\{x_0\}$. Since $\mathscr{B}$ is local, $\exists \delta > 0$ with $S_d(x_0, \delta) \in \mathscr{B}$. For each n set $T_n = S_d(x_0, \delta) \cup \bigcup_{j=1}^n B_j$. Since $\mathscr{B}$ is stable under finite unions, $\langle T_n \rangle$ is an increasing sequence in $\mathscr{B}$. We claim that $\{T_n : n \in \mathbb{N}\}$ is cofinal in $\mathscr{B}$. If this fails we can find $B \in \mathscr{B}$ not included in any T_n. For each n, pick $x_n \in B \backslash T_n$. Then for $j \geq n$, we have

$$\{x_0, x_j\} \in [T_n, d, \varepsilon_n][\{x_0\}],$$

and evidently $\{\mathscr{C}_0(X) \cap [T_n, d, \varepsilon_n][\{x_0\}] : n \in \mathbb{N}\}$ is also a local base for the relative topology at $\{x_0\}$. This yields $\{x_0\} = \tau_{\mathscr{B}, d} - \lim\{x_0, x_j\}$. But on the other hand, since $d(x_j, x_0) \geq \delta$ for each $j \in \mathbb{N}$,

$$\{x_0, x_j\} \notin [B, d, \frac{\delta}{2}][\{x_0\}],$$

and this gives the desired contradiction. $\qquad \square$

Corollary 37.3. *Let $\langle X, d \rangle$ be a complete metric space and let $\mathscr{B}$ be a bornology that is stable under small enlargements and that has a countable base. Then $(\mathscr{B}, d)$-convergence on $\mathscr{C}_0(X)$ is completely metrizable.*

Proof. The proof of the last theorem shows that we can find an equivalent metric ρ with $\min\{d, 1\} \leq \rho$ such that $(\mathscr{B}, d)$-convergence is Attouch-Wets convergence with respect to ρ. Now any ρ-Cauchy sequence in X is d-Cauchy and is hence convergent. This means that ρ is a complete metric and as a result, Attouch-Wets convergence with respect to ρ is completely metrizable (see, e.g., [8] or [17, p. 80]). □

Example 37.4. A bornology on $\langle X, d \rangle$ can have a countable base, a closed base, and an open base without $(\mathscr{B}, d)$-convergence being topological, much less metrizable. Consider $X = [0, \infty) \times [0, \infty)$ as a metric subspace of the Euclidean plane, and the bornology $\mathscr{B}$ on X having as a countable closed base all sets of the form.

$$B_n := \{(x, y) : x \geq 0, y \geq 0 \text{ and } xy \leq n\} \qquad (n \in \mathbb{N})$$

Evidently an open base for the bornology consists of

$$V_n := \{(x, y) : x \geq 0, y \geq 0 \text{ and } xy < n\} \qquad (n \in \mathbb{N})$$

For each $n \geq k, B_n$ does not shield B_k from closed sets (consider as a closed set disjoint from B_n the set B_{n+1} which is near B_k). Thus, $\mathscr{B}$ is not shielded from closed sets.

38. Ideals, Bornologies and Extensions

In this section we leave the confines of Hausdorff topological spaces and work with topological spaces $\langle X, \tau \rangle$ with no assumed separation properties. Consistent with our terminology for bornologies, if $\mathscr{B}$ is an ideal on X and $\downarrow \mathscr{B}_0 = \mathscr{B}$ we call $\mathscr{B}_0$ a *base* for $\mathscr{B}$. If $\mathscr{B}$ has a closed base, we henceforth say that the ideal $\mathscr{B}$ is *closed*.

Let I be a nonempty set disjoint from X and let σ be a topology on $X \cup I$. We call $\langle X \cup I, \sigma \rangle$ an *extension* of $\langle X, \tau \rangle$ provided X is dense in $\langle X \cup I, \sigma \rangle$ and the relative topology that X inherits from $\langle X \cup I, \sigma \rangle$ coincides with τ. The set I of new points is of course called the *remainder*. For each $i \in I$,

$$\{X \backslash W : W \in \sigma, i \in W\}$$

is a closed base for an ideal $\mathscr{B}_i(\sigma)$ on X, and

$$\mathscr{B}_i(\sigma) = \{E \subseteq X : \exists W \in \sigma \text{ such that } i \in W, \; W \cap E = \emptyset\}$$

$$= \{E \subseteq X : i \notin \mathrm{cl}_\sigma(E)\}.$$

By the denseness of X in the extension, $\mathscr{B}_i(\sigma)$ is a nontrivial ideal. It is clear that the ideal is a bornology if and only if whenever $p \in X$ there exists $W \in \sigma$ with $i \in W$ and $p \notin W$. In particular, if the extension topology is T_1 then $\{B_i(\sigma) : i \in I\}$ is a family of bornologies on X.

We will now a describe a natural way to associate an extension of a topological $\langle X, \tau \rangle$ to a family of nontrivial closed ideals on X. Let I be a nonempty set disjoint from X. Suppose $\{\mathscr{B}_i : i \in I\}$ is a family of nontrivial closed ideals on X. For each $E \in \mathscr{C}(X)$, put

$$U_E := \{i \in I : E \in \mathscr{B}_i\}, \quad V_E := X \backslash E.$$

Since each ideal contains the empty set, it is clear that $U_\emptyset \cup V_\emptyset = X \cup I$ while by the nontrivialness of each ideal, $U_X \cup V_X = \emptyset$.

Lemma 38.1. *Let $\{\mathscr{B}_i : i \in I\}$ be a family of nontrivial closed ideals on $\langle X, \tau \rangle$. Then the family $\{U_E \cup V_E : E \in \mathscr{C}(X)\}$ is closed under finite intersections.*

Proof. For E, F closed in X, one has

$$U_E \cap U_F = \{i \in I : E \in \mathscr{B}_i \text{ and } F \in \mathscr{B}_i\} = \{i \in I : E \cup F \in \mathscr{B}_i\} = U_{E \cup F}.$$

Hence,

$$(U_E \cup V_E) \cap (U_F \cup V_F) = (U_E \cap U_F) \cup (V_E \cap V_F) = U_{E \cup F} \cup V_{E \cup F}. \qquad \square$$

We have remarked that $U_\emptyset \cup V_\emptyset = X \cup I$. Since $\{U_E \cup V_E : E \in \mathscr{C}(X)\}$ is also closed under finite intersections, this family forms a base for a topology $\tau_0(\{\mathscr{B}_i : i \in I\})$ on $X \cup I$ [155, Theorem 5.3]. When no ambiguity can result, we just write τ_0 for this topology.

Proposition 38.2. *Let $\langle X, \tau \rangle$ be a topological space, and let $\{\mathscr{B}_i : i \in I\}$ be a family of nontrivial closed ideals on X. Then $\langle X \cup I, \tau_0(\{\mathscr{B}_i : i \in I\}) \rangle$ is an extension of X.*

Proof. The relative topology on X determined by τ_0 is contained in τ as the intersection of each basic open set with X evidently lies in τ. To show the reverse inclusion, let $W \in \tau$. Evidently,

$$W = (U_{X \backslash W} \cup V_{X \backslash W}) \cap X$$

which shows that W is relatively open.

To see that X is dense, suppose $i \in I$ belongs to some basic open set $U_E \cup V_E$. Then $i \in U_E$ so that $E \in \mathscr{B}_i$ which means $E \neq X$ and thus $(U_E \cup V_E) \cap X = V_E \cap X \neq \emptyset$. □

Given a family of nontrivial closed ideals $\{\mathscr{B}_i : i \in I\}$ on X, we call $\langle X \cup I, \tau_0(\{\mathscr{B}_i : i \in I\}) \rangle$ the *bornological extension* of $\langle X, \tau \rangle$ determined by $\{\mathscr{B}_i : i \in I\}$ and τ_0 the associated *bornological extension topology*. We use this terminology whether or not the ideals are bornologies. To help the reader gain some familiarity with this construct, we immediately characterize compactness of a bornological extension.

Theorem 38.3. *Let $\langle X, d \rangle$ be a topological space and let $\{\mathscr{B}_i : i \in I\}$ be a family of nontrivial closed ideals on X. Then the associated bornological extension topology is compact if and only if whenever $\{F_j : j \in J\} \subseteq \mathscr{C}(X)$ satisfies $X \cup I \subseteq \cup_{j \in J} U_{F_j} \cup V_{F_j}$, there exists $\{j_1, j_2, \ldots j_n\} \subseteq J$ with $I = \cup_{k=1}^{n} U_{F_{j_k}}$ and $\cap_{k=1}^{n} F_{j_k} = \emptyset$.*

Proof. Compactness of the extension topology can be tested using open covers taken from a fixed base for the topology. Working with the standard base $\{U_F \cup V_F : F \in \mathscr{C}(X)\}$, compactness occurs if and only if whenever $\{U_{F_j} \cup V_{F_j} : j \in J\}$ is an open cover of $X \cup I$, then we can find $\{j_1, j_2, \ldots j_n\} \subseteq J$ such that

$$X \cup I = \bigcup_{k=1}^{n} U_{F_{j_k}} \cup V_{F_{j_k}} = \bigcup_{k=1}^{n} (U_{F_{j_k}} \cup X \backslash F_{j_k}).$$

In view of DeMorgan's laws, this means that $I = \cup_{k=1}^{n} U_{F_{j_k}}$ and $\cap_{k=1}^{n} F_{j_k} = \emptyset$. □

Definition 38.4. We will say that the extension $\langle X \cup I, \sigma \rangle$ of $\langle X, \tau \rangle$ is a *bornological extension* if there exists a family $\{\mathscr{B}_i : i \in I\}$ of nontrivial closed ideals on X such that $\sigma = \tau_0(\{\mathscr{B}_i : i \in I\})$.

For an extension $\langle X \cup I, \sigma \rangle$, recall that we denoted by $\{\mathscr{B}_i(\sigma) : i \in I\}$) the family of nontrivial closed ideals naturally determined by σ. We will denote by $\tau_0(\sigma)$ the bornological extension topology $\tau_0(\{\mathscr{B}_i(\sigma) : i \in I\})$. Using the simple arithmetic of relative complements, the reader can verify that all sets of the form $\{i \in I : i \notin \text{cl}_\sigma(X \backslash W)\} \cup (W \cap X)$ where $x \in W \in \sigma$ is a local base at x for $\tau_0(\sigma)$. We will see that $\tau_0(\sigma)$ is always coarser than σ, and we give several sets of necessary and sufficient conditions for their agreement. We will see that we have agreement in any extension having a base consisting of regular open sets, a weaker condition than regularity of the extension topology. On the other, a Hausdorff extension need not be bornological.

The results and examples that follow on bornological extensions are for the most part adaptations of results from Beer and Vipera [54]. We take a naive approach to extensions, that is, we adjoin points to an initial space and consider closed ideals associated with the neighborhood systems of the points of the remainder rather than consider filter systems, as favored by most researchers in the area. The agreement of σ and $\tau_0(\sigma)$ in this parallel understanding has been described by saying that the extension $\langle X \cup I, \sigma \rangle$ is *strict* (see, e.g., [10, pp. 4–8] or [144, pp. 133–137]).

Given a bornological extension $\langle X \cup I, \tau_0(\{\mathscr{B}_i : i \in I\})\rangle$, the next result says that we can recover our family of closed ideals directly from the extension topology.

Proposition 38.5. *Let $\{\mathscr{A}_i : i \in I\}$ be a family of nontrivial closed ideals on $\langle X, \tau\rangle$ and let τ_0 be the topology on $X \cup I$ that it induces. Then $\forall i \in I$, we have $\mathscr{A}_i = \mathscr{B}_i(\tau_0)$.*

Proof. Since $\mathscr{A}_i$ and $\mathscr{B}_i(\tau_0)$ both have closed bases, we need only show that for each index i, we have $\mathscr{A}_i \cap \mathscr{C}(X) = \mathscr{B}_i(\tau_0) \cap \mathscr{C}(X)$.

Suppose first that $E \in \mathscr{B}_i(\tau_0) \cap \mathscr{C}(X)$; by definition, there exists $W \in \tau_0$ with $i \in W$ and $W \cap E = \emptyset$. In view of the standard base for τ_0, $\exists F \in \mathscr{C}(X)$ such that $i \in U_F \cup V_F$ and $(U_F \cup V_F) \cap E = \emptyset$. It follows that $E \subseteq F$, and thus, $E \in \mathscr{A}_i$ because $F \in \mathscr{A}_i$ and $\mathscr{A}_i$ is hereditary. We conclude that $E \in \mathscr{A}_i \cap \mathscr{C}(X)$ because E is τ-closed.

For the reverse inclusion, suppose $E \in \mathscr{A}_i \cap \mathscr{C}(X)$ so that $i \in U_E$. Then $U_E \cup V_E$ is a τ_0-neighborhood of i disjoint from E, and so $E \in \mathscr{B}_i(\tau_0) \cap \mathscr{C}(X)$. $\qquad\square$

Corollary 38.6. *Let $\{\mathscr{B}_i : i \in I\}$ be a family of nontrivial closed ideals on $\langle X, \tau\rangle$. Then $\tau_0(\{\mathscr{B}_i : i \in I\}) = \tau_0(\tau_0(\{\mathscr{B}_i : i \in I\}))$.*

The last corollary highlights the fact that each bornological extension $\langle X \cup I, \sigma\rangle$ must arise from the family of ideals $\{\mathscr{B}_i(\sigma) : i \in I\}$.

The next formula effectively shows that for any extension $\langle X \cup I, \sigma\rangle$ of $\langle X, \tau\rangle$, $\tau_0(\{\mathscr{B}_i(\sigma) : i \in I\}) \subseteq \sigma$.

Proposition 38.7. *Suppose $\langle X \cup I, \sigma\rangle$ is an extension of $\langle X, \tau\rangle$, and $\forall i \in I$, let $\mathscr{B}_i(\sigma)$ be the ideal generated by $\{X \backslash W : W \in \sigma, i \in W\}$. For each $E \in \mathscr{C}(X)$, put $U_E = \{i \in I : E \in \mathscr{B}_i(\sigma)\}$ and $V_E = X \backslash E$. Then*

$$(X \cup I) \backslash cl_\sigma(E) = U_E \cup V_E.$$

Proof. If $x \in X$ and $x \notin cl_\sigma(E)$, then $x \notin E$ and so $x \in X \backslash E$. On the other hand, if $x \in X \backslash E = X \backslash cl_\tau(E)$, then $x \notin cl_\sigma(E)$ because $cl_\tau(E) = X \cap cl_\sigma(E)$. Thus, the points in X that are in both sets are the same and comprise $X \backslash E = V_E$.

Moving to points in the remainder, put $W = (X \cup I) \backslash cl_\sigma(E)$, and suppose $i \in I \cap W$. Since $W \in \sigma$, $E = X \backslash W \in \mathscr{B}_i(\sigma) \cap \mathscr{C}(X)$, and hence $i \in U_E$. Conversely, let $i \in U_E$. Then $E \in \mathscr{B}_i(\sigma)$ and so there exists $W \in \sigma$ with $E \cap W = \emptyset$ and $i \in W$. Thus, $i \in (X \cup I) \backslash cl_\sigma(E)$. $\qquad\square$

Corollary 38.8. *Suppose $\langle X \cup I, \sigma\rangle$ is an extension of $\langle X, \tau\rangle$. Then*

    (1) $\tau_0(\{B_i(\sigma) : i \in I\}) \subseteq \sigma$;

    (2) $\forall E \in \mathscr{C}(X)$, $U_E \cup V_E$ is the largest element of σ whose intersection with X is $X \backslash E$.

Proof. From Proposition 38.7, σ contains a base for $\tau_0(\{B_i(\sigma) : i \in I\})$ and so it contains the entire topology. If $E \subseteq X$ is closed, then $U_E \cup V_E \cap X = V_E = X \backslash E$. If $W \in \sigma$ and $W \cap X = X \backslash E$, then $W \cap E = \emptyset$; hence $W \subseteq (X \cup I) \backslash cl_\sigma(E) = U_E \cup V_E$. $\qquad\square$

Example 38.9. We give an example of a one-point extension that is a bornological extension. Let $X = \{x_1, x_2\}$ and let $I = \{p\}$. Our extension topology is given by

$$\sigma = \{\emptyset, \{x_1, x_2, p\}, \{x_2, p\}\}.$$

This induces the relative topology $\tau = \{\emptyset, X, \{x_2\}\}$ on X, and we also have $\mathscr{B}_p(\sigma) = \{\emptyset, \{x_1\}\}$. As $\mathscr{C}(X) = \{\emptyset, X, \{x_1\}\}$, and $U_\emptyset \cup V_\emptyset = X \cup I, U_X \cup V_X = \emptyset$, and $U_{\{x_1\}} \cup V_{\{x_1\}} = \{x_2, i\}$, the extension topology σ agrees with $\tau_0(\sigma)$.

We next characterize those one-point extensions that are bornological extensions (cf. [54, Proposition 5.9]). In the case that the extensions are Alexandroff one-point compactifications of noncompact Hausdorff spaces, it is to be expected that when the base space is locally compact, then the extension will be bornological. But this is true more generally.

Proposition 38.10. *Let $\langle X \cup \{p\}, \sigma \rangle$ be a one-point extension of $\langle X, \tau \rangle$. Then $\sigma = \tau_0(\sigma)$ if and only if whenever $w \in W \in \sigma$ and $W \subseteq X$ then there exists a τ-open neighborhood V of w such that $X \backslash V \notin \mathscr{B}_p(\sigma)$.*

Proof. For sufficiency, we will show that whenever $W \in \sigma$, then W contains a $\tau_0(\sigma)$ basic neighborhood of each point of W. We consider two exhaustive cases: (a) $p \in W$; (b) $W \subseteq X$. In the first case, $W = U_{X \backslash W} \cup V_{X \backslash W}$, that is W is already a basic neighborhood. In the second case, let $w \in W$ be arbitrary and let V be as stated with respect to w. As $\mathscr{B}_p(\sigma)$ is hereditary, $X \backslash (V \cap W) \notin \mathscr{B}_p(\sigma)$, so with $E = X \backslash (V \cap W)$, we have

$$w \in V \cap W = V_E = U_E \cup V_E \subseteq W.$$

For necessity, suppose $\sigma \subseteq \tau_0(\sigma)$ and $w \in W \in \sigma$ where $W \subseteq X$. Then we can find $E \in \mathscr{C}(X)$ such that $w \in U_E \cup V_E \subseteq W$. This implies that both $w \in V_E$ and $U_E = \emptyset$, that is, $E = X \backslash V_E \notin \mathscr{B}_p(\sigma)$. $\square$

Corollary 38.11. *Let $\langle X \cup \{p\}, \sigma \rangle$ be a one-point extension of $\langle X, \tau \rangle$ such that $\forall x \in X$, x and p have disjoint neighborhoods. Then the extension is bornological. In particular, this is true if σ is Hausdorff.*

Proof. Suppose $w \in W \in \sigma$ where $W \subseteq X$. Choose $G_w \in \sigma$ and $G_p \in \sigma$ with $w \in G_w, p \in G_p$ and $G_w \cap G_p = \emptyset$. Then $G_w \cap W \in \mathscr{B}_p(\sigma)$ so that $E := X \backslash (G_w \cap W) \notin \mathscr{B}_p(\sigma)$ because the ideal is nontrivial. Clearly, $w \in X \backslash E \subseteq W$. Apply Proposition 38.10. $\square$

Later, we will produce an example of a Hausdorff extension (whose remainder is not a singleton) that fails to be a bornological extension.

Recall from an earlier section that if $\mathscr{B}$ is a nontrivial closed bornology on a Hausdorff space $\langle X, \tau \rangle$ and $p \notin X$, the notation $o(\mathscr{B}, p)$ was used for $X \cup \{p\}$ equipped with the extension topology

$$\tau \cup \{X \cup \{p\} \backslash B : B \in \mathscr{B} \cap \mathscr{C}(X)\}.$$

A prototype for such extensions is $o(\mathscr{B}, p)$ for $\mathscr{B} = \mathscr{K}(X)$ in a noncompact Hausdorff space, the so-called Alexandroff one-point compactification [107, 155].

Corollary 38.12. *Let $\langle X, \tau \rangle$ be a Hausdorff space. Then $o(\mathscr{B}, p)$ is bornological if and only if for each $x \in X$ there exists $V \in \tau$ such that $x \in V$ and $X \backslash V \notin \mathscr{B}$.*

Proof. This follows directly from the last proposition, noting that X is open in this particular extension topology. $\square$

Now if $\langle X, \tau \rangle$ is a locally compact Hausdorff space that is not compact, and $x \in X$ with $x \in V \in \mathscr{K}(X)$, then $X \backslash V$ cannot be relatively compact, else X itself would be relatively compact and closed. Thus the Alexandroff one-point compactification of such a space is a bornological extension. But this is also true for the Alexandroff one-point compactification of $\mathbb{Q}$ as a metric subspace of the real line: given $q \in \mathbb{Q}$, the complement of $S_d(q, 1)$ is not relatively compact.

We next give a concrete example of a one-point compactification of a noncompact metrizable space that fails to be a bornological extension.

Example 38.13. In the sequence space ℓ_2 with origin 0_{ℓ_2} and standard orthonormal base $\{e_n : n \in \mathbb{N}\}$ let X be this metric subspace: $X = \cup_{n=1}^{\infty} \text{co}(\{0_{\ell_2}, \frac{1}{n}e_n\})$. Then given any neighborhood V of the origin, $X \backslash V$ is a subset of a finite union of line segments and is thus (relatively) compact. By the last corollary, the one-point compactification of X cannot be a bornological extension.

Example 38.14. Next is an example of an extension with a two-point closed remainder that is not a bornological extension. Let $X = \{x\}$, let $I = \{i_1, i_2\}$ and let

$$\sigma = \{\emptyset, X \cup I, \{x, i_1\}, \{x, i_2\}, \{x\}\}.$$

Here, $\mathscr{B}_{i_1}(\sigma) = \mathscr{B}_{i_2}(\sigma) = \{\emptyset\}$, and the relative topology is $\{\emptyset, X\}$. For $\tau_0(\sigma)$, we only get the indiscrete topology $\{\emptyset, X \cup I\}$.

Example 38.15. Let $\langle X \cup I, \sigma \rangle$ be a metrizable extension of $\langle X, \tau \rangle$. We will directly show that $\langle X \cup I, \sigma \rangle$ is a bornological extension of X. Since each such extension is regular, this is to be expected.

Let d be a compatible metric for σ. For each $i \in I$, it is clear that $\mathscr{B}_i(\sigma) = \{E \in \mathscr{P}(X) : d(i, E) > 0\}$, where the distance from any point to the empty set is understood to be ∞. To show $\sigma \subseteq \tau_0(\sigma)$, let $w \in W \in \sigma$, and choose $n \in \mathbb{N}$ with

$$\left\{ u \in X \cup I : d(u, w) < \frac{1}{n} \right\} \subseteq W.$$

Put $E = \{x \in X : d(x, w) \geq \frac{1}{2n}\}$. We claim that $w \in U_E \cup V_E \subseteq W$. If $w \in X$, since $V_E = X \backslash E$ and $d(w, w) = 0 < \frac{1}{2n}$, we get $w \in V_E$. On the other hand, if $w \in I$, then $d(w, E) > 0$ gives $E \in \mathscr{B}_w(\sigma)$ which means $w \in U_E$.

For the inclusion, it is obvious that $V_E \subseteq W$: if $x \in V_E = X \backslash E$, we have $d(x, w) < \frac{1}{2n} < \frac{1}{n}$. To show $U_E \subseteq W$, we show that if $i \in I \backslash W$, then $i \notin U_E$. Since $d(i, w) \geq \frac{1}{n}$ and X is dense in the extension, we can find a sequence $\langle x_k \rangle$ in X that is convergent to i such that for all k, $d(x_k, w) > \frac{1}{2n}$. By definition, each x_k lies in E, and so $d(i, E) = 0$ which means that $i \notin U_E$, as required.

Example 38.16. The classical way to present the completion of a noncomplete metric space $\langle X, d \rangle$ is to equip equivalence classes of Cauchy sequences in X with a natural metric [155, p. 176]. Two Cauchy sequences g and h in X are declared equivalent if

$$\lim_{n \to \infty} d(h(n), g(n)) = 0,$$

in which case we write $g \sim h$. If g is convergent to a point of X, then $g \sim h_a$ for some $a \in X$ where $\forall n \in \mathbb{N}$, $h_a(n) = a$. Equivalence classes of nonconvergent Cauchy sequences correspond to points of the remainder I. The (well-defined) complete metric on pairs of equivalence classes is defined by $\hat{d}(g, h) := \lim_{n \to \infty} d(h(n), g(n))$ where g and h are representatives of each equivalence class.

We can gain a qualitative understanding of the topology σ of the completion on $X \cup I$ within our framework, using Example 38.15 as a point of departure. Now if $A \subseteq X$ and $i \in I$, we claim that $A \notin \mathscr{B}_i(\sigma)$, i.e., $i \in \text{cl}_\sigma(A)$, if and only if there exists a Cauchy sequence in A in the equivalence class of i.

For necessity, let $\langle b_k \rangle$ be a sequence in X that is a representative of i. Now $A \notin \mathscr{B}_i(\sigma)$ means that for each $n \in \mathbb{N}$ we can find $a_n \in A$ such that $\hat{d}(h_{a_n}, \langle b_k \rangle) < \frac{1}{n}$. We can find a subsequence $\langle b_{k_n} \rangle$ of $\langle b_k \rangle$ such that $\forall n \in \mathbb{N}$, $d(a_n, b_{k_n}) < \frac{1}{n}$. Clearly, $\langle a_n \rangle$ is a Cauchy sequence in A equivalent to $\langle b_k \rangle$ as $\langle b_k \rangle \sim \langle b_{k_n} \rangle$.

For sufficiency, suppose $\langle a_k \rangle$ is a Cauchy sequence in A that is equivalent to $\langle b_k \rangle$. Then $\forall n \in \mathbb{N}$ $\exists m \in \mathbb{N}$ such that for all k sufficiently large, $d(a_m, b_k) < \frac{1}{2n}$. From this estimate we conclude that $\hat{d}(h_{a_m}, \langle b_k \rangle) < \frac{1}{n}$ so that $i \in \text{cl}_\sigma(A)$.

The claim established, the family $\{\mathscr{B}_i(\sigma) : i \in I\}$ can be described by the condition: $A \in \mathscr{B}_i(\sigma)$ if and only if no Cauchy sequence in A lies in the equivalence class of i. Further, for each $A \in \mathscr{C}(X)$, the basic open set in the completion topology that $E \in \mathscr{C}(X)$ determines adjoins to $X \backslash E$ all points in I whose representatives fail to be equivalent to any Cauchy sequence in E.

For an understanding of how the Stone-Čech compactification and other Wallman-type compactifications can be described as bornological extensions, we refer the interested reader to Example 4.1 and Example 4.2 of [54].

Example 38.17. Let X be an infinite set and let I be nonempty. Equipping $X \cup I$ with the cofinite topology σ [155, p. 26] produces an extension of X, the relative topology τ being the cofinite topology on X. Note that for each $i \in I$, we have

$$\mathscr{B}_i(\sigma) = \{E \subseteq X : E \text{ is finite}\} = \mathscr{C}(X) \backslash \{X\},$$

and

$$\tau_0(\sigma) = \{\emptyset, X \cup I\} \cup \{I \cup (X \backslash E) : E \subseteq X \text{ and } E \text{ is finite}\}.$$

If I is a singleton, then $X \in \sigma \backslash \tau_0(\sigma)$. On the other hand, if I contains at least two distinct members, let I_1 be a nonempty proper subset of I that is cofinite in I. Then $X \cup I_1 \in \sigma \backslash \tau_0(\sigma)$. Thus the cofinite topology on $X \cup I$ fails to be a bornological extension.

The next result tells us when a bornological extension topology for $\langle X, \tau \rangle$ contains a particular member of τ.

Proposition 38.18. *Let $\langle X, \tau \rangle$ be a topological space, and let $\langle X \cup I, \tau_0 \rangle$ be determined by a family $\{\mathscr{B}_i : i \in I\}$ of nontrivial closed ideals on X. Then a subset W of X belongs to τ_0 if and only $\forall x \in W$, $\exists E_x \in \mathscr{C}(X)$ such that $E_x \notin \cup_{i \in I} \mathscr{B}_i$ and $x \in X \backslash E_x \subseteq W$.*

Proof. We begin with necessity. If $W = \emptyset$, the condition is satisfied vacuously. Otherwise, let $x \in W$ be arbitrary. In consideration of the standard base for the extension, there exists $E_x \in \mathscr{C}(X)$ with

$$x \in U_{E_x} \cup V_{E_x} \subseteq W.$$

Since $U_{E_x} \cap X = \emptyset$, it follows that $\forall i \in I$, $E_x \notin \mathscr{B}_i$ and $x \in X \backslash E_x \subseteq W$.

For sufficiency, simply observe that for all $x \in W$, $U_{E_x} = \emptyset$, and we get

$$W = \bigcup_{x \in W} X \backslash E_x = \bigcup_{x \in W} V_{E_x} = \bigcup_{x \in W} (U_{E_x} \cup V_{E_x}) \in \tau_0. \qquad \square$$

Using this proposition we can immediately describe when a bornological extension topology contains the initial topology τ on X.

Corollary 38.19. *Let $\langle X, \tau \rangle$ be a topological space, and let $\langle X \cup I, \tau_0 \rangle$ be determined by a family $\{\mathscr{B}_i : i \in I\}$ of nontrivial closed ideals on X. The following conditions are equivalent:*

(i) $\tau \subseteq \tau_0$;

(ii) $X \in \tau_0$, *that is the remainder I is closed in the extension;*

(iii) *there exists $\{E_j : j \in J\} \subseteq \mathscr{C}(X)$ with $\cap_{j \in J} E_j = \emptyset$ and $\forall j \in J, \forall i \in I, E_j \notin \mathscr{B}_i$.*

Proof. Conditions (i) and (ii) are equivalent because τ is the relative topology on X and so $X \in \tau_0$ is equivalent to $\tau \subseteq \tau_0$. In view of Proposition 38.18, condition (ii) implies condition (iii) taking X as our index set J. Conversely, condition (iii) implies condition (ii) because $\cap_{j \in J} E_j = \emptyset$ means $\cup_{j \in J} X \backslash E_j = X$ so that each $x \in X$ belongs to $X \backslash E_j$ for some index j. $\qquad \square$

In a bornological extension of $\langle X, \tau \rangle$, it is always the case that X must be dense. We now see when the remainder is dense (note that the remainder cannot be dense if the extension topology contains any subset of X).

Proposition 38.20. *Let $\langle X, \tau \rangle$ be a topological space, and let $\langle X \cup I, \tau_0 \rangle$ be determined by a family $\{\mathscr{B}_i : i \in I\}$ of nontrivial closed ideals on X. Then I is τ_0-dense if and only if whenever $E \in \mathscr{C}(X)$ and $E \neq X$, then $\exists i \in I$ with $E \in \mathscr{B}_i$.*

Proof. For sufficiency, suppose $E \in \mathscr{C}(X)$ and $U_E \cup V_E$ is nonempty. This means that E is a proper subset of X, and choosing i with $E \in \mathscr{B}_i$, we have $i \in U_E \subseteq U_E \cup V_E$. Thus, I is dense. For necessity, suppose I is dense and E is a closed proper subset of X. Then $U_E \cup V_E$ is nonempty as $X \backslash E$ is nonempty, so for some $i \in I$, we have $i \in U_E$, i.e., $E \in \mathscr{B}_i$. $\qquad \square$

Theorem 38.21. *Let $\langle X \cup I, \sigma \rangle$ be a topological space where X and I are disjoint dense subspaces. Then $\langle X \cup I, \sigma \rangle$ is a bornological extension of both subspaces if and only if σ has a base $\mathscr{W}$ such that whenever $W \in \mathscr{W}, \bar{x} \in W \cap X$ and $\bar{i} \in W \cap I$, there exists $\{S_1, S_2\} \subseteq \sigma$ such that $\bar{x} \in (S_1 \cap X) \subseteq W, \bar{i} \in (S_2 \cap I) \subseteq W$, and*

$$\{i \in I : i \notin \mathrm{cl}_\sigma(X \backslash S_1)\} \cup \{x \in X : x \notin \mathrm{cl}_\sigma(I \backslash S_2)\} \subseteq W.$$

Proof. We write $\tau_1(\sigma)$ for the bornological extension topology determined by $\{\mathscr{B}_x(\sigma) : x \in X\}$ where $\mathscr{B}_x(\sigma) := \{E \subseteq I : x \notin \mathrm{cl}_\sigma(E)\}$. Our proof in both directions relies on our description of a local base for $\tau_0(\sigma)$, and implicitly for $\tau_1(\sigma)$.

For necessity, we show that σ itself can serve as an appropriate base $\mathscr{W}$. Let $W \in \sigma$ be arbitrary and let $\{\bar{x}, \bar{i}\} \subseteq W$. Since $\sigma \subseteq \tau_0(\sigma)$, there exists $S_1 \in \sigma$ with $\bar{x} \in S_1 \cap X \subseteq W$ such that $\{i \in I : i \notin \mathrm{cl}_\sigma(X \backslash S_1)\} \subseteq W$. Since $\sigma \subseteq \tau_1(\sigma)$, there exists $S_2 \in \sigma$ with $\bar{i} \in S_2 \cap I \subseteq W$ such that $\{x \in X : x \notin \mathrm{cl}_\sigma(I \backslash S_2)\} \subseteq W$. From the second conditions on S_1 and S_2 alone, we get

$$\{i \in I : i \notin \mathrm{cl}_\sigma(X \backslash S_1)\} \cup \{x \in X : x \notin \mathrm{cl}_\sigma(I \backslash S_2)\} \subseteq W.$$

For sufficiency, we need only produce a $\tau_0(\sigma)$-neighborhood of $\bar{x}$ lying in W and a $\tau_1(\sigma)$-neighborhood of $\bar{i}$ lying in W. We just do the latter as the proofs are essentially the same. Since $S_2 \cap I \subseteq W$ and $\{x \in X : x \notin \mathrm{cl}_\sigma(I \backslash S_2)\} \subseteq W$, we obtain

$$\bar{i} \in \{x \in X : x \notin \mathrm{cl}_\sigma(I \backslash S_2)\} \cup (S_2 \cap I) \subseteq W$$

as required. $\qquad \square$

39. When is an Extension Bornological?

As usual, let $\langle X \cup I, \sigma \rangle$ be an extension of $\langle X, \tau \rangle$, and for each $i \in I$, let $\mathscr{B}_i(\sigma)$ be the ideal with base $\{X \backslash W : W \in \sigma, i \in W\}$. Our primary goal in this section is to produce sufficient (that may also be necessary) conditions for the extension to be a bornological extension of $\langle X, \tau \rangle$. This amounts to showing that σ agrees with $\tau_0(\sigma)$ as determined by the family of ideals $\{\mathscr{B}_i(\sigma) : i \in I\}$.

Theorem 39.1. *Let $\langle X \cup I, \sigma \rangle$ be an extension of $\langle X, \tau \rangle$. The following conditions are equivalent.*

(1) $\sigma = \tau_0(\sigma)$;

(2) $\{(X \cup I) \backslash cl_\sigma(E) : E \in \mathscr{C}(X)\}$ *is a base for σ;*

(3) *whenever $C \subseteq X \cup I$ is σ-closed, $\exists \{E_\lambda : \lambda \in \Lambda\} \subseteq \mathscr{C}(X)$ such that $C = \bigcap_{\lambda \in \Lambda} cl_\sigma(E_\lambda)$;*

(4) *whenever $C \subseteq X \cup I$ is σ-closed, $C = \bigcap \{cl_\sigma(B) : B \subseteq X$ and $C \subseteq cl_\sigma(B)\}$;*

(5) *whenever $A \subseteq I$ and $p \in (X \cup I) \backslash cl_\sigma(A)$, there exists $B \subseteq X$ with $p \notin cl_\sigma(B)$ and $cl_\sigma(A) \subseteq cl_\sigma(B)$.*

Proof. By Proposition 38.7, condition (2) says that the standard base for $\tau_0(\sigma)$ is a base for σ as well. Thus, conditions (1) and (2) are equivalent. Condition (3) says that $\{cl_\sigma(E) : E \in \mathscr{C}(X)\}$ is a closed base associated with σ, and so conditions (2) and (3) are equivalent. Clearly, condition (3) implies (4), and condition (4) implies (5). It remains to prove if condition (5) holds, then (3) holds.

To this end, let $C \subseteq X \cup I$ be closed in the extension. We have the decomposition $C = cl_\sigma(C \cap X) \cup cl_\sigma(C \cap I)$. Suppose $p \notin C$; as $p \notin cl_\sigma(C \cap I)$, by condition (5), $\exists B_p \subseteq X$ with
$$cl_\sigma(C \cap I) \subseteq cl_\sigma(B_p) \text{ and } p \notin cl_\sigma(B_p).$$

For each $p \in (X \cup I) \backslash C$, put $E_p := cl_\tau((C \cap X) \cup B_p) \in \mathscr{C}(X)$. We compute
$$cl_\sigma(E_p) = cl_\sigma(C \cap X) \cup cl_\sigma(B_p) \supseteq C,$$

and so by construction, $C = \bigcap_{p \notin C} cl_\sigma(E_p)$ as required. $\square$

Theorem 39.1 appears as Theorem 5.1 in [54]. Note that M. H. Stone [141, p. 120] originally defined strictness of an extension by a condition on the structure of its open subsets that can be viewed as a translation of our criterion (4).

We intend now to apply condition (3) to obtain an attractive proof of a classical result of M. H. Stone [141, Theorem 65] asserting that any extension having a base of regular open sets is strict in his sense. An open subset V of a topological space $\langle X, \tau \rangle$ is called a *regular open set* provided $V = int(cl(V))$; dually, a closed set C is called a *regular closed set* provided $C = cl(int(C))$. An open set is a regular open set if and only if its complement is a regular closed set. From the generally valid formula

$$\mathrm{cl}(\mathrm{int}(\mathrm{cl}(V))) = \mathrm{cl}(V) \quad (V \in \tau),$$

we see that $\mathrm{int}(\mathrm{cl}(V))$ is a regular open set for any open set V. As a result, if $\langle X, \tau \rangle$ is a regular topological space, then the space has a base of regular open sets. To see this, let $x \in W \in \tau$; by regularity we can find $V \in \tau$ such that $x \in V \subseteq \mathrm{cl}(V) \subseteq W$. From this, $\mathrm{int}(\mathrm{cl}(V))$ is a regular open neighborhood of x contained in W. A space with this property is called *semiregular*. Since semiregularity is not an hereditary property of topological spaces [155, p. 98], it is a strictly weaker property than regularity which is hereditary.

Theorem 39.2. *Let $\langle X \cup I, \sigma \rangle$ be a semiregular extension of $\langle X, \tau \rangle$. Then $\sigma = \tau_0(\sigma)$.*

Proof. Let C be σ-closed. For each $p \in (X \cup I) \backslash C$ choose by semiregularity $W_p \in \sigma$ such that $W \cap C = \emptyset$ and $W_p = \mathrm{int}_\sigma(\mathrm{cl}_\sigma(W_p))$. Since the complement of a regular open set is regular closed, we have

$$(X \cup I) \backslash W_p = \mathrm{cl}_\sigma(\mathrm{int}_\sigma((X \cup I) \backslash W_p)).$$

Using the fact that X is dense in each open subset of the extension, we can assert

$$(X \cup I) \backslash W_p = \mathrm{cl}_\sigma(X \cap \mathrm{int}_\sigma((X \cup I) \backslash W_p))$$

$$= \mathrm{cl}_\sigma(\mathrm{cl}_\tau(X \cap \mathrm{int}_\sigma((X \cup I) \backslash W_p)))$$

and $\{\mathrm{cl}_\tau(X \cap \mathrm{int}_\sigma((X \cup I) \backslash W_p)) : p \notin C\}$ is the desired subfamily of $\mathscr{C}(X)$ with respect to condition (3) of Theorem 39.1. $\qquad\square$

Semiregularity is a sufficient but not necessary condition for the extension to be bornological; Example 38.9 provides us with a bornological extension that is neither semiregular nor T_1.

It is not true in general that if $X_1 \subseteq X$ is dense in $X \cup I$ where $\langle X \cup I, \sigma \rangle$ is a bornological extension of $\langle X, \tau \rangle$, that $\langle X \cup I, \sigma \rangle$ need be a bornological extension of $\langle X_1, \mu \rangle$ where μ is the relative topology on X_1 that it inherits from σ [54, Example 5.12].

Example 39.3. Perhaps the best known example of a Hausdorff space that is not regular [155, Example 14.2] is the real line $\mathbb{R}$ equipped with the topology σ generated by the usual topology and the set $\mathbb{R} \backslash \{1/n : n \in \mathbb{N}\}$. Let $X = (-\infty, 0) \cup (0, \infty)$ equipped with the relative topology τ it inherits from σ. As $\langle \mathbb{R}, \sigma \rangle$ is a Hausdorff one-point extension of $\langle X, \tau \rangle$, by Corollary 38.11 it is a bornological extension. Of course, τ is just the relative topology that X inherits from the usual topology on $\mathbb{R}$, so that τ is a metrizable topology. As a result, $\langle X, \tau \rangle$ is a bornological extension of each of its dense subspaces.

Let $X_1 = X \backslash \{1/n : n \in \mathbb{N}\}$ equipped with relative topology μ. Viewing $\langle \mathbb{R}, \sigma \rangle$ as an extension of $\langle X_1, \mu \rangle$ as a dense subspace, it is clear that criterion (4) Theorem 39.1 fails for the σ-closed set $C = \{1/n : n \in \mathbb{N}\}$, for if $B \subseteq X_1$ with $C \subseteq \mathrm{cl}_\sigma(B)$, then $0 \in \mathrm{cl}_\sigma(B)$.

Our construction additionally shows that a bornological extension of a bornological extension need not be a bornological extension of the initial space!

A good enough reason to include criterion (5) of Theorem 39.1 is because it easily leads to a simple answer to a natural question: when is a topological space a bornological extension of each of its proper dense subspaces?

Theorem 39.4. *Let $\langle Y, \sigma \rangle$ be a topological space. Then $\langle Y, \sigma \rangle$ is a bornological extension of each of its proper dense subspaces if and only if whenever I is a nonempty subset of Y with empty interior, $A \subseteq I$ and $p \in Y \backslash cl_\sigma(A)$, there exists $B \subseteq Y \backslash I$ with $p \notin cl_\sigma(B)$ and $cl_\sigma(A) \subseteq cl_\sigma(B)$.*

Proof. Let X be a proper subset of Y and put $I = Y \backslash X \neq \emptyset$. The density of X in Y is equivalent to $\text{int}_\sigma(I) = \emptyset$. The result immediately follows from criterion (5) of Theorem 5.1 for a space to be a bornological extension of a proper dense subspace X. $\square$

Our next result is also most easily obtained from criterion (5); the routine but notationally cumbersome proof is left to the reader.

Theorem 39.5. *Let $\langle X \cup I, \sigma \rangle$ be a bornological extension of $\langle X, \tau \rangle$ and let I_1 be a nonempty subset of I. Denote the relative topology on $X \cup I_1$ by σ_1. Then $\langle X \cup I_1, \sigma_1 \rangle$ is a bornological extension of $\langle X, \tau \rangle$.*

The following result allows us to produce an example of a Hausdorff extension that is not bornological.

Proposition 39.6. *Let $\langle X \cup I, \sigma \rangle$ and $\langle X \cup I, \mu \rangle$ be extensions of $\langle X, \tau \rangle$. Suppose for each $i \in I$, the relative topology that $X \cup \{i\}$ inherits from σ coincides with the one inherited from μ. Then $\tau_0(\sigma) = \tau_0(\mu)$. Moreover, if $\langle X \cup I, \sigma \rangle$ is a bornological extension, then $\sigma \subseteq \mu$.*

Proof. Let $i \in I$ be arbitrary. By definition, a closed subset E of X is in $\mathscr{B}_i(\sigma)$ if and only if $i \notin cl_\sigma(E)$, that is, if and only if E is closed in $X \cup \{i\}$ equipped with the relative topology. Thus if σ and μ induce the same relative topologies on $X \cup \{i\}$, then $\mathscr{B}_i(\sigma) \cap \mathscr{C}(X) = \mathscr{B}_i(\mu) \cap \mathscr{C}(X)$. Therefore, under our assumption, $\tau_0(\sigma) = \tau_0(\mu)$.

If σ is a bornological extension topology, by Corollary 38.8, we obtain

$$\sigma = \tau_0(\sigma) = \tau_0(\mu) \subseteq \mu. \qquad \square$$

The next corollary speaks to the importance of one-point extensions in the study of general bornological extensions.

Corollary 39.7. *Let $\langle X \cup I, \sigma \rangle$ and $\langle X \cup I, \mu \rangle$ be bornological extensions of $\langle X, \tau \rangle$. Suppose for each $i \in I$, the relative topology that $X \cup \{i\}$ inherits from σ coincides with the one inherited from μ. Then $\sigma = \mu$.*

Example 39.8. Let X be the open upper half plane in $\mathbb{R} \times \mathbb{R}$, and let I be the x-axis. Let σ be the Euclidean topology on $X \cup I$ and let τ be the relative topology on X. For every $i \in I$, and $r \in (0, \infty)$, put

$$U_{i,r} := \{x \in X : d(x, i) < r\},$$

where d is the Euclidean metric. Then the family of subsets of $X \cup I$

$$\tau \cup \{\{i\} \cup U_{i,r} : i \in I, \ r > 0\}$$

is clearly a base for a strictly finer Hausdorff topology μ on $X \cup I$ which induces the discrete topology on I and which is also an extension topology for $\langle X, \tau \rangle$. For each $i \in I$, $\{\{i\} \cup U_{i,r} : i \in I, \ r > 0\}$ forms a local base at i for both relative topologies on $X \cup \{i\}$ and as $\tau \subseteq \sigma \cap \mu$, the one-point extensions agree. By metrizability of σ, we have $\tau_0(\sigma) = \sigma$. From Proposition 39.6 we obtain $\tau_0(\mu) = \tau_0(\sigma) = \sigma$. Hence, $\langle X \cup I, \mu \rangle$ while a Hausdorff extension is not a bornological extension.

With Theorem 39.5 in mind, Example 39.8 also shows that an extension can fail to be bornological even if all of the one-point extensions that it gives rise to as subspaces are bornological extensions.

We close this sections by displaying characteristic conditions for a bornological extension to be T_1, T_2 (Hausdorff) and T_3 (regular).

Proposition 39.9. *Let $\langle X, \tau \rangle$ be a T_1-space and let $\{\mathscr{B}_i : i \in I\}$ be a family of nontrivial closed ideals on X. Put $\tau_0 = \tau_0(\{\mathscr{B}_i : i \in I\})$. Then $\langle X \cup I, \tau_0 \rangle$ is T_1 if and only if*

(1) *each $\mathscr{B}_i$ is a bornology;*

(2) *$\mathscr{B}_i \nsubseteq \mathscr{B}_j$ for $i \neq j$;*

(3) *for every $x \in X$ and $i \in I$, there exists $F = F_{x,i} \in \mathscr{C}(X)$ such that $F \notin \mathscr{B}_i$ and $x \notin F$.*

Proof. Assume $\langle X \cup I, \tau_0 \rangle$ is T_1. Each $\mathscr{B}_i$ is already hereditary and stable under finite unions. Each $\mathscr{B}_i$ contains each singleton subset of X, because for each $x \in X$ and $i \in I$, we can find $F \in \mathscr{C}(X)$ such that $i \in U_F$ but $x \notin V_F = X\backslash F$. This means that $F \in \mathscr{B}_i$ and $x \in F$ so that $\{x\} \in \mathscr{B}_i$. To prove (2), suppose for some $i \neq j$ we have $\mathscr{B}_i \subseteq \mathscr{B}_j$. If $U_E \cup V_F$ is a basic neighborhood of i, then $E \in \mathscr{B}_j$ and so $j \in U_E \subseteq U_E \cup V_E$. Thus each neighborhood of i contains j, a contradiction. As for (3), if $x \in X$ and $i \in I$, we can find a basic neighborhood of x that does not contain i. This means that there is a closed subset F of X such that $i \notin U_F \cup V_F$ and $x \in U_F \cup V_F$, that is, $F \notin \mathscr{B}_i$ and $x \notin F$.

Conversely, assume the three conditions hold. We consider cases on distinct points of $X \cup I$. We need only look at distinct points where at least one lies in I because the base space is assumed T_1. If $x \in X$ and $i \in I$, then by condition (1), $i \in U_{\{x\}} \cup V_{\{x\}}$ which does not contain x. On the other hand, taking $F = F_{x,i}$, one has $x \in U_F \cup V_F$ which does not contain i. Finally, if $i, j \in I$ with $i \neq j$, since $\mathscr{B}_i \nsubseteq \mathscr{B}_j$ and $\mathscr{B}_i$ is closed, we can find $E \in \mathscr{C}(X)$ with $E \in \mathscr{B}_i\backslash\mathscr{B}_j$. Then $i \in U_E \cup V_E$ and $j \notin U_E \cup V_E$. Producing a neighborhood of j that does not contain i is done in the same way. □

Proposition 39.10. *Let $\langle X, \tau \rangle$ be a Hausdorff space and let $\{\mathscr{B}_i : i \in I\}$ be a family of nontrivial closed ideals on X. Put $\tau_0 = \tau_0(\{\mathscr{B}_i : i \in I\})$. The following conditions are equivalent:*

(1) *$\langle X \cup I, \tau_0 \rangle$ is Hausdorff;*

(2) *each $\mathscr{B}_i$ is a local bornology, and for distinct i, j in I, there exists $E \in \mathscr{B}_i$, $F \in \mathscr{B}_j$ with $E \cup F = X$.*

Proof. (1) $\Rightarrow$ (2). Assume that the extension is Hausdorff. By the last proposition, each $\mathscr{B}_i$ is a bornology. We now prove that each $\mathscr{B}_i$ is local. Let $x \in X$. There exist $E \in \mathscr{C}(X)$ with $x \in U_E \cup V_E$ and $F \in \mathscr{C}(X) \cap \mathscr{B}_i$ such that $(U_E \cup V_E) \cap (U_F \cup V_F) = \emptyset$. This implies $V_E \subseteq F$, and so $x \in V_E \in \mathscr{B}_i \cap \tau$. Finally, let i and j be distinct points of the remainder and let $U_E \cup V_E$ and $U_F \cup V_F$ be disjoint basic neighborhoods of i and j, respectively. Then $E \in \mathscr{B}_i$, $F \in \mathscr{B}_j$ and $V_E \cap V_F = \emptyset$, that is, $E \cup F = X$.

(2) $\Rightarrow$ (1). First, let $x, y \in X$ with $x \neq y$. Choose A and B to be disjoint open subsets of X containing x and y, respectively. Putting $E = X \backslash A$ and $F = X \backslash B$, one has $E \cup F = X$. We now obtain

$$(U_E \cup V_E) \cap (U_F \cup V_F) = (U_E \cup A) \cap (U_F \cup B)$$
$$= (U_E \cap U_F) \cup (A \cap B) = U_{E \cup F} \cup \emptyset$$
$$= U_X \cup \emptyset = \emptyset,$$

and we have produced disjoint τ_0-neighborhoods of x and y.

Next, let $x \in X$ and $i \in I$. Since $\mathscr{B}_i$ is local, we can find $W \in \tau$ with $x \in W \in \mathscr{B}_i$. Put $E = X \backslash W$ and $F = \mathrm{cl}_\tau(W)$. Since the bornology is closed, we have $F \in \mathscr{B}_i$. Since $E \cup F = X$, again one has $U_E \cap U_F = \emptyset$, and $V_E \cap V_F = X \backslash (E \cup F) = \emptyset$. Clearly $x \in U_E \cup V_E$ and $i \in U_F \cup V_F$, and these neighborhoods are disjoint.

Finally, let $i, j \in I$ with $i \neq j$. Since $\mathscr{B}_i$ and $\mathscr{B}_j$ are closed bornologies, we can choose $E \in \mathscr{C}(X) \cap \mathscr{B}_i$ and $F \in \mathscr{C}(X) \cap \mathscr{B}_j$ such that $E \cup F = X$. Then $U_E \cup V_E$ and $U_F \cup V_F$ are disjoint members of τ_0 containing i and j, respectively. $\qquad \square$

Proposition 39.11. *Let $\langle X, \tau \rangle$ be a regular space and let $\{ \mathscr{B}_i : i \in I \}$ be a family of closed nontrivial ideals on X. Then the bornological extension $\langle X \cup I, \tau_0 \rangle$ is regular if and only if the following three conditions all hold:*

(1) *for every $x \in X$ and for every $E \in \mathscr{C}(X)$ such that $x \notin E$, there is a τ-neighborhood $W_{x,E}$ of x such that $\forall i \in I$, $W_{x,E} \notin \mathscr{B}_i \Rightarrow E \in \mathscr{B}_i$;*

(2) *each ideal $\mathscr{B}_i$ has an open base;*

(3) *for every $i \in I$ and for every closed $E \in \mathscr{B}_i$, there is a family $\{ A_\lambda : \lambda \in \Lambda \}$ of closed subsets of X such that $X \setminus [\bigcap_{\lambda \in \Lambda} A_\lambda] \in \mathscr{B}_i$, and $\forall j \in I$, $\{ A_\lambda : \lambda \in \Lambda \} \cap \mathscr{B}_j = \emptyset \Rightarrow E \in \mathscr{B}_j$.*

Proof. We first prove necessity. Let $x \in X$ and $E \in \mathscr{C}(X)$ with $x \notin E$. By regularity of the extension, there exists a τ_0-neighborhood W of x such that $\mathrm{cl}_{\tau_0}(W) \cap \mathrm{cl}_{\tau_0}(E) = \emptyset$. Put $W_{x,E} = W \cap X$. If $W_{x,E} \notin \mathscr{B}_i$ then $i \in \mathrm{cl}_{\tau_0}(W_{x,E}) \subseteq \mathrm{cl}_{\tau_0}(W)$. Then $i \notin \mathrm{cl}_{\tau_0}(E)$, that is, $E \in \mathscr{B}_i$. Thus, condition (1) holds.

To prove (2), let $E \in \mathscr{B}_i$. Then there exists an open τ_0-neighborhood W of i disjoint from E. By regularity, there is $W_1 \in \tau_0$ such that $i \in W_1 \subseteq \mathrm{cl}_{\tau_0}(W_1) \subseteq W$. Then one has

$$E \subseteq X \setminus W \subseteq X \setminus \mathrm{cl}_{\tau_0}(W_1) \subseteq X \setminus W_1.$$

Clearly $X \setminus W_1 \in \mathscr{B}_i$; hence $X \setminus \mathrm{cl}_{\tau_0}(W_1)$ is an open member of $\mathscr{B}_i$ containing E.

Now we prove (3). Let $i \in I$ and let $E \in \mathscr{B}_i$. The neighborhood $U_E \cup V_E$ of i must contain a closed neighborhood. So there is a τ_0-closed subset C of $X \cup I$ and a closed member F of $\mathscr{B}_i$ such that

$$i \in U_F \cup V_F \subseteq C \subseteq U_E \cup V_E.$$

By condition (3) of Theorem 39.1, we can find $\{ A_\lambda : \lambda \in \Lambda \} \subseteq \mathscr{C}(X)$ with $C = \bigcap_{\lambda \in \Lambda} \mathrm{cl}_{\tau_0}(A_\lambda)$, and so

$$U_F \cup V_F \subseteq \bigcap_{\lambda \in \Lambda} \mathrm{cl}_{\tau_0}(A_\lambda) \subseteq U_E \cup V_E.$$

Taking the intersections with X, we obtain

$$V_F \subseteq \bigcap_{\lambda \in \Lambda} A_\lambda \subseteq V_E.$$

Taking complements, $X \setminus (\bigcap_{\lambda \in \Lambda} A_\lambda) \subseteq F$ and this implies $X \setminus (\bigcap_{\lambda \in \Lambda} A_\lambda) \in \mathscr{B}_i$.

For the last part of condition (3), suppose $\mathscr{B}_j$ does not contain A_λ for any λ. Then

$$j \in \left(\bigcap_{\lambda \in \Lambda} \mathrm{cl}_{\tau_0}(A_\lambda) \right) \cap I \subseteq U_E,$$

which means $E \in \mathscr{B}_j$ as required. We have proved (3).

We turn to sufficiency of the three conditions. Let $x \in X$ and suppose $U_E \cup V_E$ is a τ_0-basic open neighborhood of x. By regularity of $\langle X, \tau \rangle$, let W be a τ-open neighborhood of x such that $\mathrm{cl}_\tau(W) \subseteq W_{x,E} \cap V_E$, where $W_{x,E}$ is chosen to satisfy condition (1). We put $T = X \setminus W$ and $F = \mathrm{cl}_\tau(W)$. Clearly $U_T \cup V_T$ is a neighborhood of x. It suffices to prove that

$$U_T \cup V_T \subseteq \mathrm{cl}_{\tau_0}(F) \subseteq U_E \cup V_E.$$

For the inclusion $U_T \cup V_T \subseteq \mathrm{cl}_{\tau_0}(F)$, clearly, $V_T = W \subseteq F \subseteq \mathrm{cl}_{\tau_0}(F)$. If $i \in U_T$, then $T \in \mathscr{B}_i$; being that $T \cup F = X$, $F \notin \mathscr{B}_i$. This implies $i \in \mathrm{cl}_{\tau_0}(F)$. For the second inclusion, if $x \in \mathrm{cl}_{\tau_0}(F) \cap X$, then $x \in \mathrm{cl}_\tau(W) \subseteq V_E$. If $i \in \mathrm{cl}_{\tau_0}(F) \cap I$, then $F \notin \mathscr{B}_i$. Since $F \subseteq W_{x,E}$, this implies that $W_{x,E} \notin \mathscr{B}_i$ as well. Hence, by condition (1), we have $E \in \mathscr{B}_i$, that is, $i \in U_E$.

We finally consider points of the remainder which proves to be quite delicate. Let $i \in I$ and suppose $U_E \cup V_E$ is a basic open neighborhood of i. Let $\{A_\lambda\}_{\lambda \in \Lambda}$ satisfy condition (3) (with respect to i and E) and put $W = X \setminus (\bigcap_{\lambda \in \Lambda} A_\lambda)$. By condition (2), let W_1 be an open member of $\mathscr{B}_i$ containing E and put $F = \mathrm{cl}_\tau(W \cup W_1) \in \mathscr{B}_i$. By definition, $U_F \cup V_F$ is a τ_0-neighborhood of i. Put

$$C := \left[\mathrm{cl}_{\tau_0} \left(\bigcap_{\lambda \in \Lambda} A_\lambda \right) \right] \cap \mathrm{cl}_{\tau_0}(X \setminus W_1).$$

Evidently, C is τ_0-closed. We will be done if we can verify that

$$U_F \cup V_F \subseteq C \subseteq U_E \cup V_E.$$

For the first inclusion, let $x \in V_F = X \setminus F$. Clearly $x \notin W \cup W_1$; hence $x \in [\bigcap_{\lambda \in \Lambda} A_\lambda] \cap (X \setminus W_1) \subseteq C$. If $j \in U_F$, then $F \in \mathscr{B}_j$. Now by the definition of W, for each $\lambda \in \Lambda$, F contains $X \setminus A_\lambda$. As a result, $A_\lambda \cup F = X$ for every λ, and so $A_\lambda \notin \mathscr{B}_j$. This implies $\forall \lambda \in \Lambda$, $j \in \mathrm{cl}_{\tau_0}(A_\lambda)$. Similarly, $F \cup (X \setminus W_1) = X$ and $F \in \mathscr{B}_j$ jointly imply $X \setminus W_1 \notin \mathscr{B}_j$; hence, $j \in \mathrm{cl}_{\tau_0}(X \setminus W_1)$ for each $j \in U_F$.

For the inclusion $C \subseteq U_E \cup V_E$, first let $x \in C \cap X$. Then $x \notin W_1$ and as $E \subseteq W_1$, $x \in V_E$. If $j \in C \cap I$, then in particular $j \in \bigcap_{\lambda \in \Lambda} \mathrm{cl}_{\tau_0}(A_\lambda)$. This implies $\forall \lambda \in \Lambda$, $A_\lambda \notin \mathscr{B}_j$. Applying condition (3), $E \in \mathscr{B}_j$ and this implies $j \in U_E$. $\square$

References

1. M. Aggarwal, S. Hasra and S. Kundu. Finitely chainable and totally bounded metric spaces: Equivalent characterizations. Top. Appl. **216**(2017): 59–73.
2. M. Aggarwal and S. Kundu. More about the cofinally complete spaces and the Atsuji spaces. Houston J. Math **42**(2016): 1373–1395.
3. M. Aggarwal and S. Kundu. Cauchy metrizability of bornological universes. J. Convex Anal. **24**(2017): 1085–1098.
4. M. Aggarwal and S. Kundu. More on variants of complete metric spaces. Acta Math. Hungarica **151**(2017): 391–405.
5. F. Albiac and N. Kalton. Topics in Banach Spaces. Springer, New York (2006).
6. R. Arens and J. Eells. On embedding uniform and topological spaces. Pacific J. Math. **6**(1956): 397–403.
7. M. Atsuji. Uniform continuity of continuous functions of metric spaces. Pacific J. Math. **8**(1958): 11–16.
8. H. Attouch, R. Lucchetti and R. Wets. The topology of the ρ-Hausdorff distance. Ann. Mat. Pura Appl. **160**(1991): 303–320.
9. P. Alexandroff and P. Urysohn. Une condition nécéssaire et suffisante pour qu'une classe (L) soit une classe (D). C. R. Acad. Sci. Paris **177**(1923): 1274–1277.
10. B. Banaschewski. Extensions of topological spaces. Canad. Math. Bull. **7**(1964): 1–22.
11. G. Beer. Metric spaces on which continuous functions are uniformly continuous and Hausdorff distance. Proc. Amer. Math. Soc. **95**(1985): 653–658.
12. G. Beer. More about metric spaces on which continuous functions are uniformly continuous. Bull. Austral. Math. Soc. **33**(1986): 397–406.
13. G. Beer. UC spaces revisited. Amer. Math. Monthly **95**(1988): 737–739.
14. G. Beer. Conjugate convex functions and the epi-distance topology. Proc. Amer. Math. Soc. **108**(1990): 117–126.
15. G. Beer. A second look at set convergence and linear analysis. Rend. Sem. Math. Fis. Milano **59**(1990): 161–172.
16. G. Beer. A Polish topology for the closed subsets of a Polish space. Proc. Amer. Math. Soc. **113**(1991): 1123–1133.
17. G. Beer. Topologies on Closed and Closed Convex Sets. Kluwer Academic Publishers, Dordrecht, Holland (1993).
18. G. Beer. On metric boundedness structures. Set-valued Anal. **7**(1999): 195–208.
19. G. Beer. On convergence to infinity. Monat. Math. **129**(2000): 267–280.
20. G. Beer. Between compactness and completeness. Top. Appl. **155**(2008): 503–514.
21. G. Beer. Product metrics and boundedness. Appl. Gen. Top. **9**(2008): 133–142.
22. G. Beer. Operator topologies and graph convergence. J. Convex Anal. **16**(2009): 687–698.
23. G. Beer. Embeddings of bornological universes. Set-Valued Anal. **16**(2008): 477–488.

24. G. Beer. Between the cofinally complete spaces and the UC spaces. Houston J. Math. **38**(2012): 999–1015.

25. G. Beer. McShane's extension theorem revisited. Vietnam J. Math. **48**(2020): 237–246.

26. G. Beer and J. Cao. Oscillation revisited. Set-Valued Var. Anal. **20**(2017): 603–616.

27. G. Beer, A. Caserta, G. Di Maio and R. Lucchetti. Convergence of partial maps. J. Math. Anal. Appl. **419**(2014): 1274–1289.

28. G. Beer, C. Costantini and S. Levi. Total boundedness in metrizable spaces. Houston J. Math. **37**(2011): 1347–1362.

29. G. Beer, C. Costantini and S. Levi. Bornological convergence and shields. Mediterr. J. Math. **10**(2013): 529–560.

30. G. Beer and A. DiConcilio. Uniform convergence on bounded sets and the Attouch-Wets topology. Proc. Amer. Math. Soc. **112**(1991): 235–243.

31. G. Beer and G. Di Maio. Cofinal completeness of the Hausdorff metric topology. Fund. Math. **208**(2010): 75–85.

32. G. Beer and G. Di Maio. The bornology of cofinally complete subsets. Acta Math. Hungar. **134**(2012): 322–343.

33. G. Beer, L.C. García-Lirola and M.I. Garrido. Stability of Lipschitz-type functions under pointwise product and reciprocation. Rev. R. Acad. Cienc. Exactas Fis. Nat. Ser. A Mat. **114**(2020).

34. G. Beer and M.I. Garrido. Bornologies and locally Lipschitz functions. Bull. Australian. Math. Soc. **90**(2014): 257–263.

35. G. Beer and M.I. Garrido. Locally Lipschitz functions, cofinal completeness, and UC spaces. J. Math. Anal. Appl. **428**(2015): 804–816.

36. G. Beer and M.I. Garrido. On the uniform approximation of Cauchy continuous functions. Top. Appl. **208**(2016): 1–9.

37. G. Beer and M.I. Garrido. Real-valued Lipschitz functions and metric properties of functions. J. Math. Anal. Appl. **486**(2020) (no. 1): 123839.

38. G. Beer, M.I. Garrido and A.S. Meroño. Uniform continuity and a new bornology for a metric space. Set-Valued Var. Anal. **26**(2018): 49–65.

39. G. Beer, C. Himmelberg, K. Prikry and F. Van Vleck. The locally finite topology on 2^X. Proc. Amer. Math. Soc. **101**(1987): 168–171.

40. G. Beer and M. Hoffman. The Lipschitz metric for real-valued continuous functions. J. Math. Anal. Appl. **406**(2013): 229–236.

41. G. Beer, A. Lechicki, S. Levi and S. Naimpally. Distance functionals and suprema of hyperspace topologies. Ann. Mat. Pura Appl. **162**(1992): 367–381.

42. G. Beer and S. Levi. Pseudometrizable bornological convergence is Attouch-Wets convergence. J. Convex Anal. **15**(2008): 439–453.

43. G. Beer and S. Levi. Gap, excess, and bornological convergence. Set-valued Anal. **16**(2008): 489–506.

44. G. Beer and S. Levi. Total boundedness and bornologies. Top. Appl. **156**(2009): 1271–1288.

45. G. Beer and S. Levi. Strong uniform continuity. J. Math. Anal. Appl. **350**(2009): 568–589.

46. G. Beer and S. Levi. Uniform continuity, uniform convergence, and shields. Set-Valued and Variational Anal. **18**(2010): 251–275.

47. G. Beer and S. Naimpally. Uniform continuity of a product of real functions. Real Anal. Exch. **37**(2011–12): 1–8.

48. G. Beer, S. Naimpally and J. Rodríguez-López. $\mathscr{S}$-topologies and bounded convergences. J. Math. Anal. Appl. **339**(2008): 542–552.

49. G. Beer and M. Segura. Well-posedness, bornologies and the structure of metric spaces. Applied Gen. Top. **10**(2009): 131–157.

50. G. Beer and M.C. Vipera. The Alexandroff one-point compactification as a prototype for extensions. Adv. in Math. **231**(2012):1598–1618.

51. Y. Benyamini and J. Lindenstrauss. Geometric nonlinear functional analysis vol. 1. American Math. Soc. Colloquium Publications vol. 48, Providence (2000).

52. J. Borsik. Mappings that preserve Cauchy sequences. Časopis Pro Pěstování Mat. **113**(1988): 280–285.

53. J. Borwein and A. Lewis. Convex Analysis and Nonlinear Optimization. CMS Books in Mathematics, Springer, New York (2006).

54. J. Borwein and J. Vanderwerff. Epigraphical and uniform convergence of convex functions. Trans. Amer. Math. Soc. **348**(1996):1617–1631.

55. N. Bouleau. Une structure uniforme sur un espace F(E, F). Cahiers Topologie Géom. Diff. **11**(1969): 207–214.

56. N. Bourbaki. Elements of Mathematics, General Topology, Part 1. Hermann, Paris (1966).

57. A. Bouziad and E. Sukhacheva. Preservation of uniform continuity under pointwise product. Top. Appl. **254**(2019): 132–144.

58. B. Braga. Coarse and uniform embeddings. J. Functional Anal. **272**(2017): 1852–1875.

59. B. Burdick. Local compactness of hyperspaces. Ann. New York Acad. Sci. **704**(1993): 28–33.

60. B. Burdick. On linear cofinal completeness. Top. Proc. **25**(2000): 435–455.

61. J. Cabello-Sánchez. $\mathscr{U}(X)$ as a ring for metric spaces X. Filomat **31**(2017): 1981–1984.

62. A. Caserta, G. DiMaio and L. Holá. Arzela's theorem and strong uniform convergence on bornologies. J. Math. Anal. Appl. **371**(2010): 384–392.

63. A. Caterino and S. Guazzone. Extensions of unbounded topological spaces. Rend. Sem. Mat. Univ. Padova **100**(1998): 123–135.

64. A. Caterino, T. Panduri and M. Vipera. Boundedness, one-point extensions, and B-extensions. Math. Slovaca **58**(2008): 101–114.

65. S. Cobzaş, R. Miculescu and A. Nicolae. Lipschitz Functions. Springer, Cham, Switzerland (2019).

66. S. Cobzaş and C. Mustăţa. Norm preserving extensions of convex Lipschitz functions. J. Approx. Theory **24**(1978): 236–244.

67. P. Corazza. Introduction to metric-preserving functions. Amer. Math. Monthly **106**(1999): 309–323.

68. H. Corson. The determination of paracompactness by uniformities. Amer. J. Math. **80**(1958): 185–190.

69. C. Costantini. Every Wijsman topology relative to a Polish space is Polish. Proc. Amer. Math. Soc. **123**(1995): 2569–2574.

70. C. Costantini, S. Levi and J. Zieminska. Metrics that generate the same hyperspace convergence. Set-valued Anal. **1**(1993): 141–157.

71. C. Czipszer and L. Gehér. Extensions of functions satisfying a Lipschitz condition. Acta Math. Acad. Sci. Hung. **6**(1955): 213–220.

72. G. Di Maio, E. Meccariello and S. Naimpally. Decompostions of UC spaces. Questions and Answers in Gen. Top. **22**(2004): 13–22.

73. R. Doss. On uniformly continuous functions in metric spaces. Proc. Math. Phys. Soc. Egypt **3**(1947): 1–6.

74. J. Dugundji. Topology. Allyn and Bacon, Boston (1966).

75. N. Dunford and J. Schwartz. Linear Operators Part I. Wiley Interscience, New York (1988).

76. I. Ekeland and R. Témam. Convex Analysis and Variational Problems. North-Holland, Amsterdam (1976).

77. V. Efremovič. The geometry of proximity I. Math. Sbornik **31**(1952): 189–200.

78. R. Engelking. General Topology. Polish Scientific Publishers, Warsaw (1977).

79. J. Fell. A Hausdorff topology for the closed subsets of a locally compact non-Hausdorff space. Proc. Amer. Math. Soc. **13**(1962): 472–476.

80. J. Fried. Open cover of a metric space admits ℓ_∞-partition of unity. Rend. Circ. Mat. Palermo **3**(1984): 139–140.

224   Bornologies and Lipschitz Analysis

81. Z. Frolik. Existence of ℓ_∞ partitions of unity. Rend. Sem. Mat. Univ. Politech. Torino **42**(1984): 9–14.
82. A. García-Máynez and S. Romaguera. Perfect pre-images of cofinally complete metric spaces. Comment. Math. Univ. Carolinae **40**(1999): 335–342.
83. M.I. Garrido and J. Jaramillo. Homomorphisms on function lattices. Monat. Math. **141**(2004): 127–146.
84. M.I. Garrido and J. Jaramillo. Lipschitz-type functions on metric spaces. J. Math. Anal. Appl. **340**(2008): 282–290.
85. M.I. Garrido and A.S. Meroño. Uniformly metrizable bornologies. J. Convex Anal. **20**(2013): 285–299.
86. M.I. Garrido and A.S. Meroño. New types of completeness in metric spaces. Ann. Acad. Sci. Fenn. Math. **39**(2014): 733–758.
87. M.I. Garrido and A.S. Meroño. Two classes of metric spaces. Appl. Gen. Top. **17**(2016): 57–70.
88. M.I. Garrido and A.S. Meroño. The Samuel real compactification of a metric space. J. Math. Anal. Appl. **456**(2017): 1013–1039.
89. J. Giles. Introduction to the Analysis of Normed Linear Spaces. Cambridge University Press, Cambridge (2000).
90. R. Goldberg. Methods of Real Analysis. 2nd Edition, Wiley, New York (1976).
91. F. Hausdorff. Erweiterung einer Homöomorphie. Fund. Math. **16**(1930): 353–360.
92. J. Heinonen. Lectures on Analysis in Metric Spaces. Springer, New York (2001).
93. J. Hejcman. Boundedness in uniform spaces and topological groups. Czech. Math. J. **9**(1959): 544–563.
94. J. Hejcman. On simple recognizing of bounded sets. Comment. Math. Univ. Carol. **38**(1997): 149–156.
95. C. Hess. Contributions à l'étude de la mesurabilité, de la loi de probabilité, et de la convergence des multifunctions. Thèse d'état, Université Montpellier II (1986).
96. N. Hindman. Basically bounded sets and a generalized Heine-Borel theorem. Amer. Math. Monthly **80**(1973): 549–552.
97. H. Hogbe-Nlend. Bornologies and Functional Analysis. North-Holland, Amsterdam (1977).
98. A. Hohti. On uniform paracompactness. Ann. Acad. Sci. Fenn. Series A Math. Diss. **36**(1981): 1–46.
99. L. Holá. Complete metrizability of topologies of strong uniform convergence on bornologies. J. Math. Anal. Appl. **387**(2012): 770–775.
100. N. Howes. On completeness. Pacific J. Math. **38**(1971): 431–440.
101. N. Howes. Modern Analysis and Topology. Springer, New York (1995).
102. S.-T. Hu. Boundedness in a topological space. J. Math Pures Appl. **228**(1949): 287–320.
103. S.-T. Hu. Introduction to General Topology. Holden-Day, San Francisco (1966).
104. T. Jain and S. Kundu. Atsuji spaces: Equivalent conditions. Topology Proc. **30**(2006): 301–325.
105. T. Jain and S. Kundu. Atsuji completions: Equivalent characterizations. Top. Appl. **154**(2007): 28–38.
106. N. Kalton. The nonlinear geometry of Banach spaces. Rev. Mat. Complut. 21(2008): 7–60.
107. J. Kelley. General Topology. Van Nostrand, Princeton, New Jersey (1955).
108. V. Klee. Some topological properties of convex sets. Trans. Amer. Math. Soc. **78**(1955): 30–45.
109. E. Klein and A. Thompson. Theory of Correspondences. Wiley, New York (1984).
110. G. Köthe. Topological Vector Spaces I. Springer, New York (1969).
111. K. Kuratowski. Topology, vol. 1. Academic Press, New York (1966).
112. A. Lechicki and S. Levi. Wijsman convergence in the hyperspace of a metric space. Bull. Un. Math. Ital. (7) 1-B(1987): 439–452.
113. A. Lechicki, S. Levi and A. Spakowski. Bornological convergences. J. Math. Anal. Appl. **297**(2004): 751–770.

114. D. Leung and W.-K. Tang. Functions that are Lipschitz in the small. Rev. Mat. Complut. **30**(2017): 25–34.

115. E. Lowen-Colebunders. Function Classes of Cauchy Continuous Functions. Marcel Dekker, New York (1989).

116. J. Luukkainen. Rings of functions in Lipschitz topology. Ann. Acad. Scient. Fenn., Series A. I. Math. **4**(1978–1979): 119–135.

117. J. Luukkainen and J. Väisälä. Elements of Lipschitz topology. Ann. Acad. Scient. Fenn., Series A. I. Math. **3**(1977): 85–122.

118. G. Marino. When is a continuous function Lipschitzian? Extracta Math. **13**(1998): 107–110.

119. G. Marino, G. Lewicki and P. Pietramala. Finite chainability, locally Lipschitzian and uniformly continuous functions. Z. Anal. Anwend. **17**(1998): 795–803.

120. E. McShane. Extension of range of functions. Bull. Amer. Math. Soc. **40**(1934): 837–842.

121. E. Michael. Topologies on spaces of subsets. Trans. Amer. Math. Soc. **71**(1951): 151–182.

122. E. Michael. A short proof of the Arens-Eells embedding theorem. Proc. Amer. Math. Soc. **15**(1964): 415–416.

123. R. Miculescu. Approximation of continuous functions by Lipschitz functions. Real Anal. Exch. **26**(2000–2001): 449–452.

124. A.A. Monteiro and M.M. Peixoto. Le nombre de Lebesgue et la continuité uniform. Portugaliae Math. **10**(1951): 105–113.

125. S. Mrowka. On the convergence of nets of sets. Fund. Math. **45**(1958): 237–246.

126. S. Mrowka. On normal metrics. Amer. Math. Monthly **72**(1965): 998–1001.

127. S. Nadler and T. West. A note on Lebesgue spaces. Topology Proc. **6**(1981): 363–369.

128. J. Nagata. On the uniform topology of bicompactifications. J. Inst. Polytech. Osaka City Univ. **1**(1950): 28–38.

129. S. Naimpally. Proximity Approach to Problems in Topology and Analysis. Oldenbourg, Munich (2009).

130. A. O'Farrell. When uniformly continuous implies bounded. Irish Math. Soc. Bull. **53**(2004): 53–56.

131. J.-P. Penot and C. Zalinescu. Bounded (Hausdorff) convergence: Basic facts and applications. *In*: F. Giannessi and A. Maugeri (eds.). Variational Analysis and Applications. Kluwer Acad. Publ. Dordrecht (2005).

132. J. Rainwater. Spaces whose finest uniformity is metric. Pacific J. Math. **9**(1959): 567–570.

133. M. Rice. A note on uniform paracompactness. Proc. Amer. Math. Soc. **62**(1977): 359–362.

134. A.W. Roberts and D. Varberg. Convex Functions. Academic Press, New York (1973).

135. A. Robertson and W. Robertson. Topological Vector Spaces. Cambridge Univ. Press, Cambridge (1973).

136. R. Rockafellar and R. Wets. Variational Analysis. Springer, Berlin (1998).

137. S. Romaguera. On cofinally complete metric spaces. Q & A in Gen. Top. **16**(1998): 165–170.

138. D. Sherbert. Banach algebras of Lipschitz functions. Pacific J. Math **13**(1963): 1387–1399.

139. D. Sherbert. The structure of ideals and point derivations in Banach algebras of Lipschitz functions. Trans. Amer. Math. Soc. **111**(1964): 240–272.

140. R. Snipes. Functions that preserve Cauchy sequences. Nieuw Archief Voor Wiskunde **25**(1977): 409–422.

141. A. Stone. Paracompactness and product spaces. Bull. Amer. Math Soc. **54**(1948): 977–982.

142. M.H. Stone. Applications of the theory of Boolean rings to general topology. Trans. Amer. Math. Soc. 41(1937): 374–481.

143. A. Taylor and D. Lay. Introduction to Functional Analysis. 2nd Edition, Wiley, New York (1980).

144. W. Thron. Topological Structures. Holt, Rinehart and Winston, New York (1966).

145. G. Toader. On a problem of Nagata. Mathematica (Cluj) **20**(1978): 78–79.

146. H. Torunczyk. A short proof of Hausdorff's theorem on extending metrics. Fund. Math. **77**(1972): 191–193.

147. F. Valentine. On the extension of a vector function so as to preserve a Lipschitz condition. Bull. Amer. Math. Soc. **49**(1943): 100–108.

148. F. Valentine. Convex Sets. McGraw-Hill, New York (1964).

149. H. Vaughan. On locally compact metrizable spaces. Bull. Amer. Math. Soc. **43**(1937): 532–535.

150. L. Vietoris. Bereiche zweiter ordnung. Monatsh. Math. Phys. **32**(1922): 258–280.

151. M. Vipera. Some results on sequentially compact extensions. Comment. Math. Univ. Carolinae **39**(1998): 819–831.

152. T. Vroegrijk. Uniformizable and realcompact bornological universes. Appl. Gen. Top. **10**(2009): 277–287.

153. N. Weaver. Lipschitz Algebras. World Scientific Publishers, Singapore (1999).

154. R. Wijsman. Convergence of sequences of convex sets, cones and functions, II. Trans. Amer. Math. Soc. **123**(1966): 32–45.

155. S. Willard. General Topology. Addison-Wesley, Reading, Massachusetts (1970).

Index

Printed in the United States
by Baker & Taylor Publisher Services